OXFORD WORLD'S CLASSICS

ANECDOTES AND ANTIDOTES

AN ABRIDGED TRANSLATION OF

THE BEST ACCOUNTS OF THE CLASSES OF PHYSICIANS

Ibn Abī Usaybiʿah was born at Damascus. The son of a physician, he studied medicine at Damascus and in 1234 he was appointed physician to a new hospital in Cairo. His only surviving work is *The Best Accounts of the Classes of Physicians*, which covers 1,850 years of medical practice, from the mythological beginnings of medicine with Asclepius through Greece, Rome, and India, down to the author's day. Written as much to entertain as to inform, it is not only the earliest comprehensive history of medicine but the most important and ambitious of the medieval period, incorporating accounts of over 432 physicians, their training, practice, and medical compositions, all interlaced with amusing poetry and anecdotes illustrating their life and character. The 'Herodotean' breadth of the book reflects the geographical and cultural reach of the Islamic empire. Written by a man who was a medic and a poet, this highly readable history reflects considerable medical experience and lies at the interface of the serious medical practice of the day with society's interest in biography and gossip. He died in 1270, having created a book that is one of the most delightful productions of Classical Arabic.

Henrietta Sharp Cockrell worked as a specialist for Christie's Islamic Dept in London for several years. Now freelance, her consultancy work has included contributing to the Nasser D. Khalili catalogue, *Gems and Jewels of Mughal India*, and assisting on the *New Catalogue of Arabic Manuscripts* at the Bodleian Library, where she devised the Sharp Scale for quantifying paper translucency. She also worked in Kuwait for UNESCO after the Iraqi invasion and is an occasional writer for *The Art Newspaper*.

Contributors: Geert Jan van Gelder, FBA, was Laudian Professor of Arabic, University of Oxford; Emilie Savage-Smith, FBA, was Professor of the History of Islamic Science, University of Oxford; Simon Swain, FBA, is Professor of Classics, University of Warwick; Ignacio Sánchez is a Senior Research Fellow, Department of Classics & Ancient History, University of Warwick; N. Peter Joosse was Senior Research Fellow at Leiden University and Department of Classics & Ancient History, University of Warwick; Alasdair Watson is Bahari Curator of Persian Collections, Bodleian Library, University of Oxford; Bruce Inksetter was for many years an Arabic translator with UNESCO; and Franak Hilloowala wrote her doctoral thesis on Ibn Abī Usaybiʿah. Credit for the maps goes to Daniel Burt.

OXFORD WORLD'S CLASSICS

For over 100 years Oxford World's Classics have brought readers closer to the world's great literature. Now with over 700 titles—from the 4,000-year-old myths of Mesopotamia to the twentieth century's greatest novels—the series makes available lesser-known as well as celebrated writing.

The pocket-sized hardbacks of the early years contained introductions by Virginia Woolf, T. S. Eliot, Graham Greene, and other literary figures which enriched the experience of reading. Today the series is recognized for its fine scholarship and reliability in texts that span world literature, drama and poetry, religion, philosophy, and politics. Each edition includes perceptive commentary and essential background information to meet the changing needs of readers.

OXFORD WORLD'S CLASSICS

IBN ABĪ USAYBIʿAH

Anecdotes & Antidotes

A Medieval Arabic History of Physicians

A NEW TRANSLATION
ABRIDGED EDITION

Translated by
EMILIE SAVAGE-SMITH, SIMON SWAIN, AND GEERT JAN VAN GELDER (POETRY), WITH IGNACIO SÁNCHEZ, N. PETER JOOSSE, ALASDAIR WATSON, BRUCE INKSETTER, AND FRANAK HILLOOWALA

Selected and Edited with Notes by
HENRIETTA SHARP COCKRELL

With Introduction by
GEERT JAN VAN GELDER

OXFORD
UNIVERSITY PRESS

OXFORD
UNIVERSITY PRESS

Great Clarendon Street, Oxford, OX2 6DP,
United Kingdom

Oxford University Press is a department of the University of Oxford.
It furthers the University's objective of excellence in research, scholarship,
and education by publishing worldwide. Oxford is a registered trade mark of
Oxford University Press in the UK and in certain other countries

First published as an Oxford World's Classics paperback 2020

Impression: 1

Published in the United States of America by Oxford University Press
198 Madison Avenue, New York, NY 10016, United States of America

British Library Cataloguing in Publication Data
Data available

Library of Congress Control Number: 2020933194

ISBN 978–0–19–882792–4

Printed and bound in Great Britain by
Clays Ltd, Elcograf S.p.A.

CONTENTS

INTRODUCTION

Geert Jan van Gelder

THIS introduction discusses the extraordinary biographical world history of medics and medical culture—indeed, the first world history of medicine ever attempted—from which the present volume's selected passages are taken.[1] This extraordinary work was written by a Syrian physician called Ibn Abī Usaybiʿah, who died in 1270, having created a book that is one of the most delightful productions of Classical Arabic.

Ibn Abī Usaybiʿah was a Muslim, a practising physician, and a man with a brilliant gift for describing fellow-doctors and the societies they served without prejudice to their religion or background. His book is somewhat 'Herodotean' in scope: a highly readable account of over 430 physicians, featuring their lives and training as well as their medical and other works, including four 'autobiographies' to which he had access.

The rhyming Arabic title of his book, *ʿUyūn al-anbāʾ fī tabaqāt al-atibbāʾ*, translates as *The best accounts: on the classes of physicians*—a very Arabic formulation. The 'classes' in question are the different groups and categories—chronological, geographical, ethnic, religious—who were encountered by Ibn Abī Usaybiʿah in his tour through some 1,850 years of medics, patients, and patrons, from ancient Greece and Rome to India and the lands in between. The whole is interlaced with amusing poems and tales illustrating the characters and achievements of the author's subjects, embellished with witty aphorisms and plentiful quotations from their writings.

This is a world where science was shared between Muslims, Jews, and Christians and where physicians had their surgeries and hospitals as they do today. Our author's grandfather, father, and uncle were all well-known physicians of their generations and served in the

[1] The following is a description of the work as a whole and not every physician or fact mentioned here is included in the present selection; interested readers are referred to the complete scholarly edition and translation (see below). Those physicians included in this selection are given numbered entries.

courts of various sultans and princes, including the famous ruler Salāh al-Dīn (Saladin). Ibn Abī Usaybiʿah himself trained in ophthalmology at the renowned Nāsirī hospital founded in Cairo by Saladin in 1171. Indeed, his remarkable history of medicine contains our most important accounts of medical activity administered in medieval hospitals.

But this is also a world in which physicians were firmly part of the general intellectual and cultural scene. It was a culture where the ability to write stylishly and entertain one's peers in prose and verse was the basis of social credibility. Through this book a window opens onto the truly multi-cultural, multi-religious world of thirteenth-century Syria and insight is gained into the origins of the medical profession.

Ibn Abī Usaybiʿah was eager to earn his place in history. He was a competent physician, who was employed at the courts of princes of the reigning Ayyubid dynasty. Like many physicians he combined practising medicine with writing books or treatises on medicine. There are no indications that he was a particularly outstanding physician, either practical or theoretical. His fame rests entirely on the only book of his that is preserved. Instead of producing yet another of the many medical compendia or the countless specialized treatises on diseases and ailments, hygiene, therapeutics, and prophylactics, he decided to write what may be called a literary history of physicians. It was an ambitious venture on which he spent considerable time, revising it extensively. It was worth it, for with it he made his mark in history.

The subtitle chosen for this volume, *A Medieval Arabic History of Physicians*, is an accurate characterization of the book. One should be aware, however, that the equivalents of three of the four main words—medieval, Arabic, history—do not appear in the Arabic original title that was given above. More will be said on this title below, but it gives already an adequate indication of the author's intention: his principal interest is in persons and stories. Nevertheless, he also presents the history and progress of medical knowledge and practice through the ages, even though his love of a good story may override his recording of medical fact. Another way of rendering the title in English would be *The Book of Physicians, by One of Themselves* (to use a variation of Thackeray's description of his *Book of Snobs*).

The Author, His Name, and His Life

Like many pre-modern Arabs or writers in Arabic[2] of any standing, our author has a name that tends to sprawl. The very short entry on him in the massive biographical encyclopaedia by al-Safadī (d. 1363)[3] calls him 'Ahmad ibn al-Qāsim ibn Khalīfah al-Khazrajī Muwaffaq al-Dīn Abū l-ʿAbbās, known as Ibn Abī Usaybiʿah'. It cannot be called his 'full' name, for it could easily be extended by including more ancestors; some are known from the entry on his paternal uncle Rashīd al-Dīn ʿAlī ibn Khalīfah ibn Yūnus ibn al-Qāsim ibn Khalīfah. Ahmad is our author's given name, al-Qāsim and Khalīfah are his father and paternal grandfather. Al-Khazrajī is an adjective derived from the Arab tribe of al-Khazraj, one of the two tribes that dwelled in Medina at the time of the prophet Muhammad. Ibn Abī Usaybiʿah's family was descended from Saʿd ibn ʿUbādah al-Khazrajī (d. 633 or 634), a famous 'companion' of the prophet Muhammad who asked him to mediate between them after his conversion to Islam and who, it is said, might have been the first caliph after the prophet's death but for the intervention of Abū Bakr and ʿUmar.

Muwaffaq al-Dīn (approx. 'Given Success in Religion') is an honorific, and Abū l-ʿAbbās ('Father of al-ʿAbbās') is a 'paedonymic', a name mentioning one child, usually one's oldest son. A person could be generally known by any of these elements, or by a nickname or agnomen. This, confusingly, may begin with Ibn or Abū. In the present case one finds both, for Ibn Abī Usaybiʿah literally means 'Son of the Father of the Little Finger' (*abī* being the genitive of *abū*, and *usaybiʿah* being the diminutive form of *isbaʿ*, 'finger' or 'toe'), but 'Father of' should be taken figuratively, as 'the man with'.[4] Ibn Abī Usaybiʿah says that his grandfather Khalīfah was also known as Ibn Abī Usaybiʿah, confusingly; the name seems to have served as a kind of family name. It appears that this grandfather, or some earlier

[2] Many of the most famous writers in Arabic—the philosopher and physician Ibn Sīnā, the theologian al-Ghazālī, the poet Bashshār ibn Burd, to name but three—were not ethnically Arab. Ibn Usaybiʿah wrote in Arabic, spoke it as his mother tongue, and traced his origin, in the paternal line, to an ancient Arab tribe.

[3] al-Safadī, *al-Wāfī bi-l-wafayāt*, 30 vols. Beirut, Wiesbaden, and Berlin: Franz Steiner / Klaus Schwarz, 1931–2005, vol. 7, p. 295.

[4] A contemporary of Ibn Abī Usaybiʿah with a very similar name is the Egyptian literary critic Ibn Abī l-Isbaʿ (d. 1256).

ancestor, was known for something digital: perhaps he had a deformed finger, or a supernumerary one; it is even possible that, paradoxically, he missed a finger. As a result of this slight misfortune, the reader who is unaccustomed to Arabic will repeatedly be confronted with a name that even in its short form does not trip easily off the tongue. It cannot be helped.

Unfortunately, Ibn Abī Usaybiʿah did not include an entry on himself in his book, but biographical details can be gleaned from the entries on his relatives and teachers. He was born into a family of physicians. His grandfather, born in Damascus, had followed Saladin when he moved to Egypt in 1171; his two sons, Sadīd al-Dīn al-Qāsim (the author's father), born in Cairo, and Rashīd al-Dīn (Ch. 15, no. 18), born in Aleppo, became physicians too. The family moved frequently between Egypt and Syria in the service of Ayyubid princes and other local rulers and it is even not certain whether our author was born in Damascus, as most sources say, or in Cairo. Nor is his year of birth certain: it may have been *c.*1200 or some years earlier. He grew up in Damascus and was trained as a physician. We know the names of several of his teachers, among them his uncle, the famous botanist and pharmacologist Ibn al-Baytār (d. 1248) and al-Dakhwār (Ch. 15, no. 17), the founder of the first school in Damascus devoted to the study of medicine (al-Madrasah al-Dakhwāriyyah). After a brief spell in Cairo he was asked, in 1237, to become the private physician of the emir ʿIzz al-Dīn al-Muʿazzamī, the ruler of Sarkhad under the Ayyubids. He accepted, perhaps lured by higher pay, and left the great metropolis. Sarkhad, now better known as Salkhad, was a fortified town situated on a trade route in southern Syria and it was there, around the year 1242, that Ibn Abī Usaybiʿah wrote the first version of his book, revising it and adding to it over the following decades. There are indications that he thought Sarkhad was a backwater. A colleague, Sharaf al-Dīn al-Rahbī, addressed him in a letter with a long poem in which he urges him to return to the 'Paradise on Earth', i.e. Damascus:

Muwaffaq al-Dīn! What's this mindlessness of yours,
despite the rank you have earned in knowledge and erudition?
Have you sold your soul for something trifling and paltry?
You sold it cheap, after being serious and assiduous for so long!
You have been staying in a town that mocks its inhabitants;
no sensible person of standing would be content with it.

It is remote from all that is good; barren; nothing is there
except rocks and blazing heat.
. . .
If where you are now were a place to acquire riches,
it would not compensate for spending your life in hardship.
So how is it what with the little regular pay and its meanness,
and being so far from all virtuous and erudite people?
Come back, then, to the Paradise on Earth. . . .

Ibn Abī Usaybiʿah responded with a poem that has some unkind words about his patron, who started his life as a *mamlūk*, a non-Arab slave and member of the military ruling class, and who in other sources is praised for his benevolent patronage of the arts and his erudition:

. . . How can life be enjoyed by someone whom Time has allotted
to people who are firewood![5]
In their ignorance they do not know the worth of a scholar,
which is not surprising in the case of ignorant people.
I came to someone in whose courtyard my merit was wasted. Would the
stupidity of the non-Arabs be aware of the intelligence of the Arabs?

He expresses his resolve to return to Damascus, but it was not to be. He died in Sarkhad in 1270.

Egypt and Syria in the Thirteenth Century

Even by Middle Eastern standards, Ibn Abī Usaybiʿah lived in turbulent times. Syria and Egypt suffered attacks from Crusaders; the several branches of the Ayyubid Dynasty, in Egypt, Damascus, Aleppo, Homs, Hama, and elsewhere, were mostly in decline after the glorious time of its founder Saladin, who had died in 1193. The Crusaders' invasions and conquests were dwarfed by the massive upheaval of the onslaught of the Mongols, who, led by Hülegü, the grandson of Genghis Khan, took Baghdad in 1258, putting an end to the Abbasid Dynasty that had reigned for just over five hundred years. The advance of the Mongols, or Tatars (*Tatar*) as Ibn Abī Usaybiʿah called them, was stopped in 1260 in Syria by the Egyptian Mamluks, who had supplanted the Ayyubids in Cairo ten years earlier.

[5] Presumably an allusion to Qur'an *Jinn* 72:15: «*As for the unjust, they are firewood for Hell.*»

In spite of the disruptions, cultural life still flourished in many places. The thirteenth century is generally considered one of the most brilliant epochs in Damascus, which was a political, commercial, and industrial hub as well as a centre of learning, some evidence of this being still visible in the many architectural monuments dating from that time. The 'new' city of Cairo (as opposed to al-Fusṭāṭ or Old Cairo) meanwhile, though only founded and made great in the tenth century by the Fatimids, rose during this time to become the most important city, politically and culturally, in the Arab countries.

In the influential but now somewhat dated *Arabic Literature: An Introduction* by H. A. R. Gibb,[6] the Golden Age lasted from the establishment of the Abbasid Dynasty in 750 until 1055, when the Turkish Seljuqs took power in Baghdad and the central Islamic lands. The Silver Age ended with the Mongol conquest of Baghdad in 1258 and was followed by the 'Age of the Mamlūks'. Ibn Abī Usaybiʿah, then, seems to have been partly silver but assigned in his old age to a metal unspecified but of less value. In recent decades it has been made clear, however, that this paradigm of a relentlessly downward slope is no longer valid. There were great Arabic poets and prose writers in the fourteenth and fifteenth centuries, philosophy did not end with Ibn Rushd (Averroes, d. 1198) (Ch. 13, no. 13), and al-Ghazālī (d. 1111) was not the last great Muslim theologian. Nor should it be thought that religion was dominant everywhere and in every field. It is true that in general the majority of titles in the libraries of scholars were on religious subjects (Qur'an, Hadith, Islamic Law, polemics, homiletics) but there were exceptions. The catalogue of a medium-sized, privately endowed library founded in the thirteenth century and attached to a mosque in Damascus has been preserved and has been studied in detail by Konrad Hirschler.[7] Of its more than two thousand titles, volumes of poetry (secular, not religious) and literary works greatly outnumber all other categories, and even medicine is better represented than religion.

Among Ibn Abī Usaybiʿah's older contemporaries were Ibn al-Fāriḍ (d. 1235), the greatest Arabic mystical poet; the controversial, hugely

[6] H. A. R. Gibb, *Arabic Literature: An Introduction*, Oxford: Oxford University Press, 1926, 1964[2].

[7] K. Hirschler, *Medieval Damascus: Plurality and Diversity in an Arabic Library. The Ashrafiyya Library Catalogue*, Edinburgh: Edinburgh University Press, 2016.

influential, and extraordinarily prolific mystic Ibn al-ʿArabī (d. 1240); the polymath ʿAbd al-Laṭīf al-Baghdādī (d. 1231), who has an entry in Ibn Abī Usaybiʿah's book (Ch. 15, no. 12); the historian ʿIzz al-Dīn Ibn al-Athīr (d. 1233); and his brother Diyāʾ al-Dīn Ibn al-Athīr, literary critic and epistolographer. Yāqūt (d. 1229) compiled two great and useful encyclopaedic works, one on place names and one on men of letters. Ibn Khallikān (d. 1282) wrote a very often quoted encyclopaedia of famous people from Islamic times. Not long after Ibn Abī Usaybiʿah's death, Ibn Manẓūr (d. 1311) completed his celebrated dictionary, appropriately entitled *Lisān al-ʿArab* (*The language of the Arabs*). In the *madrasah*s ('schools'), Islamic studies, especially Islamic law, remained central but Arabic language and rhetoric were much studied as ancillary disciplines; speculative theology (*kalām*) was to some extent compatible with and merged with philosophy (*falsafah*).

Medicine

The medical communities of Ibn Abī Usaybiʿah's day were a mix of Christian, Jewish, and Muslim physicians. The constant interchange and occasional rivalry of these three communities is one of the prominent themes of this work.

In works on the history of medicine in the lands of Islam one often finds the phrase 'Islamic medicine'. As so often, the word 'Islamic' is misleading. It suggests, wrongly, that the medical theory and practice of the Muslims are inspired by and thoroughly imbued with religion, a counterpart of 'Christian medicine', if such a thing existed. Rather, it should be taken to mean 'medicine in the world of Islam'. The useful but inelegant word 'Islamicate' was introduced by Marshall G. S. Hodgson to refer to non-religious aspects, as distinct from 'Islamic';[8] but the word has not found currency except among a few specialists and it will not be used here.

Many Muslims, however, will argue that there is something that could rightly be called 'Islamic medicine'. It is what in Arabic is

[8] M. G. S. Hodgson, *The Venture of Islam: Conscience and History in a World Civilization*, 3 vols (Chicago: University of Chicago Press, 1974), see vol. 1, 56–60. Though somewhat outdated especially for the early history of Islam, it is still an extremely thoughtful and comprehensive presentation of Islamic culture (relevant for the present purpose are parts of vol. 1, *The Classical Age of Islam*, and vol. 2, *The Expansion of Islam in the Middle Periods*).

termed *al-tibb al-nabawī*, 'Prophetic medicine', or medicinal knowledge allegedly going back to the prophet Muhammad. This was devised by Islamic religious scholars, who collected a mixture of traditional folk medicine and magic, with Greek ideas in disguise, skilfully combined with sayings rightly or wrongly attributed to the Prophet and the early pious Muslims, all this to serve as a counter to the more scholarly medicine that was so heavily influenced by Greek theory and practice and therefore suspect as being of heathen origin in the eyes of the orthodox. It was and still is popular among the pious, but regular physicians, whether Muslim, Christian, or Jewish, rather than openly criticizing it, usually ignored it. Significantly, Ibn Abī Usaybiʿah does not explicitly mention it.

Medicine in the central Arab lands in Islamic times was largely a continuation and development of Greek medical practice and theory; other traditions, Persian, Indian, or Arabian, are marginal by comparison. This is reflected in Ibn Abī Usaybiʿah's book, which after an introductory chapter begins with sections on Greeks, from the mythical Asclepius (Ch. 2) to Hippocrates (Ch. 4, no. 1) and many others, and above all Galen (d. 216, Galenus in Latin, Jālīnūs in Arabic) (Ch. 5), whose towering influence enveloped Islamic and Western medicine for many centuries. It is not surprising that Ibn Abī Usaybiʿah's entry on him is the longest in the entire book. The prominence of Greek medicine in the book is one of the many results of the translation movement in the eighth, ninth, and tenth centuries, one of the most remarkable feats of cultural transmission in world history, when large quantities of Greek works were translated into Arabic, either via Syriac (the form of Aramaic used by Eastern Christians) as an intermediate language or directly from Greek. Not just any Greek works: the Arabs had little interest in literary works, considering their own poetry far superior, but above all works on philosophy, logic, mathematics, natural sciences, and medicine. It is no coincidence that several of the translators were physicians themselves, among them the star translator Hunayn ibn Is'hāq (d. 873 or 877) (Ch. 8, no. 12), a Christian who was at home in Greek, Syriac, and Arabic, and on whose life we are well-informed, partly thanks to the long and eminently readable chapter on him in Ibn Abī Usaybiʿah's book, which includes a lengthy autobiographical passage. He translated over one hundred Greek medical works, many of them by Galen.

With time, particularly in the twelfth and thirteenth centuries, medical theory and practice in Islamic lands developed distinctive and innovative techniques and practices of its own, especially in the fields of ophthalmology and pharmacology. These are reflected in many of the accounts given by Ibn Abī Usaybiʿah. The underlying medical theory, however, remained Galenic, just as it did in European medicine for many centuries.[9] One understands its appeal and its popularity. Until the development of technologies such as microscopy, X-rays, and general anaesthesia allowing extensive and minute examination of the interior of the body, the structure and function of the human body as outlined in Galenic physiology seemed adequate. In the Galenic system, the concept of four humours (blood, phlegm, yellow bile, and black bile)—what historians today call humoral pathology—dominated the theory of disease. The soul was distributed through the main organs, brain, liver, heart, which in turn were the origins of the body's major functions. Galen and other physicians of antiquity had made only limited progress in aligning the health of the body and the mind. Their key treatment was in the area of therapeutics, which was directed towards restoring a general balance to the system, countering plethora with evacuation, including bloodletting. Thanks to the ninth-century translators a new Arabic technical terminology was developed, making use of the morphological potential and the lexical plenty of the Arabic language. Notwithstanding this, one comes across Greek words in Arabic guise, such as *mālīkhūliyā*, 'melancholia', as well as literal or loan-translations such as *al-mirrah al-sawdāʾ* ('black bile').

'Philosophy' (*falsafah*) included the natural sciences and was often a close companion to medicine in the lives and works of physicians. A prominent illustration of this is provided by the Persian Ibn Sīnā, known in the West as Avicenna (d. 1037) (Ch. 11, no. 2), whose works on philosophy and medicine were extremely influential, in Islamic countries and subsequently in Europe too. Naturally, Ibn Abī Usaybiʿah devotes a lengthy entry to him. Ibn Sīnā's commentaries on Aristotelian philosophy (including logic and the natural sciences) are

[9] Most of the introductions to Islamic medicine listed in the Select Bibliography will provide a summary of the Galenic system; a good introduction to Galen and his writings is his *Selected Works*, translated with an Introduction and Notes by P. M. Singer (Oxford: Oxford University Press, 1997). See also V. Nutton, *Ancient Medicine* (London: Taylor & Francis, 2004).

collectively known as *al-Shifāʾ* (*The cure*)—a metaphor taken from medicine. His great medical compendium is entitled *al-Qānūn fī l-ṭibb* (*The canon: on medicine*), the loan-word *qānūn*[10] being a reminder of the lasting influence of the ancient Greeks on the development of 'Arab' or 'Islamic' civilization. The close relationship between medicine and philosophy is clearly visible also in Ibn Abī Usaybiʿah's book, which has lengthy sections on Plato (Ch. 4, no. 6) and Aristotle (Ch. 4, no. 7), whose names do not spring to mind as being physicians. Galen thought of himself, naturally, as a philosopher as well as a doctor. In Arabic the word *hakīm*, literally 'sage', often means 'philosopher' as well as 'physician', a synonym of *tabīb*.

Many physicians, especially court physicians of caliphs and viziers, were Christians; some were Jewish. Some limited themselves to practising medicine but numerous physicians wrote medical treatises, works of theory or practice, or even large compendia. An important and valuable part of Ibn Abī Usaybiʿah's book are the extensive booklists he provides for individual authors. They have for the most part had to be omitted or shortened for the present selection even though they offer intriguing reading especially to lovers of lists and curiosa, for there are numerous titillating and tantalizing titles mostly of books and treatises now lost. These lists also show a very large number of works on a wide range of topics that are only tangentially related to medicine, or not at all. So, for such list-loving people, here is a small taste of the breadth of subject matter, without authors' names: *A treatise on vertigo; On the finite and the infinite; A discourse on the vacuum; A treatise on hunchbacks; A treatise on rhubarb and its beneficial properties; An epistle on why it is difficult to find a competent physician and why ignoramuses are numerous; A treatise on why women become fat when past their youth; A treatise on the fact that ignorant physicians are too strict in preventing the sick from indulging in their sensual cravings; A treatise on the mouse; The main plants and trees of al-Andalus; An epistle on buttocks and pains affecting them; The book of hallucinations; An epistle on why wild beasts and lions were created; On the fact that those who have no experience with demonstrations cannot grasp the fact that the Earth is round and the people are around it; The wisdom of backgammon; An*

[10] In modern Arabic, *qānūn* (from Greek *kanōn*, 'rule') means '(secular) law' (as distinct from Islamic law); it is also a popular zither-like musical instrument.

epistle on why people use ambiguities in their speech; Why seawater is salty; A treatise on tickling.

Apart from booklists, which are valuable mainly to scholars, the book also has much information on medicine as a craft and a practice, with many anecdotes about cures, on the daily routine of physicians at court or in hospitals, on their relationships with patients, patrons, and one another. The book is, in fact, our major source for the structure and function of hospitals in Damascus and Cairo in the twelfth and thirteenth centuries. The fact that the author's father, uncle, and grandfather all worked in them, and that he was trained in one, gives us some first-hand accounts of what was done, even if the details are woefully lacking.

Ibn Abī Usaybiʿah's Medieval Arabic History of Physicians: Genre and Title

Obviously, the author himself did not use the word 'medieval', of which there is no pre-modern Arabic equivalent. Some scholars[11] rightly dislike using 'medieval' when speaking of Arab-Islamic history because the word was coined specifically for European history and also because to many it has, unfortunately, some negative associations. The Arabic title of Ibn Abī Usaybiʿah's book provides clues to its origin and the genre or genres to which it aspires to belong. *ʿUyūn al-anbāʾ fī tabaqāt al-atibbāʾ* literally translates as '*The best accounts* (or *Choice reports*): *on the classes of physicians*'. It stands in a long tradition of works about 'classes' or categories of people, a genre that started when Islamic scholars, keen to establish the names, dates, and reliability of the transmitters of Hadith ('traditions, reported sayings') from the time of the prophet Muhammad and his Companions, began to compile lists of shorter or longer biographies. The Arab fondness of tribal genealogy, often with political implications, may also have played a part. But besides religion and politics, it was a widely shared love of personal stories, anecdotes, and gossip that lies behind the proliferation

[11] A forceful plea against using the word is T. Bauer, *Warum es kein islamisches Mittelalter gab. Das Erbe der Antike und der Orient* (Munich: C. H. Beck, 2011). Nevertheless, it was decided to use the term because it is customary, and one could argue that the author indeed lived in the Islamic 'Middle Ages' between late antiquity and the modern period. It is hoped, moreover, that the present selection will help to dispel any associations of 'medieval' with backwardness and barbarity.

of biographical works. Similar works were written on poets, jurisprudents, theologians, Qur'an commentators, ascetics and mystics, philologists, and grammarians. The organization of these works varies and the term 'class' (*tabaqah*) is usually not clearly defined nor does it necessarily imply a ranking in quality; they may be arranged roughly chronologically (in which case *tabaqah* could be translated as 'generation') or geographically.

Ibn Abī Usaybiʿah had predecessors who wrote on the history of medicine. Is'hāq ibn Hunayn (d. *c.*910) (Ch. 8, no. 13), eminent translator of Greek works, is the author of a *History of Physicians*, preserved in fragments, in which he writes on the origin of medicine and mentions a few ancient philosophers and physicians. The Egyptian Yūsuf ibn Ibrāhīm ibn al-Dāyah (d. between 941 and 951) compiled a work on physicians, now lost but one of Ibn Abī Usaybiʿah's sources. Another is the Andalusian Ibn Juljul (d. after 994), who collected biographies of philosophers and physicians in his *Tabaqāt al-atibbāʾ wa-l-hukamāʾ* (*The classes of physicians and sages*), a work much smaller than that of Ibn Abī Usaybiʿah and with a more limited scope. Only the last of its seven 'classes' deals with physicians in the Islamic period. Yet another predecessor, also from Spain, is Sāʿid al-Andalusī (d. 1070). His slim but important volume entitled *Tabaqāt al-umam* (*The classes of nations*) deals with the scientists and scholars of various 'civilized' nations, which to him are the Indians, Persians, Chaldeans, the ancient Greeks, *Rūm* (Romans or Byzantines), (pre-Islamic) Egyptians, Arabs, and Jews. One of Ibn Abī Usaybiʿah's sources is not a work of 'classes'; it is the celebrated *al-Fihrist* (*The catalogue*), compiled *c.*988 by Ibn al-Nadīm, an exceptionally erudite bookseller in Baghdad, who listed all the Arabic books known to him and their authors by subject area. An older contemporary of Ibn Abī Usaybiʿah, the Egyptian Ibn al-Qiftī (d. 1248), wrote an alphabetically arranged book on physicians and philosophers, *Ikhbār al-ʿulamāʾ bi-akhbār al-hukamāʾ* (*Informing scholars of the reports on sages*). When Ibn Abī Usaybiʿah, after more than twenty years, revised and expanded his first version he incorporated material from Ibn al-Qiftī's work.

It is worth stressing that, in contrast to what is often found in Western medieval treatises, Ibn Abī Usaybiʿah regularly and scrupulously mentions his sources, oral as well as written, the former kind especially in the entries on his contemporaries in Egypt and Syria. In this outstanding scholarly practice, he was no exception, for Arabic and Islamic writers very often made a point of conscientiously quoting and acknowledging their sources.

Ibn Abī Usaybiʿah's book, however, surpasses his predecessors in scope and volume. Moreover, the first half of its title hints at its character as being entertaining and part of *adab*, a term that variously, even in modern Arabic, means 'good manners', 'proper conduct', 'erudition', or 'belles lettres', and is also used for works that contain a mixture of literary entertainment, moral education, instruction, and information. The words *ʿUyūn al-anbāʾ* (*Best accounts* or *Choice reports*) are reminiscent of a seminal literary anthology by the polymath Ibn Qutaybah (d. 889), entitled *ʿUyūn al-akhbār* (which also translates as *Best accounts* or *Choice reports*). Pre-modern Arabic book titles very often consist of two rhyming halves, the first being somewhat flowery and figurative while the second, mostly introduced by *fī* (a preposition that normally means 'in', but here rather 'on, about', or the equivalent of the colon preceding an English subtitle) indicates more clearly the topic of the book. The word *anbāʾ* ('reports', plural of *nabaʾ*) would have readily suggested itself: it provides a rhyme, even a rich rhyme, with *atibbāʾ*, 'physicians' (the plural of *tabīb*). Ibn Abī Usaybiʿah, by his inclusion of so much poetry and countless entertaining anecdotes, maxims, and wise sayings, clearly wished his work to be seen as standing in the same tradition of *adab*.

Many anthological works of *adab* and biographical encyclopaedias are rather impersonal, with the compiler being almost wholly without a personal voice. It is very different with Ibn Abī Usaybiʿah's book, for he is very much present in the text, as transmitter and participant in anecdotes, personal observations, opinions, and comments. He is himself part of a network described in detail. The word *aqūl*, 'I say', at the end of a quotation or the introduction of a personal comment, is one of the more common words in the text. He does not seem to have been a very prominent physician himself and we have no stories from him or others about his medical skills. Al-Safadī, the fourteenth-century biographer quoted above, says that he composed a fine 'history of physicians' and calls him 'a man of letters, a physician, and a poet', perhaps in order of proficiency.

The author considered his book as an ongoing project, revising it over the years until his death, while earlier versions were already circulating. He mentions that a colleague found his book in Cairo and bought it for his own library (Ch. 14, no. 16). Another colleague, looking through a copy in the author's presence, complained that he himself had not been included. He has, however, an entry in the final version (Ch. 15, no. 6). Many manuscripts of the book survive, some of them copied from drafts in the author's hand, but no autograph has been preserved. He mentions some of his other works, but they are lost. [See Appendix 5, Figure 1 for the colophon of a copy completed in Syria in 1372, approximately one hundred years after Ibn Abī Usaybiʿah wrote his final version.]

A Medieval Arabic History of Physicians: Contents

For an overview of the contents of *A Medieval Arabic History of Physicians* one is referred to the notes at the beginning of the fifteen individual chapters. These chapters are of widely varying length. The roughly chronological order of Chapters 2 to 9 gives way to an arrangement based on geography, with chapters on Iraq, the lands of the Persians, India, the Muslim West, Egypt, and Syria; the entries within each chapter are arranged more or less chronologically. With the exception of the short section on India these chapters are lengthy. The last chapter especially is full of long entries on physicians close to the author, and as well as containing many first-hand accounts, it also provides evidence of the poetic skills of many physicians, including the author himself. Poetry is abundant in other chapters too, in such quantities that it deserves a separate heading.

POETRY

The presence in the book of some 3,600 lines of poetry—and an Arabic line is often as long as an English couplet—is proof of its importance to the author and his intended readership. Many modern readers had considerably less patience with it. August Müller, the first to edit the text, wrote that the author 'unfortunately devoted himself so persistently to *adab* and poetry that the *qasīdah*s and shorter poems composed by himself and others drove to well-justified despair the otherwise quite patient copyist of the excellent model of the Brit. Mus. manuscript Add. 7340'.[12] Juan Vernet, in his entry on the author

[12] In *Actes du 6ème Congrès International des Orientalistes tenu en 1883 à Leide, Deuxième partie, Section 1: Sémitique*, Leiden, 1885, 262 (my translation from the German). A *qasīdah* is a longer poem, often translated as 'ode'.

in the second edition of *The Encyclopaedia of Islam*, says disparagingly that the book contains 'some long series of verses which have nothing to do with the main theme'. It is true: there are relatively few poems that could be called 'medical' and whereas Ibn Sīnā's poetry is well represented, his famous medical versification *al-Urjūzah fī l-tibb* (translated into Latin in the thirteenth century) is not there. There are no 'didactic' poems, for instance, for memorizing the bones in the body. One does find, however, short pieces with general advice or poetic exchanges between patient and doctor: a vizier, writing to Ibn Sīnā, complains of pustules on his forehead; the Syrian nobleman Usāmah ibn Munqidh wrote to Ibn al-Naqqāsh about his bad knees; another vizier asks the physician and translator Is'hāq ibn Hunayn about the latter's bowel movements after taking a purgative.

There are many poems, some of them lengthy, in praise of physicians, which refer to medical matters. The addressee is said to be 'the successor of Hippocrates' or he would be able to teach Galen if he were alive, and so forth. Although many of these poems are full of clever conceits, striking imagery, learned allusions, and ingenious wordplay, they are not among the more appealing poems to a modern readership, especially when much of their rhetorical brilliance is lost in translation. The customary hyperbole, the unrelenting reference to their 'lofty qualities' and 'excellence', the endless series of words for generosity, munificence, benefaction, liberality, beneficence, bounteousness, and the constant need of copious annotation for the sake of the modern reader: all this can make for tediousness, and they have mostly not been included in the present selection. To the author and his colleagues, however, they were relevant, serving as tokens of friendship and esteem, an essential part of polite social intercourse between educated equals, between patron and dependent, or between subject and ruler. It is no coincidence that there are many eulogies and congratulatory poems, especially in the last chapter, the one on Syria, where the author writes about his teachers and colleagues, who are likely to read his book. More readable, however, are the countless epigrams: lampoons, complaints, wisdom, love, wine, poetic descriptions of objects, and riddles. There are humorous poems, such as those by Abū l-Hakam al-Maghribī (Ch. 15, no. 2; also see Ch. 15, no. 4). He is one of several whose entries consist mainly of poetry and one has the impression that Ibn Abī Usaybiʿah included him and some others because they excelled at poetry, rather than being eminent physicians.

APHORISMS

Many short, epigrammatic poems belong to what in Arabic is called *hikmah*, 'wisdom', a word that may also mean 'philosophy' and is related to the word *hakīm*, 'sage' and also 'physician', as mentioned above. Wisdom may also be expressed in short, apodictic prose: in an aphorism, in short. Aphorisms and short admonitions or precepts are another major form in the book, called *hikam* (plural of *hikmah*) or *ādāb* (approximately 'rules for good life and conduct'). The ancient Greeks excelled in them and the Arabs eagerly collected wise sayings of the philosophers. Often quoting from one such collection by the eleventh-century Egyptian al-Mubashshir ibn Fātik, who himself has a short entry in the book (Ch. 14, no. 6), Ibn Abī Usaybiʿah incorporated hundreds of them, by Hippocrates, Pythagoras, Socrates, Plato, Aristotle, Galen, and others. Although to the Arabs poetry was the preferred literary form for wisdom, they also tried their hand at the prose maxim. Among its practitioners was the author's uncle, Rashīd al-Dīn ibn Khalīfah.

Earlier Editions and Translation

The German Arabist August Müller (1848–92) published a critical edition of the Arabic text of Ibn Abī Usaybiʿah's book in 1882. His was an outstanding achievement but he had every reason to be unhappy with the result. As he writes in his German preface:

In more than one aspect the form in which the present book sees the light corresponds with neither my original intention nor my subsequent expectations. This is due to the history of its publication, a short account of which, as far as is possible and I hope as far as is necessary, will serve as justification and apology for those shortcomings that to me are most frustrating.[13]

This 'short account' takes some ten pages and is followed by scores of pages with critical remarks, variant readings, and corrections. Müller had been persuaded by his friend Spitta Bey (Wilhelm Spitta, 1853–83) to have the Arabic text printed in Cairo. To his chagrin, the result was a dog's dinner. The Arabic title page of the first of its two volumes

13 My translation from the German (GJvG).

looks impressive, with its ornate border and prolix text in traditional style; it translates as[14]

The Book of
Choice Reports on the Classes of Physicians
Composed by the Eminent Physician and Erudite Scholar
Muwaffaq al-Dīn Abū l-ʿAbbās Ahmad ibn al-Qāsim ibn Khalīfah
ibn Yūnus al-Saʿdī al-Khazrajī
Known as Ibn Abī Usaybiʿah
God rest his Soul
Transcribed and Edited from the Manuscripts Found in Several Libraries
by the Humble Man Needing God's Help and Mercy[15]
Imruʾ al-Qays ibn al-Tahhān
****
First Edition, at the Printing House of Wahb[16]
in the Year 1299 of the Hijra Corresponding to the Year 1882

One supposes August Müller is responsible for the whimsical 'translation' of his own name, for Imruʾ al-Qays, a famous pre-Islamic name, has few sounds in common with August, but ibn al-Tahhān, at least, means 'the Miller's (Müller's) Son'. [See Appendix 5, Figure 2 for the title page of August Müller's 1882 edition.]

For all its deficiencies Müller's edition remained the standard edition of the text. The often-used edition by Nizār Ridā (Beirut, 1965) is wholly dependent on Müller's; it is lightly annotated and by no means a critical edition. The edition by Muhammad Bāsil ʿUyūn al-Sūd (Beirut, 1998) is similar. The attempt by ʿĀmir al-Najjār to make a new critical edition, which appeared in Cairo between 1996 and 2004 in six volumes with extensive introduction, critical apparatus, and indexes, is unfortunately marred by errors, typos, and inaccuracies to such an extent that it cannot be deemed to be a proper scholarly edition.

There exist some partial, incomplete translations of the book into European languages.[17] Some years ago an incomplete English translation by Lothar Kopf was placed on-line without the (deceased) translator's

[14] Although there are no capitals in the Arabic alphabet, their use in English seems appropriate here.

[15] A customary formula of humility.

[16] This refers to the publisher, Mustafā Wahbī.

[17] See the entry 'Ibn Abī Uṣaybiʿah' (F. Hilloowala) in *The Encyclopaedia of Islam Three* (Leiden: Brill, 2000 [available on-line]).

or library's permission, using a typed copy or draft partial translation prepared before 1969 and placed on deposit in the National Library of Medicine, Bethesda, MD. While it was a monumental one-man undertaking, it was never finished, lacks most of the poetry, and contains many errors and infelicities.

NOTE ON TRANSLATION AND SELECTION

A NEW edition, based on all the important available manuscripts, including some copies unavailable to earlier editors, along with a full and richly annotated English translation, has been prepared by a team of scholars who collectively call themselves the ALHOM (A Literary History of Medicine) group.[1] Each of them was the principal editor and translator of one or several of the fifteen chapters. It was published early in 2020 by E. J. Brill (Leiden) in five volumes and also in an electronic form available as open access.[2] The project was financially supported by the Wellcome Trust. The reader is encouraged to consult the full edition and translation, which also includes a volume with introductory essays.

Since it was thought that the book deserves a wider readership than is envisaged for the full scholarly edition and translation, the present selection was prepared; again, it was in some ways a collective effort. Normally, in the case of selections for a series such as Oxford World's Classics, the one who selects is the translator, who would also provide an introduction and annotations. For the present volume, however, the selection was made by Henrietta Sharp Cockrell, who used and adapted the translations and annotations made by the ALHOM team for the full version. This was done in consultation with the translators.

[1] They are Emilie Savage-Smith and Simon Swain (both of whom were 'principal investigators' of the project as underwritten and supported by the Wellcome Trust), with Geert Jan van Gelder, Ignacio Sánchez, N. Peter Joosse, Alasdair Watson, Bruce Inksetter, and Franak Hilloowala.

[2] For the print version, see *A Literary History of Medicine: The ʿUyūn al-anbāʾ fī ṭabaqāt al-aṭibbāʾ of Ibn Abī Uṣaybiʿah*, edited and translated by E. Savage-Smith, S. Swain, and G. J. van Gelder, with I. Sánchez, N. P. Joosse, A. Watson, B. Inksetter, and F. Hilloowala, 5 vols (Leiden: Brill, 2020). An open access version is available at https://dh.brill.com/scholarlyeditions/library/urn:cts:arabicLit:0668IbnAbiUsaibia/.

EDITORIAL NOTE

THE original book has close to 326,000 words in Arabic and over 560,000 in English translation. In order to reduce it to the approximately 100,000 words of a manageable paperback, four-fifths, or 80 per cent, of the original has had to be removed. While the selection for this edition has been slightly weighted towards the later chapters that reveal most about life and medicine in the medieval Islamic World, still only a quarter of the 368 entries in Chapter 7 and onward could be included. Poetry, lists of aphorisms or sayings, and book lists have had to be drastically reduced, though effort has been made with the latter to include a selection of both important works and some of the more obscure and surprising titles. It is hoped that the feel and balance of the original has been retained to some degree, but inevitably the present edition is only a small window into the whole.

Literary Arabic of Ibn Abī Usaybiʿah's time tends to be flowery, to some extent, formulaic, and, to an unaccustomed reader, can feel slightly long-winded. Many passages have been streamlined to make them clearer and more accessible to the expected audience. The threads of narration can also be confusing: '*X* told me he heard it from his friend, whose servant used to work for the caliph's chamberlain; in the chamberlain's words . . .'. For clarity, such threads have generally been reduced or restructured. Long chains of narration, often included to confer authenticity, have also been omitted for the most part.

Not only are Arabic names often long and complex, as mentioned in the introduction, but Ibn Abī Usaybiʿah also refers to people at various times by different variations of their name. To avoid confusion, one version has been chosen and maintained throughout.

In the section titles where 'Ibn' meaning 'son of' refers to a person who is the son of the previous entry, it has been translated to emphasize the connection.

Some characters, especially non-Arabs, are referred to by their anglicized name if they have one, such as Maimonides and John the Grammarian. Others, such as Ibn Sīnā, have the name they were known by in Europe (Avicenna) in brackets.

Some personages, such as Saladin, were felt to be so well known that their Arab name (Salāh al-Dīn) has not been used.

Ibn Abī Usaybiʿah rarely explains who people are since the majority of their names would have been well known in his time. Where appropriate their role or title (sultan, scribe, vizier, etc) has been inserted to help the reader keep track of unfamiliar characters. Readers may refer to the appendix for further details on more obscure figures.

As explained in the Introduction, the vast majority of Ibn Abī Usaybiʿah's work was from conscientiously and scrupulously cited source material. He used a wide range, both verbal and written, from many different religious and ethnic communities. Nearly all are important, and some preserve the only passages from, or give us the only known reference to, a work now lost in the original. Nevertheless, after much discussion, it was decided that including the details of all of these was beyond the scope of this edition and therefore many source names have been omitted. The majority of the sources that remain are those who appear frequently or who are also entries in the book itself. Full details of the network of sources used by Ibn Abī Usaybiʿah will be found in the complete print and online texts published by Brill.

Unsurprisingly, in a medieval history of more formal, scholarly medicine, mentions of women in general, and especially in roles other than as patients, are uncommon. Almost all have been included in this volume, but even with this disproportionate favour, they remain very few and generally very brief.

Regarding the poetry, one of the many conventions of a good verse was a form of word play or punning, which was used extensively. Much of this has obviously been lost in translation, but occasionally words have been inserted in brackets to give a very small idea of how this worked.

Dates in the original text were given in Hijri dating—the lunar calendar used in Islam which begins in 622 when the Prophet and his followers journeyed from Mecca to Medina and founded the first Muslim community. The year has twelve months in 354 or 355 days. This shorter year means that a particular Hijri month does not always fall in the same season and that a Hijri year nearly always spans part of two Gregorian years, for example: 445 runs from 23 April 1053 to 13 April 1054.

In this edition, the first date given is the original Hijri date and the second is the Gregorian equivalent, for example: Monday 15 Safar 628/22 December 1230. If only the Hijri year is given (without month or day), then only the first year of its Gregorian equivalent span is listed, for example: 445/1053, unless the event is known to have taken place in the latter year.

Many thanks are due to Emilie Savage-Smith, Geert Jan van Gelder, and Simon Swain for their help and collaboration on this edition.

HSC

NOTE ON TRANSLITERATION AND PRONUNCIATION

For Arabic names and terms a simplified version of the full scholarly system of transliterating Arabic is used. Subscript dots to distinguish between phonemes (*d/ḍ, h/ḥ, s/ṣ, t/ṭ, z/ẓ*) are omitted, but macrons to indicate long vowels (*ā, ī, ū*) are given, as are the signs to represent the letter *'ayn* (a voiced pharyngeal fricative) and the *hamzah* (ʾ, the glottal stop). A divider is used to distinguish *s-h* as in 'mishap' (e.g. Is'hāq) from *sh* as in 'ship'. Names and terms current in English appear in their customary forms: Mecca, Qur'an, emir, rather than Makkah, Qurʾān, amīr. An exception to this is 'shaykh' which was felt to better convey the meaning of 'authority', 'teacher', and sometimes 'old man' than the anglicized 'sheikh', which is commonly used to refer to royal, tribal, or religious leaders. In bibliographical references the transliteration of the original has been preserved.

Arabic vowels (short: *a, i, u*, long: *ā, ī, ū*) are pronounced roughly as in Italian but each with a wider range, depending on the adjacent consonants, which explains common non-scholarly spellings such as Ahmed (instead of Ahmad) and Omar (instead of ʿUmar). There are two diphthongs, *ay* and *aw* (as in 'eye' and 'now' but shorter and 'tenser'). *Th* is as in 'think', *dh* as in 'this'. Consonants not found in standard English are the notorious ʿ, described above as a voiced pharyngeal fricative and sometimes, a bit unfairly, as a 'vomiting sound'; non-specialists are advised to pronounce it as ʾ, the glottal stop. *Kh* is as in 'Bach' or 'loch', *gh* is a bit like the uvular Parisian *r*, but not rolled, *q* is like *k* but further back in the mouth; it is *not* followed by *w* as in English *qu*. Unlike English *k* and *t*, aspirated when preceding a stressed vowel, Arabic *k* and *t* are never aspirated. *R* is rolled as in Scottish. *H* is always pronounced as *h*, even when followed by a consonant as in Ahmad, *qahwah* ('coffee'), but in the common ending *-ah* it is heard only weakly or not at all. Any consonant may be doubled in transliteration, i.e. lengthened in pronunciation. When the article *(a)l-* is followed by a consonant made with the tip of the tongue, such as *d, t, r, s, n*, the *l* assimilates to this consonant, which is lengthened (*al-dīn* is pronounced *addīn*, like the beginning of Italian *addio*), but this assimilation is not represented in the spelling used here.

Stress falls on a final syllable if this ends in a long vowel plus consonant (Hārūn, *atibbāʾ*) or in two consonants (Dimáshq, Hunáyn; note that a digraph such as *sh* or *th* counts as one consonant). The penultimate syllable is stressed if this does not apply and the penultimate vowel is long (*khalīfah*) or followed by two consonants (Muwáffaq), or when the word has only two syllables (Ábī, Ásad). In other cases the stress will normally fall on the antepenultimate syllable (al-Sáfadī, Káladah, Usáybiʿah, *mádrasah*). But in the Egyptian pronunciation of standard Arabic one may hear Usaybíʿah, *madrásah*, stressed on the penultimate.

SELECT BIBLIOGRAPHY

A Literary History of Medicine: The ʿUyūn al-anbāʾ fī ṭabaqāt al-aṭibbāʾ of Ibn Abī Uṣaybiʿah, edited and translated by E. Savage-Smith, S. Swain, and G. J. van Gelder, with I. Sánchez, N. P. Joosse, A. Watson, B. Inksetter, and F. Hilloowala, 5 vols (Leiden: Brill, 2020).

https://dh.brill.com/scholarlyeditions/library/urn:cts:arabicLit:0668IbnAbiUsaibia/

Medicine in the pre-modern Islamic world

Amar, Z., and E. Lev, *Arabian Drugs in Early Medieval Mediterranean Medicine* (Edinburgh: Edinburgh University Press, 2017).

Bürgel, J. C., *Ärztliches Leben und Denken im arabischen Mittelalter*. Bearbeitet von Fabian Käs (Leiden: Brill, 2016).

Conrad, L. I., 'The Arab-Islamic Medical Tradition', in L. I. Conrad *et al.*, *The Western Medical Tradition: 800 BC to AD 1800* (Cambridge: Cambridge University Press, 1994), 93–138.

Dols, M. W., *Majnūn: The Madman in Medieval Islamic Society*, ed. Diana E. Immisch (Oxford: Clarendon Press, 1992).

Giladi, A., *Muslim Midwives: The Craft of Birthing in the Premodern Middle East* (Cambridge: Cambridge University Press, 2015).

Isaacs, H. D., 'Arabic medical literature', in M. J. L. Young, J. D. Latham, and R. B. Serjeant (eds), *Religion, Learning and Science in the ʿAbbasid Period* (Cambridge: Cambridge University Press, 1990), 342–63.

Lindberg, D. C., and M. H. Shank (eds), *The Cambridge History of Science*, vol. 2: *Medieval Science* (Cambridge: Cambridge University Press, 2013), see section 1: 'Islamic Culture and the Natural Sciences' (F. J. Ragep), 27–82; section 4: 'Medicine in Medieval Islam' (E. Savage-Smith), 139–67; section 6: 'Science in the Jewish Communities' (Y. T. Langermann), 168–89.

Mitchell, P., *Medicine in the Crusades: Warfare, Wounds and the Medieval Surgeon* (Cambridge: Cambridge University Press, 2004).

Nutton, V., *Ancient Medicine*, 2nd edn (London: Routledge, 2012).

Perho, I., *The Prophet's Medicine: A Creation of the Muslim Traditionalist Scholars* (Helsinki: Finnish Oriental Society, 1995).

Pormann, P., and E. Savage-Smith, *Medieval Islamic Medicine* (Edinburgh: Edinburgh University Press, 2007).

Ragab, A., *The Medieval Islamic Hospital: Medicine, Religion, and Charity* (Cambridge: Cambridge University Press, 2015).

Savage-Smith, E., 'Ṭibb', in *The Encyclopaedia of Islam*, new [= 2nd] edn, vol. 10 (Leiden: Brill, 2000), 452–60.

Ullmann, M., *Islamic Medicine* (Edinburgh: Edinburgh University Press, 1978).

On the translations from Greek into Arabic

Gutas, D., *Greek Thought, Arabic Culture: The Graeco-Arabic Translation Movement in Baghdad and Early ʿAbbāsid Society (2nd–4th/8th–10th centuries)* (London: Routledge, 1998).

Rosenthal, F., *The Classical Heritage in Islam*, trans. by E. and J. Marmorstein (London: Routledge, 1992) [orig. title: *Das Fortleben der Antike im Islam*, 1965].

Some individual physicians and treatises in translation

Galen, *Selected Works*, trans. with an introduction and notes by P. M. Singer (Oxford: Oxford University Press, 1997).

Ibn Butlān, *Le Taqwīm al-ṣiḥḥa (Tacuini sanitatis) d'Ibn Buṭlān: un traité médical du XIe siècle, histoire du texte, édition critique, traduction, commentaire*, par H. Elkhadem (Leuven: Peeters, 1990).

Ibn Butlān, *Das Ärztebankett*, trans. by F. Klein-Franke (Stuttgart: Hippokrates verlag, 1984) [an edition with English translation by P. Kennedy is in preparation].

Ibn Hindū, Abū l-Faraj ʿAlī ibn al-Ḥusayn, *The Key to Medicine and a Guide to Students (Miftāḥ al-ṭibb wa-minhāj al-ṭullāb)*, trans. by A. Tibi, reviewed by E. Savage-Smith (Reading: Garnett, 2011).

Ibn al-Jazzār, *Ibn al-Jazzar on Sexual Diseases and their Treatment: A Critical Edition, English Translation and Introduction of Zād al-musāfir wa-qūt al-ḥāḍir (Provision for the Traveller and the Nourishment of the Sedentary)*, ed. and trans. by G. Bos (London: Kegan Paul, 1997).

Ibn Ridwān, *Medieval Islamic Medicine: Ibn Riḍwān's Treatise 'On the Prevention of Bodily Ills in Egypt'*, trans. with an introduction by M. W. Dols, Arabic text ed. by A. S. Gamal (Berkeley: University of California Press, 1984).

Maimonides, *On Rules Regarding the Practical Part of the Medical Art*, ed. and trans. by G. Bos and Y. T. Langermann (Provo, Utah: Brigham Young University Press, 2014).

Qustā Ibn Lūqā, *Qusṭā Ibn Lūqā's Medical Regime for the Pilgrims to Mecca (The Risāla fī tadbīr safar al-ḥajj)*, ed. and trans. by G. Bos, with commentary (Leiden: Brill, 1992).

Yaʿqūb ibn Is'hāq al-Isrāʾīlī, *Yaʿqūb ibn Isḥāq al-Isrāʾīlī's 'Treatise on the Errors of the Physicians in Damascus': A Critical Edition of the Arabic Text together with an Annotated English Translation*, ed. by O. Kahl, *Journal of Semitic Studies* Suppl. 10 (Oxford: Oxford University Press, 2000).

Related topics

Bosworth, C. E., *The New Islamic Dynasties: A Chronological and Genealogical Manual* (Edinburgh: Edinburgh University Press, 2004).

Chipman, L., *The World of Pharmacy and Pharmacists in Mamluk Cairo* (Leiden: Brill, 2010).

Hill, D. R., *Islamic Science and Engineering* (Edinburgh: Edinburgh University Press, 1993).

Hinz, W., *Islamische Masse und Gewichte: umgerechnet ins metrische System* (Leiden: Brill, 1970); Engl. tr. by M. I. Marcinkowski, as *Measures and Weights in the Islamic World:* (Kuala Lumpur: IIUM, ISTAC, 2003).

Montgomery, S. L., and A. Kumar, *A History of Science in World Cultures: Voices of Knowledge* (London: Routledge, 2016).

CHRONOLOGY

*c.*490 BC The Buddha is probably active in present-day Nepal and India.

*c.*430 BC Hippocrates (Ch. 4, no. 1) is active in Greece.

399 BC Socrates (Ch. 4, no. 5) dies in Athens.

347 BC Plato (Ch. 4, no. 6) dies in Athens.

323 BC Alexander the Great dies at Babylon following his conquest of the Persian Empire.

322 BC Aristotle (Ch. 4, no. 7) dies at Chalcis, Greece.

202 BC Han dynasty begins in China, lasting until 220.
The Silk Road routes are formally established for trade between China and the West.

AD 14 Augustus, founder of the Roman Empire, dies at Nola, Italy.

*c.*65 Dioscorides of Anazarbus (Ch. 4, no. 3) is active.

*c.*100 Rufus of Ephesus (Ch. 4, no. 2) is active.

*c.*216 Galen (Ch. 5) dies in Rome.

224 Ardashir I founds the Sasanid dynasty in Persia, lasting until 651.

*c.*330 The kingdom of Aksum (*c.* 100–940, in present-day Eritrea and Northern Ethiopia) adopts Christianity, the foundation of the Ethiopian Orthodox Church.

*c.*500–750 School of Alexandria (Ch. 6) is active.

*c.*522 Arabs and Byzantines obtain silk worm eggs, beginning sericulture and silk manufacture.

565 Justinian I, Roman emperor, dies in Constantinople.

579 Khusraw (Chosroes/Kisrā) Anūshirwān, twenty-first and most famous Sasanid ruler, dies at Ctesiphon, Iraq.

603–28 The great Roman-Persian War occurs.

618 Tang dynasty begins in China, lasting until 906.

622 The Prophet Muhammad emigrates (the Hijrah) from Mecca to Medina, the start of the Islamic calendar.

632 The Prophet Muhammad dies at Medina.
The reigns of the four 'rightly guided' caliphs start.

634–5 al-Hārith ibn Kaladah (Ch. 7, no. 1) dies.

642 Alexandria is captured by Arab forces.

661 Umayyad caliphate is established in Damascus, lasting until 749.

711 Islamic conquest of Spain is underway.
Arab armies reach the Indus.

732 Muslims are defeated at the Battle of Tours by Charles Martel.

749 Abbasid caliphate established, centred in Iraq, lasting until 945 (nominally until 1258).

751 Paper-making is said to be introduced to Islamic lands when Muslim soldiers capture Chinese paper-makers at the Battle of Talas, on the border of present-day Kazakhstan and Kyrgyzstan.

762 Baghdad (Madīnat al-Salām, 'City of Peace') is founded by the caliph al-Mansūr.

769 Jurjīs (ibn Jibrīl ibn Bukhtīshūʿ) (Ch. 8, no. 1) dies in Gondeshapur.

786–809 Reign of Hārūn al-Rashīd.
Sālih ibn Bahlah al-Hindī (Ch. 12, no. 3) is active.

Late eighth–tenth century Greek-to-Arabic translation movement is underway in Baghdad.

800 Charlemagne is crowned emperor in Rome.

813–33 Reign of Hārūn al-Rashīd's son, al-Maʾmūn, an important promoter of the study of philosophy and the sciences, and sponsor of translations.

857 Yūhannā son of Māsawayh (Ch. 8, no. 11) dies in Baghdad.

859 The madrasah, later university, of al-Qarawiyyīn is founded in Fez, Morocco.

870 Bukhtīshūʿ son of Jibrīl (Ch. 8, no. 3) dies in Iraq.
Yaʿqūb al-Kindī (Ch. 10, no. 1) dies shortly thereafter in Iraq.

873 Hunayn ibn Is'hāq (Ch. 8, no. 12) dies 873 or 877 in Iraq.

*c.*878 Ibn al-Dāyah, author of *Accounts of physicians*, dies in Egypt.

901 Thābit ibn Qurrah (Ch. 10, no. 2) dies in Baghdad.

909 Fatimid caliphate is established, lasting until 1171.

925 Abū Bakr al-Rāzī (Rhazes) (Ch. 11, no. 1) dies in Rayy, Persia.

*c.*932 Is'hāq al-Isrāʾīlī (Ch. 13, no. 2) dies in Egypt.

942 Sinān ibn Thābit (Ch. 10, no. 3) dies in Baghdad.

949 The Būyid ruler ʿAdud al-Dawlah, founder of the ʿAdudī hospital, begins his forty-year rule in Baghdad.

960 Song dynasty begins in China, lasting until 1279.

969 New Cairo is founded by the Fatimids close to Fustat (Old Cairo).

*c.*970 al-Azhar madrasah, later a university, is founded in Cairo.

*c.*994 Ibn Juljul dies in Spain.

996 al-Hākim, Fatimid ruler of North Africa and Egypt, begins twenty-five-year reign.

1037 Ibn Sīnā (Avicenna) (Ch. 11, no. 2) dies in Persia.

1039 Ibn al-Haytham (Alhazen) (Ch. 14, no. 5) dies in Egypt.

*c.*1048 al-Mubashshir ibn Fātik is active.

1057 The Nizāmiyyah law college is founded in Baghdad.

1058 ʿUbayd Allāh son of Jibrīl (Ch. 8, no. 6) dies after 1058 in Baghdad.

1066 Ibn Butlān (Ch. 10, no. 6) dies in Syria.

William I 'the Conqueror' begins his rule in England.

1068 Ibn Ridwān (Ch. 14, no. 7) dies in Egypt.

1085 Christians capture Toledo.

1088 University of Bologna, considered the oldest existing European university, is founded.

1096 Teaching is known to exist at Oxford, from which the university later develops and is founded in the mid-twelfth century.

1099 Crusaders capture Jerusalem during the First Crusade (1096–9).

1121 Almohad caliphate is established in the Maghrib, lasting until 1269.

1131 Abū l-ʿAlāʾ ibn Zuhr (Ch. 13, no. 9) dies in Cordova.

1154 The great Nūrī hospital in Damascus is founded by Nūr al-Dīn ibn Zangī.

Ibn al-Tilmīdh (Ch. 10, no. 9) dies in Baghdad 1154 or 1165.

1162 Abū Marwān ibn Zuhr (Avenzoar) (Ch. 13, no. 10) dies in Seville.

1169 Saladin, founder of the Ayyubid dynasty and ruler of Egypt, begins his twenty-five-year reign.

1179 Ibn Abī Usaybiʿah's father, Sadīd al-Dīn al-Qāsim, is born in Cairo.†

† It should be noted that some dates given by Ibn Abī Usaybiʿah, particularly birth dates, are probably 'best guesses' and should not be considered definitive.

1183 Ibn Abī Usaybiʿah's uncle, Rashīd al-Dīn ʿAlī ibn Khalīfah (Ch. 15, no. 18) is born in Aleppo.

1187 Jerusalem is re-captured by Muslims.

1198 Ibn Rushd (Averroes) (Ch. 13, no. 13) dies in Marrakesh.

Ibn Jumayʿ (Ch. 14, no. 9) dies in Egypt.

*c.*1200 Ibn Abī Usaybiʿah is born, in Damascus or Cairo. His grandfather returns to Damascus from Cairo with Ibn Abī Usaybiʿah's father and uncle.

1204 Maimonides (al-Raʾīs Mūsā ibn Maymūn) (Ch. 14, no. 11) dies in Fustat (Old Cairo).

1213 Ibn Abī Usaybiʿah's father enrols in the service of the Ayyubid sultan, al-Malik al-ʿĀdil, in Damascus.

1219 Ibn Abī Usaybiʿah's uncle dies in Damascus, aged 36.

1227 Genghis Khan, Mongol conqueror, dies in Yinchuan, China.

1230 al-Dakhwār (Ch. 15, no. 17) dies in Damascus.

Approximate start of the Mali Empire in West Africa, capital at Timbuktu.

1231 ʿAbd al-Latīf al-Baghdādī (Ch. 15, no. 12) dies in Baghdad.

1233 Ibn Abī Usaybiʿah first meets Asʿad al-Dīn (Ch. 14, no. 17) in Damascus.

He meets Ibrāhīm son of Maimonides (Ch. 14, no. 12) in Cairo at the Nāsiri hospital in 1233 or 1234.

1236 Cordova falls to Christians.

1237 Ibn Abī Usaybiʿah meets the religious scholar, Abū Marwān al-Bājī, in Damascus, acquiring much information about Andalusian physicians.

He also becomes private physician to the emir ʿIzz al-Dīn al-Muʿazzamī, the ruler of Sarkhad (modern Salkhad), Syria, under the Ayyubids.

*c.*1243 Ibn Abī Usaybiʿah writes the first version of his book, in Sarkhad.

1245 Ibn Abī Usaybiʿah gives a dedicated copy of his book to (and composes a poem for) Amīn al-Dawlah (Ch. 15, no. 16), at the vizier's request.

1248 Ibn al-Qiftī, author of a book on physicians and philosophers, *Informing scholars of the reports on sages*, dies in Aleppo.

Seville falls to Christians.

1250 Mamluk rulers of Egypt and Syria begin rule, lasting until 1517.

Frederick II dies in Italy.

1250–1 Amīn al-Dawlah (Ch. 15, no. 16) is executed in Cairo.

1251 Ibn Abī Usaybiʿah's father dies in Damascus, aged 72.

1258 Baghdad falls to the Mongols, end of the Abbasid dynasty in Baghdad.

1260 Mongol advance is halted by Egyptian Mamluks.

1269/70 Ibn Abī Usaybiʿah produces the final version of his book.

1270 Ibn Abī Usaybiʿah dies in Sarkhad, aged about 70.

Eighth Crusade, 1270–2.

Louis IX of France dies in Tunis.

1281 Ottoman dynasty is established in Anatolia, lasting until 1924.

1291 Crusaders lose Acre and abandon Crusader States.

c. 1325 The Renaissance begins in Italy.

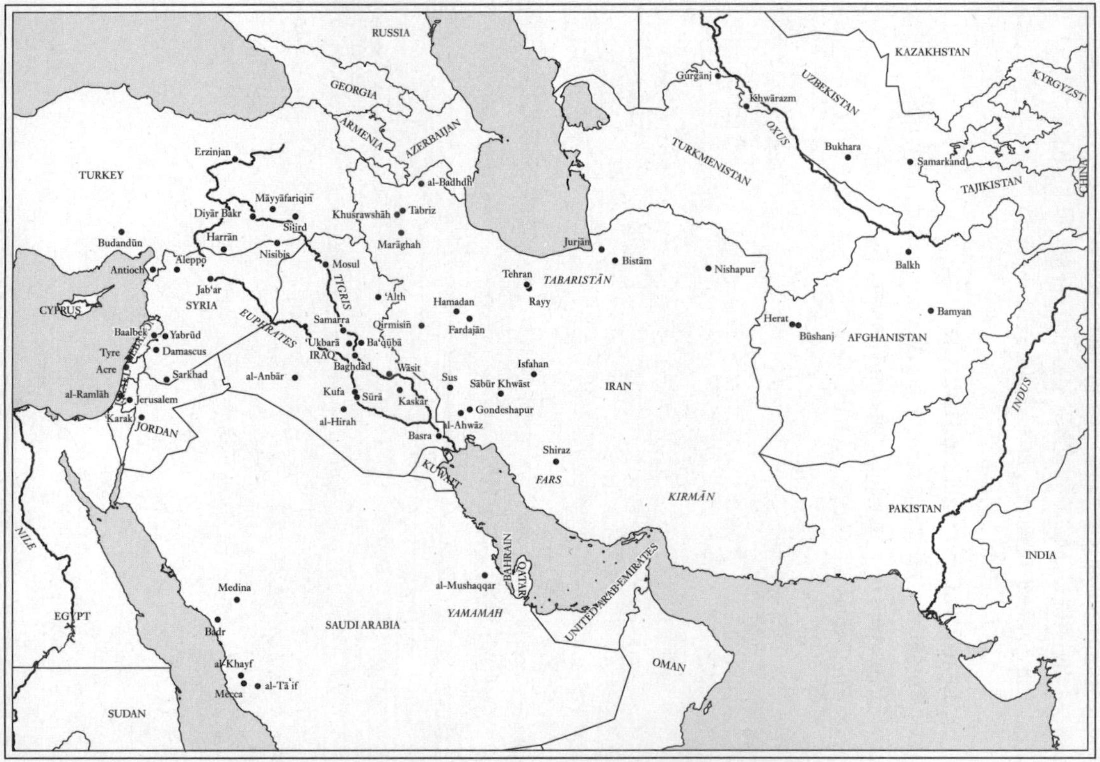

MAP 1 Map of the Islamic World circa 1200

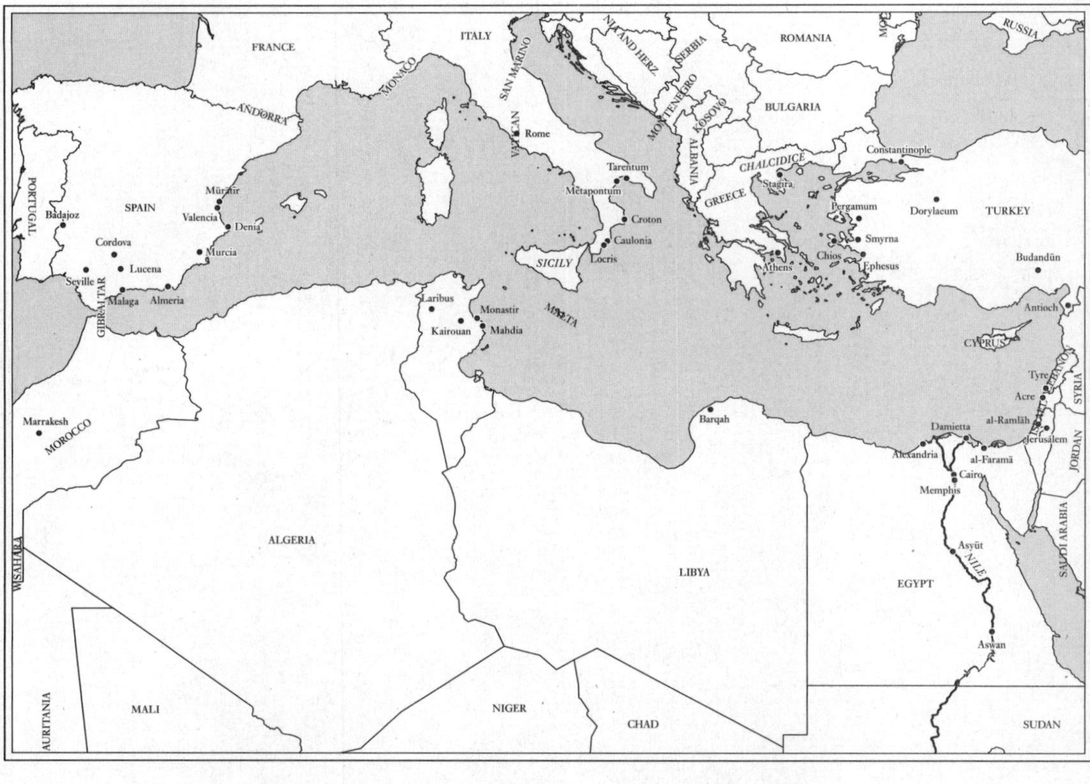

MAP 2 Map of the Islamic World circa 1200

ANECDOTES AND ANTIDOTES. A MEDIEVAL ARABIC HISTORY OF PHYSICIANS. A NEW TRANSLATION

PREFACE OF IBN ABĪ USAYBIʿAH

*The preface is included in its entirety. Here Ibn Abī Usaybiʿah sets out his reasons for writing the book and what he hopes to achieve. He points out the importance of good health and thus of the art of medicine, reminding the reader that knowledge of human bodies (*ʿilm al-abdān*) is second only to religious knowledge (*ʿilm al-adyān*).* He also explains that he was prompted to compile the book because, despite the importance of the art of medicine through the ages, no one had previously written a comprehensive history of its many great practitioners. He states that he will include material, both academic and entertaining, and emphasizes the close relationship with the learning and teaching of past physicians felt by himself and his contemporaries.*

Though short, this is the only section where Ibn Abī Usaybiʿah does not quote from any other source, and in some ways his voice can be heard most clearly here, speaking directly to the reader.

In the name of God, merciful and compassionate. None but He can grant me success.

Praised be God, He who has dispersed the world's peoples and will gather them at the Resurrection, the creator of life and healer of the sick, He who graciously bestows abundant blessings but warns those who disobey Him that painful punishment and retribution await them, He who with consummate art brings creatures into being out of nothingness, who ordains diseases and reveals remedies with the most perfect skill and utmost sagacity. I testify that God is one, speaking in all sincerity, as one who pays what is due, avoiding the sins that come of words lightly spoken and subsequently regretted. And I testify that Muhammad is His servant and His apostle whom He sent to bring the divine message to all, Arab and non-Arab alike, and who dispelled the gloom of darkness with the shining light of his mission, destroyed the proud and tyrannical with his weapon of the miraculous Qur'an, and extirpated the disease of polytheism with the incontrovertible evidence of his prophethood. May God bless him and grant him peace forever, as long as the lightning flashes and the rain falls, together with all his exalted and noble house, his companions, who strove to follow his example, and his wives, the mothers of the believers, who are free from all blemish. May God favour them all with honour and dignity.

Turning to my subject, the art of medicine is among the noblest of professions and the most profitable of occupations. Its excellence is attested in scripture and the provisions of law, to such an extent that knowledge of the human body is deemed to be second in importance only to knowledge of religion.

Wise men have said that there are two types of aspiration: to what is good, and to what is pleasant. Now, a man cannot attain either of these unless he is in good health, for what is pleasant is obtained from this world, while what is good leads ultimately to our hope of Paradise; in both cases continuing good health and a robust constitution are essential, and those can be secured only through the art of medicine, for it maintains health and restores it when it has been lost.

Inevitably, then, since the art of medicine is such an exceptionally noble profession, and since there is such need for it everywhere and at all times, particularly great attention has always been paid to it, and there has always been the keenest and most intense interest in acquiring an understanding of its rules, both general and particular.

Consequently, from the time when the art of medicine first arose down to our own day, there have been many who have practised it and sought to investigate its fundamental principles and learn about it. These have included a number of very great physicians, men of the utmost skill and renown, whose excellence is proverbial and whose outstanding capacity and nobility have been recognized by successive generations, and whose compositions testify to their ability. To my knowledge, however, none of these masters of the art of medicine and no one with a thorough knowledge of that art has ever written a comprehensive book dealing with physicians through the ages and recounting their history in a coherent fashion.

I have therefore determined to relate in this volume a number of reports and accounts that are told of the most distinguished of the physicians, both in antiquity and in more recent times, with the information regarding them arranged sequentially in chronological order. I also propose to record some of the remarks attributed to them and stories about them that have come down to us, together with entertaining anecdotes about them and the discussions in which they engaged. Lastly, I shall list the titles of some of their books. I hope by this means to enable the reader to glean some understanding of the learning that Almighty God enabled these men to acquire and the copious natural talent and intelligence that He bestowed upon them. Many of them, though they lived long ago and their days are past, stand in much the same relation to us as a teacher does to his pupil, or a recipient of kindness to his benefactor, owing to the advantages that we derive from their written works and the benefits found in the material they gathered together in their books.

I have also included accounts of a number of wise men and philosophers who were interested in the art of medicine and gave attention to it. I have related the main events of their lives, giving anecdotes about them and listing the titles of their works. Each of these accounts appears in its appropriate place, depending on their time period and their eminence.

After dealing with the art of medicine, I propose to discuss great scholars in other fields of knowledge at some length, God willing, in another book to be entitled *The outstanding personalities of all nations and reports of those endowed with wisdom.*

For the time being, however, I determined that it would be desirable to compose the present work, which is entitled *The best accounts of the classes of physicians* and is divided into fifteen chapters. It is a contribution to the library of the great, learned, and just minister, the paragon of masters, lord of viziers, wisest of wise men, chief of learned men, glorious sun of religion, Amīn al-Dawlah, may God prolong his happiness and grant him his wishes in this world and the next.* I pray to Almighty God for assistance and for success, for these are in His hand and are His to grant if He will.

CHAPTER 1

THE ORIGIN AND FIRST APPEARANCE OF THE ART OF MEDICINE

Ibn Abī Usaybiʿah discusses the various theories on how the art of medicine may have arisen, including divine revelation, dreams, coincidence, observation, and instinct, both animal and human. Many specific examples are given, some touching on the early development of pharmacology. Ibn Abī Usaybiʿah refers to various authors who have already addressed this subject but also interjects his own opinions, concluding that divinely inspired human ingenuity, combined with both experience and chance, have given rise to the art of medicine. He finishes by briefly discussing the transmission of medical knowledge, and mentions Asclepius, the first known physician, who will be discussed in the following chapter.

THE investigation of this subject is difficult, for a number of reasons. In the first place, it is remote in time, and the study of antiquity—especially this particular aspect of it—is no easy matter. In the second place, we find that the ancients, eminent authorities, and persons of sound judgement, have not always agreed with one another in the matter, so that we cannot look to them for clear-cut, definitive conclusions on which we can rely. In the third place, those who have discussed the art of medicine have belonged to various sects and have differed markedly in their accounts, depending on the particular bias of each of them, and this complicates the task of determining which version is the true one.

Galen, in his commentary on Hippocrates' *Book of oaths*,* stated that the task of determining which of the ancients first discovered the art of medicine was not an easy one. Let us begin by reviewing Galen's account, supplemented with my own remarks, in an effort to identify these various divergent views.

There are two basic schools of thought about the art of medicine: one that holds it to have existed from all eternity, and a second that holds it to have been created.

The proponents of the theory of the creation of medicine are divided in their turn into two schools of thought. One of these holds that the art of medicine must have been created at the same time as mankind, in view of the fact that it is beneficial to the human race.

The other school, which has the majority of adherents, holds that the art of medicine was invented later. Those who take this position are further subdivided into two factions. One of them considers that it was an inspiration from Almighty God to mankind. The second faction considers that medicine is a human invention. However, they disagree on the question of where and how it was invented.

Among those who hold that the art of medicine comes from God, there are some who consider that it was inspired in dreams, alleging as evidence the fact that some people have dreamed of medicines which they subsequently used while awake and found that they cured serious ailments.

There is a school of thought that maintains that God inspired the art of medicine by practical experience, and that it grew and developed further as time went on. By way of evidence, they relate that there was once a woman in Egypt who suffered from great sadness and anxiety, with outbursts of red-faced rage, and in addition was afflicted with weakness of the stomach, bad humours in the chest, and retained menstruation. But it happened that she consumed elecampane on numerous occasions, feeling a craving for it, and soon she found that all her ills had vanished and that she had been restored to health. It subsequently appeared that patients with any of those symptoms also found relief by using elecampane.* This led people to experiment with other remedies.

Those who hold that God created the art of medicine argue that human reason cannot possibly have devised so sublime a science. Galen held this view:

'For my own part, I consider it most plausible and most fitting to hold that it was God, the blessed and exalted, who created the art of medicine and inspired man with it. For such a majestic science cannot have been conceived by the mind of man; it can only have been God, blessed and exalted is He, who was its author, for none but He had the power to create it. I take this view because I do not regard medicine as being inferior to philosophy, and it is generally accepted that philosophy originated from God, the blessed and exalted, who then inspired man with it.'

I have read in a work by Ibn al-Mutrān,* entitled *The gardens of physicians and meadows of the intelligent*, as follows:

To maintain that the art of medicine could not have been devised by human intelligence is an error, for in that case, what of the intelligence of those who devised arts more lofty than medicine?

We may reasonably assume that at the very beginning of human history there was one person who was in need of the art of medicine, just as people are in need of it in our swarming, crowded world of the present day. Perhaps he felt bloated, his eyes were red, and, in general, he was experiencing the symptoms of an excess of blood. He did not know what to do, but something he ate brought on a nosebleed, whereupon he felt better. He took note of this, and when the condition recurred he gave himself a blow on the nose, brought on a flow of blood, and again was relieved. Having definitely established the efficacy of the treatment, he taught it to his offspring and their children. In the course of time, as the art of medicine became further refined, it came to be standard practice to let blood by opening a vein.

It is told that there was once a man who bought some fresh liver from a butcher and took it home. Being obliged to leave for some errand, he placed the liver on some plant leaves that were spread out on the floor. When he returned for the liver after having completed his errand, he found that it had deliquesced and liquefied into blood. He took the leaves and identified the plant, and then began to sell it as a lethal drug, until he was found out and sentenced to death.

I—Ibn Abī Usaybiʿah—have been informed that many kinds of herbs grow at the foot of the mountain on the other side of which the town of Siʿird stands, near the esplanade. A poor elderly man from the town once came to that place and lay down amid the vegetation to have a nap. As he lay sleeping, a group of people came by and found him there. They noticed that there was blood under him, seeping from his nose and from the direction of his anus, and so they awakened him. All concerned were puzzled, until they realized that the bleeding was due to a plant on which the old man had been lying.

My informant told me that he had gone to the place himself and seen the plant in question. He described it as similar to chicory,* but taller, and bitter to the taste. 'I have often seen a person put it to his nose and sniff several times,' he said, 'and it produces bleeding from the nose immediately.' Such was his account, but I have been unable to determine whether the plant in this case is the one referred to by Galen or a different one.

Ibn al-Mutrān's narrative concludes:

Galen wrote a work on the invention of the various arts* and in essence says nothing more than I have.

I—Ibn Abī Usaybiʿah—continue:

As we can see, there is a very large measure of divergence and disagreement, making our search for the origin of the art of medicine difficult in the extreme. However, if an intelligent man applies his wits to the matter, he will undoubtedly conclude that the origins of the art of medicine must have been, for the most part, broadly consistent with the above discussion.

At all events, I hold that the art of medicine is essential to human beings and is invariably a feature of their society wherever they are found. However, there may be differences among different societies.

At this point, I propose to proceed with a discussion of the origins of the art of medicine to the best of my ability.

In the first place, humanity may have obtained something of it from the prophets and friends of God, peace be upon them, owing to divine assistance granted to them by Almighty God.

Ibn ʿAbbās* relates that the Prophet, God bless him and keep him, said, 'When Solomon, the son of David, prayed, he would see a tree growing in front of him, and would ask it, "What is your name?" If it was a crop tree, it would be planted, and if it was a medicinal tree, the fact would be noted.'

The Zoroastrians* allege that Zarathustra, whom they call their prophet, gave them twelve thousand volumes bound in buffalo hide dealing with four sciences, including one thousand that were devoted to the art of medicine.

In the second place, some part of the art of medicine may have come to mankind through true dreams. Galen relates in his work *On blood-letting** that he learned to open an artery from a dream. 'Twice,' he says, 'I dreamed of being told to open the artery* between the index finger and the thumb of my right hand. When I awoke, I did so, letting the blood flow until it stopped naturally, for so I had been instructed in the dream. The quantity of blood lost was less than a *ratl*, and as a result, a long-standing pain that I had first felt as a youth, at the point where the liver touches the diaphragm, vanished at once.'

Another example of a true dream of that kind is to be found in a story about one of the caliphs of the West, who had been afflicted with a long illness for which he had been treated with many kinds of medication, without success. But then one night he dreamed of the Prophet, God bless him and keep him. The caliph told him of his ailment, and the Prophet, God bless him and keep him, said to him, 'Anoint yourself with neither and eat of nor, and you will be cured.' When the caliph awoke, he recalled his dream with astonishment, but could not understand what it meant. He consulted experts in the interpretation of dreams, but none of them could explain it, except one, who said, 'O Commander of the Faithful,* the Prophet, God bless him and keep him, has instructed you to anoint yourself with olive oil and to eat olive oil in order to be cured.' When the caliph asked him, 'How did you know that?' he replied, 'From the words of God mighty and glorious in the Qur'an: «*From a blessed tree, an olive neither of the east nor of the west whose oil would well-nigh give light even though no fire had touched it*».'*

The following account is taken from a copy of the commentary by Ibn Ridwān,* written in the author's own hand:

I had suffered for years from an excruciating headache resulting from an excess of blood in the blood vessels of my head. I bled myself, but with no result, even after I had repeated the treatment a number of times. Then I dreamed that Galen came to me and told me to read his *Method of healing* to him, so I read seven chapters of it aloud. When I reached the end of the seventh chapter, he said, 'You have forgotten your headache,' and instructed me to have the back of my head cupped. Then I awoke, and when I had myself cupped at that point, I was cured of my headache forthwith.

ʿAbd al-Malik ibn Zuhr says:

My vision had been strained as a result of a critical bout of vomiting, and subsequently my pupils suddenly became dilated. I was distressed because of this, but then I dreamed of a man who had been a medical practitioner during his life. He told me, in my dream, to put rose syrup into my eyes. At that time I was a student, and while I had learned a good deal about the art of medicine, I had had little practical experience. Accordingly, I consulted my father about the matter. He thought for some time, and finally said, 'Do as you were told to do in your dream.' I did so, and the result was so satisfactory that as of the time of writing this book, I have continued to use that treatment as a means of strengthening patients' eyesight.

There have been many such examples of medical treatments that have been discovered in true dreams.

In the third place, medical knowledge has sometimes been acquired by mere chance or serendipity. This was the case with Andromachus the Younger, who discovered the virtues of viper's flesh as an ingredient in theriac.* This, in his own words, is one of the events that put the idea into his mind and thus led him, quite accidentally, to that discovery:

The first experience involved some ploughmen who worked on one of my estates in the place known as Būrnūs. The estate in question was about two parasangs distant from the place where I lived, and I would go out there every morning to oversee the work, returning home when the men had finished for the day. I made a habit of loading some food and drink for the men on the donkey that my slave-boy rode, in order to keep them in good heart and fit for their work. One day, going out there as usual, I brought them some wine in a big green earthenware jar, unopened, with its top sealed with clay, as well as some food. They ate the food, and then they turned to the jar of wine and opened it. One of them took a mug and dipped it into the wine to take a drink, and there in the jar was a viper the flesh of which had become macerated. The men would not drink the wine after that, 'but,' they said, 'here in this village there is a man who is a leper, whose state is so terrible that he wishes to die. We shall give him some of this wine to drink, and he will die. That will be a meritorious thing to do, for we shall have released him from his suffering.' Accordingly, they went to his home with food and drink, and gave him some of the wine, quite sure that he would die before the day was out. At nightfall, his whole body became greatly swollen, remaining so until the morning. The outer skin then sloughed off, so that the red inner skin appeared; in due course it hardened, and he was cured. He lived for a long time after that, in excellent health, ultimately dying a natural death from cessation of innate heat. This showed me that viper's flesh is useful in treating cases of severe illnesses and long-standing bodily disorders.

The following account is a further illustration of the same phenomenon—that is, the acquisition of medical knowledge by mere chance.

There was once a man in Basra who suffered from such a severe case of dropsy that his family despaired of his life. They had consulted doctors who had recommended many remedies, but to no avail, until finally they became convinced that his case was hopeless. He overheard them saying as much, and said to them, 'In that case, you may as well let me indulge myself in the pleasures of life while I still can and eat whatever I like, instead of killing me with these diets!' 'As you wish,' they replied. Accordingly, he took to sitting by the door of the house, and whenever a food vendor came by, he would purchase and eat some of his wares.

One day a man selling fried locusts passed that way, and the patient bought a large quantity of them. When he had eaten them, he began to pass such copious quantities of yellow fluid that for three days he was near death. Finally, however, the flow stopped, and it was evident that the illness had left him. His strength returned, and he was cured, and he went about attending to his business, perfectly well.

One of the physicians who had been consulted in the case happened to see him and, surprised at his recovery, asked him about it, so the former patient told him what had happened. 'Locusts on their own could not have had that effect,' said the physician, 'from whom did you buy them?' The man told him, and the doctor went to see the locust vendor and asked him where he had caught the locusts. The vendor took him to the place, and there they found locusts feeding amid a dense growth of the shrub known as mezereon.* This plant is an effective remedy for dropsy: the administration of one dirham's weight to the patient causes a copious flow of fluid that is virtually uncontrollable. However, the treatment is a risky one, and consequently it is seldom prescribed. Evidently the digestion of mezereon within the bodies of the locusts, combined with the subsequent cooking, had attenuated its effect, so that when the patient ate the fried locusts, he recovered.*

In the fourth place, medical knowledge has sometimes been acquired from the observation of animals and imitation of them. Abū Bakr al-Rāzī* gives as an example the swallow, which, upon finding its nestlings affected with jaundice, goes to look for a so-called 'jaundice stone',* a type of small white stone that it knows of. The bird brings back one of these stones and places it in its nest, whereupon the young birds recover. According to al-Rāzī, anyone requiring treatment for jaundice need only find some swallow nestlings and daub them with saffron. The parent bird, thinking its young are suffering from jaundice, flies off and returns with a stone. The patient then takes the stone and hangs it around his neck, thereby obtaining relief.

Another example is afforded by the female eagle, which sometimes has great difficulty laying her eggs, so much so that her life may be endangered. The male, seeing this, flies off in search of a stone known as the 'rattle', because when shaken it produces a rattling sound. He places it in the nest, where it facilitates the emergence of his mate's eggs. These stones serve the same purpose for human women in cases of difficult childbirth. When one of them is broken, nothing is found inside it, but each fragment, when shaken, produces the same rattling sound as the original stone. They are commonly referred to as 'eagle stones', because their beneficial effect was inferred from the behaviour of eagles.*

In the fifth place, some elements of medical knowledge may have been acquired through divinely implanted instinct, as we see in the case of many animals. We observe that cats eat 'catnip' in the springtime, and if they cannot find any, they will nibble the palm fronds of a broom. As everyone knows, that sort of thing is not what they usually eat, but they do it instinctively, for God has made this behaviour a source of good health for them. When they eat green stuff, they vomit up the various humours that have accumulated within their bodies. They go on doing this until they feel that their accustomed natural good health has returned, and then they revert to their usual diet.

Oleander* is said to be injurious to livestock that browse on it in the spring. An animal that has done this loses no time in grazing on a herb which is an antidote* to oleander and will counteract its toxicity. The truth of this is to be seen from something that happened quite recently. A man reported that while on his way to Karak he had stopped for the night at a place where many oleanders grew. He and another traveller had made camp there, with oleanders all around them. The other traveller's slaves had tethered their pack-animals nearby, and they began to browse on whatever they could reach, including the oleanders. The narrator's slaves, however, had neglected to tie up their party's animals who wandered about, whereas the other animals were unable to stray and had to stay where they were. In the morning, the narrator's animals were found to be in good health, while all the others were dead.

Dioscorides says in his *De materia medica* that when one of the wild goats of Crete has been shot with an arrow that remains stuck in its body, it eats a variety of mint, which causes the arrow to work its way out of the flesh, leaving the animal unhurt.

Many illustrative examples of this kind could be mentioned. Now, if animals, which lack the ability to reason, know instinctively what is beneficial and useful to them, how much better equipped to do so are human beings, who are endowed with reason and intelligence and are capable of distinguishing between right and wrong, and as such stand at the apex of the animal kingdom. Here we have the most telling evidence in support of those who believe that the art of medicine was inspired by God, glorified be He, as a gift to His creation. In brief, it may well be that experience and chance together account for much of the knowledge of the art of medicine that humanity originally acquired. As time went on, that knowledge proliferated, buttressed by analogy and observation as interpreted by keen minds. Scholars in that way came to formulate general rules and principles that could serve for purposes of investigation and instruction.

In addition, I maintain, as stated earlier, that the origins of the art of medicine need not necessarily be ascribed to one specific place and no other. Nor can any one people claim exclusive credit for it. A wide variety of treatments have been devised, and any given people naturally tend to use treatments with which they are familiar.

Tradition transmitted through the generations, including in particular the accounts we possess from Galen and others, informs us that when Hippocrates saw that the art of medicine was about to perish and that knowledge of it was fading among the descendants of Asclepius (of whom Hippocrates himself was one), he took measures to prevent that from happening by disclosing and propagating the art among other peoples, strengthening it, promoting it, and making it known by recording it in books. That is why many have taken Hippocrates to be the inventor of the art of medicine and the first to compose works on the subject. This is inaccurate, as we know from generations of tradition. In fact, he was the first of the descendants of Asclepius to write such books with the intent of teaching the art to anyone capable of learning it. Physicians since his time have done the same, for the tradition has been maintained to this day.

Asclepius was the first to discuss some aspects of medicine, as we shall see in the following pages.

CHAPTER 2

PHYSICIANS WHO PERCEIVED THE RUDIMENTS OF THE ART OF MEDICINE AND INITIATED THE PRACTICE OF THAT ART

This chapter covers the semi-divine Asclepius, the originator of medicine and considered by Ibn Abī Usaybiʿah to have been a real historical figure. He is important as he bridges the gap between the divine and human, and his place in the hierarchy of physicians is clearly stated. He is the founder of the empirical method, believed by Galen to be key to being a physician, and he is the ancestor of a number of philosophers. He is also the pupil of Hermes—the first of several mythological or long-dead figures to be given a fascinatingly detailed physical description.

ASCLEPIUS

AMONG the philosophers and physicians of antiquity, there is a wide measure of agreement that Asclepius* was the first known physician. He is credited with having been the first to discuss aspects of medicine according to the empirical method. Asclepius was Greek.

We read in the *Annotations** of the eminent scholar, al-Sijistānī al-Mantiqī:

Asclepius, son of Zeus,* was said to have been of supernatural birth. He was the founder of the art of medicine and the father of most philosophers. Euclid was descended from him, as were Plato, Aristotle, Hippocrates, and most of the Greeks. Hippocrates was a sixteenth-generation offspring from his line.

I—Ibn Abī Usaybiʿah—add here:

The Arabic translation of the name Asclepius is 'prevention of dryness', but some have asserted that the name is derived from Greek words meaning 'beauty' and 'light'.* As far as can be determined, Asclepius was gifted, quick of understanding, deeply interested in the art of medicine, and assiduous in pursuit of it. In addition, there were a number of occasions when good luck helped him on his way, and he obtained insight thanks to inspiration from God, mighty and exalted is He.

Asclepius is said to have discovered the science of medicine in one of the temples in Rome, known as the Temple of Apollo, which was dedicated to the sun. According to some accounts, it was Asclepius himself who had that temple built, and it was known as the Temple of Asclepius.

It is said that the Temple of Asclepius was a shrine in the city of Rome in which there was an image that answered questions put to it by petitioners. The image was reputed to have been the work of Asclepius in remote antiquity.

There are numerous passages in the works of Galen stating that the art of medicine as practised by Asclepius was divine in nature. 'It was as far superior to our own medicine,' he says, 'as ours is to that of the remedies you find in the streets.'*

Other authorities have asserted that Asclepius enjoyed such esteem among the Greeks that people afflicted with illness would visit his grave in the hope of being restored to health.

Plato states that Asclepius was willing to treat anyone with an ailment that could be cured. In a case of a fatal disorder, however, he would not prolong a life that was of no use either to the patient himself or to others; he would not treat such a patient.

In the *Choicest maxims and best sayings* of al-Mubashshir ibn Fātik* we read:

This Asclepius was a pupil of Hermes, and travelled with him. This was Hermes the First. The initial letter is not pronounced; it is thus 'Ermes', which is the name of Mercury. Among the Greeks he was known as Trismegistus, the Arabs know him as Idrīs, while to the Hebrews he is Enoch. He was born in Egypt, in the city of Memphis, and (al-Mubashshir adds) he lived here on earth for eighty-two years.

Hermes, peace be upon him, had a dark complexion. He was tall, bald, handsome, thick-bearded, a well-made man, with long arms and broad shoulders; he was heavy-boned, carrying no spare flesh, and had flashing black eyes. His speech was slow and infrequent. He was controlled in his movements, and tended to look at the ground when he walked. He was often pensive. At the same time, he could be short-tempered, with a frowning demeanour, and would shake his forefinger when talking.

Another source tells us that Asclepius was a pupil of the Egyptian Hermes the Third who lived in Egypt following the Flood, and was a physician and philosopher. He was knowledgeable about lethal drugs and harmful animals: he composed a work devoted exclusively to venomous animals. He liked to travel in the country, and came to know his way around the various cities, becoming familiar with their features, as well as the natures of the people who lived there. He is the author of a number of excellent aphorisms on the art of alchemy, also relevant for many other arts, such as glassblowing, the piercing of beads, ceramics, and the like. It was Hermes the Third who had a pupil named Asclepius, who lived in the land of Syria.

Returning now to the subject of Asclepius: he is said to have been able to cure patients who had been despaired of. So impressive were his feats that the people came to think that he could raise the dead. He was celebrated in the works of the Greek poets, who claimed that God, exalted is He, had taken Asclepius up* to Himself as a means of exalting and honouring him, and had made him one of the heavenly host.

John the Grammarian states that Asclepius' life lasted for ninety years, fifty before the divine power was disclosed to him, and forty devoted to learning and teaching.

He was survived by two sons, both of whom were expert in the art of medicine. He forbade them to teach medicine to anyone apart from their own offspring and other relatives; they were not to initiate any outsider into that art. Successors were to be subject to the same restrictions. Specifically, he left instructions that they were to go to live in the midst of the populated regions of the land of the Greeks (by which he meant three islands, one of them being Cos, the island where Hippocrates was born), and were not to reveal knowledge of the art of medicine to outsiders; rather, fathers should teach it to their sons.

Asclepius' two sons* were with Agamemnon when he sailed to conquer Troy. Because of their extensive learning, he held them in high regard and showed them great honour.

We read in a work by Thābit ibn Qurrah:*

'It is said that there were pupils of Asclepius in every region of the world, numbering twelve thousand in all, and that Asclepius taught the art of medicine exclusively through the spoken word, and that men of his line inherited their knowledge of it through successive generations, down to the time of Hippocrates, until degeneration set in. Hippocrates perceived that his family and clan had become reduced in numbers, and there could be no certainty that that knowledge would not be lost forever. Accordingly, he began to set it down in written form, concisely expressed.'

The following remarks are taken from Galen:*

Asclepius was shown as bearded, and his appearance was enhanced by the presence of a thick fringe of hair covering the front part of his head.

Asclepius stands with his garments tucked up, as if to show that every physician must always be a philosopher. Those parts that must be covered for modesty's sake are covered, while the limbs that are used for the art of medicine are revealed.

He holds a staff, and this is to show that by the art of medicine, one may live to be so old that one will need a staff for support. Alternatively, it may indicate that where God, blessed and exalted is He, grants a man benefits, He deems him worthy of bearing a staff. Asclepius' staff is crooked and branched to denote the several disciplines and subdivisions found within the art of medicine.

Carved upon it is a serpent. The serpent has a number of affinities with Asclepius: its vision is keen, and it is constantly awake, never sleeping for an instant. He who pursues the study of the art of medicine must not be distracted by falling asleep; he must be incessantly alert. Or the reason is that the serpent is very long-lived, and those who resort to the art of medicine may also live to a venerable age.

Statues of Asclepius show him with a crown of laurel leaves, for the laurel dispels sorrow, as do physicians. Alternatively, the reason may be that the arts of both medicine and divination were symbolized by a laurel wreath. Also, the laurel tree* possesses curative properties. Wherever one is planted, venomous creatures that may have infested the site will go elsewhere. The laurel seed exerts an action like that of castoreum* when used as an embrocation.

The sculptors of antiquity also put an egg in Asclepius' hand, which represents the universe, as the whole world needs the art of medicine.

It may be appropriate to discuss the animals that are sacrificed in Asclepius' name. It appears that no one has ever offered up a goat, because goat hair is harder to spin than wool, and because frequent consumption of goat meat generates poor chyme* and may cause epilepsy. Moreover, it has a desiccant effect, is coarse in texture, acrid upon the palate, and fosters melancholic blood.

The following are a number of aphorisms and sage remarks uttered by Asclepius and preserved in the work of al-Mubashshir ibn Fātik:

> Have you not been known to curse your current situation, then, in altered circumstances, look back on your former state with regret? Many an event may enrage you when it befalls, yet leave you weeping once it is past.
>
> It is better to do without something than to try to obtain it from the wrong person.
>
> It is surprising to me that a man who will refrain from eating rotten food from fear of being made ill will yet not refrain from sinful behaviour for fear of the life to come.
>
> It becomes a man of religion and virtue to expend effort and resources for the sake of his friend, to display a cheerful countenance and amiability towards a person with whom he is acquainted, to act justly towards his enemy, and to refrain always from unseemly behaviour of any kind.

CHAPTER 3

THE GREEK PHYSICIANS DESCENDED FROM ASCLEPIUS

The text of this short chapter has not been included in this edition. In the chapter Ibn Abī Usaybiʿah follows the chain of learning that descended from Asclepius and his six sons, describing the five (of eight) 'celebrated physicians of antiquity' who followed Asclepius—the first one. They are Ghurus, Minas, Parmenides, Plato the Physician, and Asclepius the Second, a fictional list the first two members of which may hide the philosophers Anaxagoras and Anaximenes. The seventh and eighth, celebrated physicians of antiquity, Hippocrates and Galen, follow in Chapters Four and Five.

The chapter is a rather dry chronology of numbers and names that fill the period between Asclepius and Hippocrates. Ibn Abī Usaybiʿah notes how long each of the five physicians lived, how long he was an apprentice, how long a physician, and how long elapsed between the death of one and the birth of the next. A very brief summary of his philosophy and practice is given, with, sometimes, the names of other physicians who flourished during that time. Finally, the sons or kinsmen are listed, to whom the art of medicine was transmitted, and who transmitted it to their sons or kinsmen, until the time of the next celebrated physician.

This leads on to mention of Hippocrates, who is the first to change the tradition by teaching outsiders, i.e. non-relatives, the art of medicine, and draws the reader into the next chapter.

CHAPTER 4

GREEK PHYSICIANS TO WHOM HIPPOCRATES TRANSMITTED THE ART OF MEDICINE

Ibn Abī Usaybiʿah continues with the Greeks, starting with Hippocrates and moving on to others more famous as philosophers, such as Pythagoras, Socrates, Plato, and Aristotle. These are included both because of the strong link between philosophy and medicine and to satisfy the expectations of his readers, the equivalent of our 'chattering classes' today. Ibn Abī Usaybiʿah believed, as Galen had stated, that every physician must be a philosopher and that a physician must know the works of Aristotle in particular, to be able to use Aristotelian logic to understand and practise medicine properly. Where Asclepius brought empiricism to medicine, Hippocrates and others add rational thought and theory.

1. HIPPOCRATES (*d. c.*375 BC)

LET us begin by reporting some information about Hippocrates himself.

Hippocrates was the seventh of the celebrated great Greek physicians, whose forefather was Asclepius. He was one of the most noble and eminent men of his family. Hippocrates learned medicine from his father and his grandfather. He lived ninety-five years, sixteen of them as a child and student, seventy-six as scholar and master. When Hippocrates looked at the state of medicine, he feared for the survival of this art, for he had seen it disappearing from the majority of places. Only Kos, the city in which Hippocrates lived, kept the art alive thanks to a small number of Asclepius' descendants there.

Hippocrates resolved to spread the art of medicine to all lands and to teach it to all those worthy of learning it lest it disappear. He claimed that the gift of doing good should be given to anyone worthy, relative or not. Hippocrates assembled some strangers, taught them this precious art, and made them hold true to the vow he composed, swear the celebrated oaths to promise not to offend against the rules imposed upon them, and not to teach this art to anyone who had not previously taken this vow.

What follows is the text of the covenant composed by Hippocrates, who said:

'I swear by God, that I will honour this covenant and this pledge, and that I will hold my teacher in this art as a member of my own family: I will support him with my own livelihood, and my wealth will provide assistance if he is in need. I will hold the people of his family equal to my own, and I will teach them this art unconditionally if they need to learn it, and without asking for any fee. My sons and the sons of my master will be equals for me, and also the pupils for whom this agreement was written, and who have sworn to follow the medical law. But I will not do this with those who do not observe the oath.

'I will treat the sick and seek their benefit according to that which is in my power, and I shall prevent things that do them harm. I will not administer any medicine that may kill, even if they ask for it, nor will I provide any suggestions in this regard. I will not treat women with pessaries to abort foetuses. I will keep myself pure and clean for treatment and the practice of this art, and I will not extract stones from the bladders, for I will leave this for those who carry out this task. I shall enter the houses exclusively to treat the patients, and only when I am free of wrong and without intention of damaging anything there, or having intercourse with women and men, be they free or slaves. As for things I witness during treatment sessions or matters that are not discussed publicly, I will refrain from talking about them and mentioning things of this kind.

'Those who fulfil this oath without contravening any of its dispositions shall excel in their conduct and in their art in the most perfect and noble way, and will forever be praised by everybody. Those who contravene this will deserve the opposite.'

This is the text of the well-known *Testament* of Hippocrates, also called *Etiquette of medicine*:

'The student of medicine ought to be a free man with a good nature, young, and of regular stature and proportionate members. He should have excellent discernment and oral skills, demonstrate good judgement when taking part in deliberations, be chaste and brave, not infatuated with money, able to restrain himself when angered; but he does not hold back in extremes, nor is vacuous. He should sympathize and be compassionate with the patients, and keep their secrets because many of the sick people whom we visit suffer diseases about which they do not want anyone to know. If he is insulted, he should be forbearing because we experience this with people affected by delirium and melancholy; we should be patient with them and know that this behaviour is not intentional, but rather the result of non-natural factors.

'The hair of the student should be proportionately and evenly trimmed, not completely shaved or let to grow in excess. He should not clip his fingernails too short nor let them grow beyond the tip of the fingers. His clothes should be white, clean, and soft. The physician should not walk hastily, since this indicates fickleness, nor ploddingly, for this is a sign of a weak spirit. When attending the sick, he should sit cross-legged to hear about their state in silence, calmly and without any kind of disturbance. These forms of etiquette and conduct are better for me than any other.'

Hippocrates always showed great interest in helping and tending the sick. It is said that he was the first one who conceived of and established a hospital. He used to work in a garden close to his house, a section of which was reserved for the sick; then he provided servants to take care of their treatment and called it *xenodokheion*, i.e. 'a place to host the sick'. This is also the meaning of the Arabic term for hospital which comes from Persian: *bīmār* means sick and *stān* means place, thus *bīmāristān* means 'a place for the sick'. Hippocrates did not abandon himself to this comfortable life for the rest of his days: he continued researching medicine and treating the sick, never coveting riches by serving kings; the needs of the destitute were for him more important than his own wealth.

Ibn Juljul said:*

There is a fine story about Hippocrates that we will gladly tell as an example of his excellence. Polemon, the author of the *Physiognomy*,* claimed in his book that he could know a man's character through his physical traits. One day, when the students of Hippocrates were together, one of them said: 'Do you know of anyone better in our days than this exceptional man?' Some answered that they did not, but others said: 'Wait, let us examine him by applying Polemon's claims about physiognomic science.' Thus, they drew a picture of Hippocrates and handed it to Polemon. He looked at it, made a comparison of his body parts and then he pronounced his opinion, saying:

'This is a man who loves fornication.' They replied: 'You are lying! This is the picture of Hippocrates, the sage.' But he said: 'My theory cannot be wrong.' They went back to Hippocrates, and when they told him the opinion of Polemon, he replied: 'Polemon was right, I certainly like fornication, but I restrain myself.' This illustrates the excellence of Hippocrates, his self-control, and how he exercised his soul with virtue.

Al-Mubashshir ibn Fātik says in his *The choicest maxims and best sayings*:

Hippocrates was tall, white-skinned, and handsome; he had pale blue eyes, big bones and prominent tendons, an evenly trimmed white beard, and a hunched back. He was tall and moved slowly, and when he looked back he used to turn his body fully around. Hippocrates bowed his head constantly, always had the right words and spoke calmly, repeating his words to his listeners. When he was seated, his shoes were always in front of him. When asked he replied; when no one talked to him, he asked; he used to stare at the ground while sitting. He was playful, frugal, and prone to fasting. He always carried a scalpel or a little instrument for applying eye-salve.

These are some of the wise sayings of Hippocrates on medicine and some of his unique witticisms:*

We eat to live, not live to eat.

Wine is a friend for the body, and apples are friends for the soul and the spirit.

The sperm in your back is like water in a well: if you drain it, it gushes out, if you leave it, it oozes away.*

Someone asked: 'How often should coitus be performed?' And Hippocrates replied: 'Once a year.' 'And if this cannot be achieved?' they said. And Hippocrates replied: 'Then once a month.' 'And if this is also impossible?' they asked. To which Hippocrates replied: 'Then once a week.' 'And if this is not possible?' they insisted. And Hippocrates said: 'It is his soul, he may let it go whenever he wants.'

The most pleasurable things in this world are four: food, drink, sexual intercourse, and music. Three of these pleasures—or that which is related to them—cannot be attained but with great effort and difficulty, and they are harmful when enjoyed in excess. But listening to music, in dearth or in excess, is an undemanding and effortless pleasure.

Hippocrates also said:

Having few people dependent on oneself is a great fortune.

When Hippocrates was asked: 'What is the best life?' He replied: 'Poverty in safety is better than richness in fear.'

Whoever befriends the ruler should not worry about his severity, like the pearl diver does not worry about the salt in the water of the sea.

Intelligent persons fall in love when they share their rational abilities, but it does not occur between idiots when they share their idiocy, because reason flows in an organized stream and two persons can agree in the same way, while idiocy flows in a confused way and two persons cannot coincide.

There are five ways to treat the whole body: the head by gargling, the stomach by vomiting, the body with purgatives for the bowels, the parts between the two skins by sweating, and the insides and the interior of the veins by bloodletting.

Al-Mubashshir ibn Fātik reports the following sayings by Hippocrates:

In order to preserve health one should not be too lazy to exercise, and stop over-eating and drinking.

Wine should be served to wise men, hellebore* to the stupid.

The man should behave in his worldly existence like the guest at a banquet, who takes a cup only when he is offered and, if the cup passes him over, does not stare at it or reach for it. This is how he should conduct himself with people, wealth and descendants.

I—Ibn Abī Usaybiʿah—continue:

Hippocrates was the first to set down the art of medicine in writing. As we said before, he divulged it and made it well known. The methods of instruction he adopted in these books took three forms:

the first was the use of enigmatic language, the second was to write as succinctly as possible, and the third was to write in an introductory and explanatory way. Reports or books genuinely ascribed to Hippocrates amount to thirty works, including his twelve, most famous books used by students of medicine.

These are: *On embryos*, *The nature of man*, *On airs*, *Waters and lands*, *Aphorisms*, *Prognosis*, *On acute diseases*, *On ailments of women*, *On visiting diseases*, *On humours*, *On nourishment*, *The physician's establishment*, and *On fractures and bone setting*.

Though some of them are falsely attributed to him, Hippocrates wrote many other books including:

On ailments of virgins.

On eyesight.

On madness.

On the extraction of arrowheads.

THE FOLLOWERS OF HIPPOCRATES

When Hippocrates died he was succeeded by his direct descendants and students. There were fourteen of them, four descendants and ten students, including relatives and foreigners. This is what John the Grammarian says. Others claim that Hippocrates had exactly twelve disciples, and that he would not take a new student unless one died. This custom was followed for some time in the porticos where the Greeks used to teach. I also found in some books that Hippocrates had a daughter named Mālānā Ārsā,* who was even more skilled than her brothers in the art of medicine.

2. RUFUS OF EPHESUS (active *c.*100)

Leaving aside the sons and direct disciples of Hippocrates, the physicians worthy of mention from the period between Hippocrates and Galen include Rufus the Great, originally from the city of Ephesus, unmatched in the art of medicine in his time, and mentioned by Galen in his books, where he praises and quotes him.* Among Rufus' fifty-eight works are:

On melancholy, written in two sections. This is one of his finest works.

Jaundice and yellow bile.

On virgins.

On halitosis.

On raising children.

Vertigo.

On urine.

The drug named liquorice.

Intestinal obstruction.

3. DIOSCORIDES (*d. c.*90)

Another physician who lived between Hippocrates and Galen was Dioscorides of Anazarbus, a man of pure soul and greatest benefactor of mankind. He wore himself out travelling the world collecting knowledge about simple drugs. He assessed and verified both their beneficial properties and their theoretical effects, to corroborate and establish these as principles. His work is the basis of every single drug, and all those who came later rely on him. Blessings for this virtuous soul who worked himself to exhaustion to bring good to humankind!

Proof that Dioscorides travelled through lands in search of knowledge about plants comes in the introductory words to his book where he says: 'As you know, an unfathomable passion for acquiring knowledge of *materia medica* took hold of us from a very early age; and we travelled to many lands with this purpose, living the life of one who does not settle in a single place.' [See Appendix 5, Figure 3 for an illustration to an Arabic translation of the treatise on medicinal substances by Dioscorides, copied in Ibn Abī Usaybiʿah's lifetime.]

Also there were the great philosophers. Sāʿid al-Andalusī says:

The Greek philosophers occupy the most pre-eminent position among all people of knowledge, because of their sound commitment to all branches of knowledge. The most distinguished philosophers among the Greeks were five: in chronological order Empedocles, Pythagoras, Socrates, Plato, and Aristotle son of Nicomachus.

4. PYTHAGORAS (*d. c.*485 BC)

The qadi or judge, Sāʿid al-Andalusī says:

Pythagoras left Syria for Egypt to learn from Solomon, the son of David. When he learned geometry with the Egyptians he went back to Greece, where he introduced the study of this discipline along with natural philosophy and religion. With no other help than his own intellect, Pythagoras deduced the science of melodies and musical composition, and their relationship with arithmetical proportions.

He admonished his disciples to love one another and to instruct themselves by engaging in the discussion of elevated matters, to oppose sin, to protect and edify the soul for its struggles, to fast regularly, and to sit in their chairs and read books continually. He ordained that men should teach men, and women women; and urged them to excel in eloquence and learn how to admonish kings.

When he was in charge of the temples, he used to satisfy hunger using poppy seeds, sesame, peeled squill* well washed until its fibres stood out clearly, mallow leaves,* asphodel,* barley flour, chickpeas, and barley; he would crush them and make a paste with honey. To satisfy thirst he made drinks with cucumber seeds, plump seedless raisins, coriander flowers, mallow and *asūfā* seeds,* purslane,* a kind of bread and fine flour, all mixed with honey.

Pythagoras lived a well-balanced life, and it was not the case that he was sometimes well and sometimes ill or sometimes thin and sometimes fat. He possessed a pleasant personality, not given to excessive joy or grief, and was never seen laughing or crying. He always gave precedence to his friends, that is why he is considered to be the first to say 'The wealth of the friends is collective and undivided.'

It is said that Pythagoras used to convey his knowledge using symbolic language to keep it secret. Examples of his use of riddles are:

> 'Do not stir the fire with a knife, if it is already hot.' This meant: do not use offensive words when someone is already angered.
>
> And: 'Do not sit on a bushel,' which means: do not live idly.
>
> And: 'Do not walk by the dens of lions,' which means: do not follow the ideas of the rebellious.

And: 'Swallows do not live in houses,' which means: do not be like those arrogant and garrulous men who cannot control their own tongues.

And also: 'Do not wear ring-stones with images of angels,' which means do not reveal your religion nor disclose divine secrets to the ignorant.

Al-Mubashshir ibn Fātik says:

Pythagoras had a daughter who settled in Croton and instructed the virgins of the city in the religious laws and duties. His wife also used to instruct the other women. When he died, the devout Demetrius took care of the wise man's house and turned it into a temple.

Pythagoras' diet consisted of honey and clarified butter for lunch and millet-bread and raw or cooked legumes for dinner; he would not eat meat, except that of the sacrificial victims that were offered up to God Almighty. When he became the chief-priest of the temples he began to eat food that would relieve him from hunger and thirst.

Once he had the need to travel to various places, and before leaving he wished to meet only his closest friends. He gathered them in the house of Milon (of Croton) and while they were there, a man named Cylon (of Croton) entered without permission. He was a man of noble rank and enormous wealth and very arrogant, insolent, and inclined to injustice. He accosted Pythagoras and began to praise himself, but Pythagoras rejected him in front of all the others and admonished him to work for the salvation of his soul.

Cylon gathered his friends, reviled Pythagoras, accusing him of being an apostate, and agreed with them to murder him and his friends. When they attacked, forty of Pythagoras' companions were killed. Some were able to escape and remained in hiding while the calumnies and persecution continued.

Since they feared for the life of Pythagoras, a group of them helped to slip Pythagoras out of that town during the night. Some of them accompanied him to Caulonia, and from there to Locris. When the calumnies against Pythagoras reached the ears of the people of that city, the elders of the town told him, 'O Pythagoras, for what we have seen, you are a wise man; but the calumnies cast against you are extremely grave. Leave our land if you want to be safe.'

Pythagoras left their town and went to Tarentum but to his surprise he met there a group of people from Croton who almost killed him and his friends. Then he moved to Metapontum but disturbances due to his presence were so frequent that this period was remembered by the people of this town for many years. Then he fled to the Temple of al-Asnān known as the Temple of the Muses. Pythagoras and his friends found protection there and stayed in the temple forty days. They were not forced to flee, but the temple was set on fire. The friends of Pythagoras ran to him, surrounded him so that their own bodies would protect him against the fire. As the flames grew, Pythagoras was seized by the pains of the burning heat and fell dead. The catastrophe affected the whole group and every single one of them perished in the fire. Such was the cause of Pythagoras' death.

I copy here some of Pythagoras' instructive sayings and pieces of advice from the *Choicest maxims and best sayings* of al-Mubashshir ibn Fātik.

> Constant talk about God—may He be exalted—proves a man's incapacity to know Him.
>
> Beware of being involved in anything reprehensible, alone or in company; and be more ashamed of yourself than of anyone else.
>
> Earn money by rightful means and spend it accordingly.
>
> If you hear a lie let your soul bear it with patience.
>
> Do not neglect the health of your body; you ought to eat, drink, practise sex, and exercise with moderation.
>
> If you know a man personally and you find that he is not good enough to be a righteous and loyal friend, beware of making him your enemy.
>
> Nothing is more righteous for a man than doing what he has to do and not what he wants to do.
>
> Most disasters befall animals because they cannot speak and humans because they can.
>
> Pythagoras once saw a man wearing splendid clothes who suffered from a speech impediment and told him, 'Either use a language that matches the quality of your clothes or wear clothes that match your language.'

Once he said to an old man who wanted to learn and was ashamed to be seen as a student, 'Are you ashamed of being better at the end of your life than at the beginning?'

When death struck Pythagoras' wife in a foreign land, his friends offered their condolences for her death, lamenting the fact of her dying abroad, but he said, 'O friends! There is no difference between dying at home or abroad, for the path to the next world is the same from every land.'

It is said that the books of Pythagoras that are undoubtedly authentic come to 280.

5. SOCRATES (469–399 BC)

The qadi Sāʿid al-Andalusī says in his *Book on the categories of the nations*:

Socrates was one of the students of Pythagoras. In philosophy, he focused on metaphysics. He shunned and rejected material pleasures, and spoke openly against the Greek custom of worshipping idols. When he disputed with their leaders, Socrates used clear arguments and proofs, but they roused the masses against him and compelled their ruler to kill him. The king had him imprisoned and ordered that he be given poison to appease the people and keep himself safe from the threat of the populace. Socrates left behind noble counsels, excellent moral instruction, and famous wise sayings. His opinions on the Hereafter, however, are deficient, fall beyond the boundaries of pure philosophy, and deviate from the correct doctrines.

Al-Mubashshir ibn Fātik says in *The choicest maxims and best sayings*:

The name Socrates means in Greek 'maintaining justice'. He was born and raised in Athens. He lived there and left behind three sons. It was customary among them to marry the most excellent women so that their virtues would be passed on to their progeny. Socrates, however, sought to marry a stupid woman whose impudence had no parallel in that town; and so, by habituating himself to her stupidity and patiently bearing with her bad nature, he was able to endure the ignorance of both commoners and elites.

Socrates' unmeasured reverence for wisdom hurt those who sought knowledge after him because he considered it too noble to be confided to leaves and scrolls. Thus his saying, 'Knowledge is something pure and holy; we ought not to write it down, but treasure it in our living souls, away from the skins of dead animals, and protected from the insolent.' He never wrote a single book nor did he dictate anything for his pupils to record.

Rather, Socrates instructed them orally, as he had learned from his teacher Timotheus, who said: 'How can you trust in the skins of dead animals and reject the thoughts of living minds? Imagine that a man meets you in the street and poses a question to you. Would it be proper for you to return to your home with him in order to look for the answer in your books? That would not be right—so exercise your memory.'

Socrates used to speak in symbolic language in the same way as Pythagoras. Some of his symbolic sayings are:

> 'Speak at night where there are no bats' nests,' which means: you ought to be in dialogue with your soul in moments of solitude and prevent your soul from concerning itself with material preoccupations.
>
> 'Cover all five windows to shed light on a place of illness,' which means: shut your five senses off from wandering in whatever is useless, in order to illuminate your soul.
>
> 'Fill the vessel with perfume,' which means: nurture your intellect with clarity, discernment, and wisdom.
>
> 'At the moment of death, do not be an ant,' which means: do not hoard the treasures of the senses when you need to secure the salvation of your soul.

Al-Mubashshir ibn Fātik also says:

Socrates was a man of pale skin, reddish hair, and blue eyes. He was large-boned, with an ugly face, and narrow shoulders. His movements were slow, but he was quick in his answers. He had a dishevelled beard and was not particularly tall. When asked he would bow his head for a moment before answering with persuasive words. He was often alone, did not eat and drink but little, and often mused about death. He travelled little but exercised his body persistently, his clothes were humble but dignified, his speech eloquent and flawless.

He died by poison when he was one hundred and some years. It is said that Socrates left behind twelve thousand disciples among his students and their students.

These are some of the many sayings of Socrates that al-Mubashshir ibn Fātik has compiled:

> If ignorant people would remain silent, disagreement would disappear.
>
> If you need to ask someone to keep a secret of yours, do not confide it to him.
>
> There is nothing more harmful than ignorance and nothing worse than women.
>
> Once he looked at a little girl who was learning to write and said, 'Do not add harm to harm.'
>
> When Socrates was asked about his opinion on women, he replied, 'Women are like an oleander tree: it has beauty and splendour, but is lethal for the inexperienced man who eats from it.'
>
> Someone once told him, 'How do you dare to criticize women, when you, like any other wise men, would not exist without them?' He replied, 'Women are like a palm tree full of prickles that might wound the body of a person but is loaded with dates to harvest.'
>
> Prefer legitimate poverty to forbidden wealth.

6. PLATO (*d.* 347 BC)

Ibn Juljul says that Plato was a wise man originally from the city of Athens. He was a Greek-speaking philosopher, an ancient Greek, and a physician with good knowledge of geometry and of the nature of numbers. Plato composed a book on medicine and books and poems on philosophy.

In *The choicest maxims and best sayings* al-Mubashshir ibn Fātik says:

In their language Plato means 'wide and broad'. Plato's father was

Ariston of Collytus and all his ancestors were noble Greeks of the kin of Asclepius. His mother was an aristocratic descendant of Solon, the lawmaker. As a child, Plato studied poetry and language and excelled in these disciplines, until one day he met Socrates, who rebuked the art of poetry in a way that impressed Plato and resulted in him rejecting all that he had learned about it. He joined Socrates and studied with him for five years until he died. Then Plato heard about a group of followers of Pythagoras in Egypt and travelled there to learn from them.

Then Plato returned to Athens and founded two centres of learning where he instructed the people. Afterwards he went to Sicily, where he encountered Dionysius (II of Syracuse), the tyrant who reigned there and who subjected Plato to severe trials until he was able to escape and return to Athens. In Athens he lived a virtuous life and made good by helping the destitute.

Plato lived eighty-one years. He was a man of good morals and virtuous deeds who did plenty of good for both friends and strangers; he was measured, forbearing, and patient.

Plato used to wrap his wisdom in parables and talk in riddles in such a way that its meaning was only grasped by the learned. He composed numerous books, the titles of fifty-six of which have come down to us.

Plato had tawny skin, an average stature, was beautiful in appearance with perfect features, had a beautiful beard, and little hair on his cheeks. He was silent and soft-speaking, had quick and bright eyes, a black mole at the bottom of his chin, a long arm span, and gentle speech. He was fond of empty areas, of wild places, and of solitude, and most of the time his whereabouts were revealed by the sound of his weeping, which could be heard from miles away deep in the desert and the wilderness.

These are some of the many witty sayings and exhortations of Plato collected by al-Mubashshir ibn Fātik:

> Habit governs all things.
>
> If the wise man avoids people, search for him; if he seeks them, avoid him.
>
> He who in wealth is not generous with his friends will be abandoned by them in poverty.

A king is like a powerful river from which smaller rivers derive: if sweet-watered they will have sweet water, if salty they will be salty as well.

My soul aches for three things only: a rich man who becomes poor, a powerful man who falls in disgrace, and a wise man who is mocked by the ignorant.

Do not try to work fast but rather to work well: people will not ask about the time it took you to complete your work, but only about its quality.

When he was asked, 'What should one be wary of?' he replied, 'Of the powerful enemy, the troublesome friend, and the wrathful lord.'

He was once asked, 'What is the most profitable thing for men?'; and he replied, 'To care for his own rectitude more than that of others.'

7. ARISTOTLE (384–322 BC)

The name Aristotle means 'of perfect virtue'.

Ibn Juljul, states in his book:

Aristotle was the philosopher of the Greeks par excellence, their wisest man. He excelled in the art of medicine, but then philosophy took hold of him.

In his *Epistle to Gallus on the Life of Aristotle*,* Ptolemy says:

Aristotle was originally from a city named Stagira in Chalcidice. His father was a physician. Aristotle was still a young boy when his father died and he was entrusted to Plato. It is said that Aristotle studied with Plato for twenty years. When Plato returned from Sicily, Aristotle began a school dedicated to Peripatetic philosophy. After Plato had died, Aristotle went back to Athens and taught there at the Lyceum until one of the priests accused him of impiety because he did not worship the idols that were revered at that time. When Aristotle heard about it, he left for his homeland in Chalcidice because he hated the thought that the Athenians would cause him as much trouble as they had for Socrates. He settled and lived there until he died at the age of 68 years.

When Philip (II of Macedon) died and was succeeded by Alexander (the Great), Aristotle decided to adopt a life of retirement and to abandon all his relationships and dealings with kings. He went back to Athens, where he founded the aforementioned centre of learning which was dedicated to Peripatetic philosophy. He was eager to improve the lives of the people and to help the destitute and the needy, to marry the widows and provide sustenance to the orphans, whom he was committed to educating and helping, as well as anybody who wished to acquire knowledge and instruction in any kind of discipline. Aristotle helped them to achieve this and set them on their feet, he gave alms to the poor, brought amenities to the cities, and restored the buildings of his home town, Stagira. He was extremely pleasant and respectful, and received everybody cordially, the humble and the great, the powerful and the weak; there are no words to describe his commitment to his friends. The stories written by his biographers bear testimony to that, and all the accounts on Aristotle's life are unanimous on this matter.

Al-Mubashshir ibn Fātik states:

When Aristotle reached his eighth year, he was taken by his father to Athens and brought to the Lyceum. His father introduced him to poets, rhetoricians, and grammarians, under whom he studied for nine years. They called this teaching system 'comprehensive', on account of the fact that everyone needs it, because they are the means and route to every wisdom and virtue, and the way of presenting the findings of every form of learning. Aristotle claimed that no discipline can dispense with the knowledge of grammarians, rhetoricians, and poets, because articulate speech is a tool proper to their arts, and the superiority of men over beasts rests on their ability to speak. Those who most properly may be considered to be human are those with the greatest ability to speak, those most capable of expressing their own thoughts, and those who best choose the words to express themselves in the most concise and appealing manner. Wisdom is the most precious thing and the expression of it rests on proficiency in argument, command of idiom, concision in phraseology, and freedom from the ugliness of stuttering and stammering. Once Aristotle had perfected this art, he pursued other disciplines. At 18, he joined Plato and became his student and pupil.

Al-Mubashshir ibn Fātik also says:

Aristotle was a pale-skinned man, partly bald, of a fair stature, with large bones, small blue eyes, thick beard, aquiline nose, small mouth, and a broad chest. He used to walk fast when he was alone and slow when he was in the company of friends. He studied books constantly, he never talked drivel and he pondered about every single word. He used to bow his head in silence for a while when he had to reply to a question, and his answers were laconic. During the day he used to retire to the wilderness and to walk beside rivers. He loved listening to music, and meeting mathematicians and dialecticians. He was fair-minded when he debated with opponents, and always acknowledged right and wrong. He was modest in his clothing, and moderate when eating, drinking, having sexual intercourse, and moving around. In his hand he always held an instrument to measure the stars and hours.

Abū Nasr al-Fārābī* says:

Aristotle divided the study of logic into eight parts, each one of them treated in a book:

> The first book contains the rules applied to single intelligibles and the terms that denote them.
>
> The second book deals with the rules of compound propositions, which are compound intelligibles formed by two single intelligibles.
>
> The third book discusses the rules for statements whereby the syllogisms proper to all five arts are distinguished.
>
> The fourth book deals with the rules whereby the demonstrative statements are examined, and the rules that make philosophy coherent so that all that it undertakes is most complete, virtuous, and perfect.
>
> The fifth is focused on the rules whereby dialectical statements are examined.
>
> The sixth book deals with rules which serve distortion and deviation from the truth. It enumerates the devices used by those whose aim is misrepresentation and deceit, then lists what is needed to refute misleading statements, and how one can be on one's guard, including from making false statements in regard to one's own claims.
>
> The seventh book is focused on the rules whereby the rhetorical statements may be examined, on the kinds of rhetoric, and the discourses of the orators and rhetoricians.

The eighth book deals with the rules that define poetry, and all the elements in each genre. It lists all the things that give the art of poetry coherence, and when each should be used in order to be most eloquent and extraordinary.

These are all the divisions of logic. The fourth part enjoys pre-eminence on account of its nobility and importance. Only this fourth part is required to achieve the primary goal of logic.

The three parts that precede it are introductory and pave the way to the fourth part. The four parts that follow have two aims: first, they support and assist the fourth part; second, they foster awareness to distinguish these arts from one another. Otherwise, no one could be sure whether they were using the elements of dialectic when seeking truth without recognizing them as such, and thus might abandon the realm of certainty for the realm of strong speculation; or they might inadvertently use the features of rhetoric and slide into the realm of persuasion; or perhaps, they might use the tools of sophistry and mistake untruth for truth and believe it; or they might use the features of poetry without recognizing them and then base their beliefs on fictions.

Ptolemy says:

On his deathbed, Aristotle left a testament, parts of which are as follows:

'Upon my daughter reaching puberty, my executor, Nicanor shall take care of her and if she dies without issue, her possessions shall revert to Nicanor. I stipulate that Nicanor shall take care of the administration of the affairs of my daughter and son, Nicomachus, as if he were their father or brother.

'The testamentary executors are to take care of my maid, Herpyllis,* for she deserves that in view of her diligent services and her efforts to please me. They should satisfy all her needs and, if she should wish to marry, not to consent to her marrying anyone but a virtuous man, and to pay her one silver talent, and three maidservants in addition to her own maid and her male servant. Should she desire to reside in Chalcis, she shall receive accommodation in my estate, in the house that is next to the garden. If she desires to stay in Stagira, she should be provided with accommodation in the house of her choice from among the houses of my family.

'As for the members of my household and my children, Nicanor should take care of Myrmex, the boy, and let him go back to his homeland with all his possessions whenever he desires. My maidservant Ambracis should be emancipated and if she is still at my daughter's service when she marries, she should receive 500 drachmas and keep the maidservant that she now has. Simon shall be given the price of a slave so that he can buy him for himself, in addition to the slave for whom he has already paid. When my daughter marries, my slaves Tycho, Philo, and Olympius ought to be manumitted. Olympius' son shall not be sold but shall remain at the service of the house until he achieves a proper age, then ought to be manumitted and rewarded according to his merits.'

Hunayn ibn Is'hāq says in his *Stories of the philosophers and the sages*:

The assemblies of philosophers originated because the kings of the Greeks and elsewhere used to teach their children wisdom and philosophy and different kinds of literature. For this reason, they painted their temples with all kinds of images so that their hearts would be inspired by contemplating them. The Jewish synagogues were painted like this, and also the Christian churches, and the Muslims adorned their mosques so that their hearts would find pleasure in contemplating them and their souls would reflect upon them.

I—Ibn Abī Usaybiʿah—continue with a saying of Aristotle, which epitomizes the fundamentals of the preservation of health, stating: 'I would be surprised to see someone falling ill if he drinks "water of the vine" and eats bread and meat; moves, rests, sleeps, and stays awake in the right proportion; regulates his sexual life correctly; and keeps his humours balanced.' Eighty-eight other wise sayings of Aristotle have been collected by al-Mubashshir ibn Fātik.

At his deathbed, Aristotle disposed that he would be buried in an octagonal mausoleum. In each of the eight sides some words comprising the principles on which the well-being of people reside were to be written. [See Appendix 5, Figures 4 and 5 for a fourteenth-century manuscript and a modern diagram illustrating the eight sayings that Aristotle directed be placed on the sides of his octagonal mausoleum.]

Among Aristotle's famous works Ptolemy lists ninety-six works, and there are more that he found in the library of Apellicon.* Ptolemy states: 'These are all the works by Aristotle that I have seen, but others have seen additional books.'

I—Ibn Abī Usaybiʿah—continue: I have come across at least thirty-nine other books by Aristotle that were not listed by Ptolemy.

CHAPTER 5

PHYSICIANS FROM OR AFTER THE TIME OF GALEN

Galen is perhaps the most important figure in the book for Ibn Abī Usaybiʿah and recurs as a presence and model for all physicians throughout. Devoted almost entirely to him, this chapter comprises the longest single biography in the original text. It is also the only known life of Galen, and Ibn Abī Usaybiʿah tries hard to do justice to this great figure. He rejects John the Grammarian's sectarianizing idea that Galen was a contemporary of Christ by using the chronological researches of a trusted Christian writer, ʿUbayd Allāh ibn Bukhtīshūʿ, and includes a long anecdote cementing Galen's superior status even in comparison with a physician as wealthy and successful as Jibrīl ibn Bukhtīshūʿ.

Regarding book lists, Ibn Abī Usaybiʿah outdoes himself recording of all Galen's works. Quoting heavily from Hunayn ibn Is'hāq (another model figure to Ibn Abī Usaybiʿah), we get some insight into the effort and determination of medieval Arab scholars to gather, quantify, record, and translate the works of classical Greece.

GALEN (*d. c.*216)

LET us begin our account with a complete discussion of the evidence for Galen and his activities. We shall follow this with a summary report of the physicians who lived after Galen's time or near to it.

What we know about the status of Galen and the fame he earned across many nations is that he was the 'Seal'* of the great master physicians and was the eighth of them and that no one came close to him in the art of medicine let alone equalled him. For when he began his career he found the art of medicine was dominated by the teaching of charlatan physicians and its excellent characteristics were being erased. He took the problem in hand and showed that the doctrines of people like that were useless. He championed and invigorated the arguments and teachings of Hippocrates and his successors. It was as a result of his abilities that this was the medicine that triumphed. He composed many treatises on it and through these he clarified many of the obscurities of the art and gave eloquent expression to its truths, allowing medicine's real language to prevail. Physicians who lived after him were inferior and dependent on him for their learning.

The length of Galen's life was: 'eighty-seven years. He was a youth and a student for seventeen years and a scholar and a teacher for seventy years.' So runs the account of John the Grammarian. The only way to check what he says is to use the reports of Galen himself and Galen's words are clearly different from what is recorded about him.

There is no truth in the claim that Galen was Christ's contemporary, visited him, and believed in him since Galen mentions Moses and Christ in various passages in his works and it is clear that he lived after Christ. I have found that ʿUbayd Allāh ibn Bukhtīshūʿ, son of Jibrīl, made a careful enquiry into this matter and concluded that the chronographers show very clear contradictions in their works.*

For example, he notes that:

Trajan's deputy in Palestine wrote to him saying, 'The more Christians I kill, they more grow to love their religion,' and so Trajan gave orders to stay the sword from them.* In the tenth year of his reign Galen was born. The next king was Hadrian and the next king was Antoninus Caesar.* Galen's career began in the days of this king and it was Antoninus who took Galen into his service. Proof comes from the statement at the very start of the first section of the *Anatomical procedures*. In the passage concerned he says, 'I had previously written a book on anatomical procedures during my first visit to the city of Rome at the start of the reign of King Antoninus, who is still king in our time.'

ʿUbayd Allāh also observes:

I am mystified as to how errors occur despite the clarity of the passages from Galen and the texts of the accurate chronographers. The copyists are responsible for this type of error and it persists so that people come up with arguments that mislead those who fail to investigate the truth of the matter.

Galen's own statements in his interpretation of Plato's *Republic** testify to the fact that Christ preceded him by a long time. We learn there that, in the time of Christ, Christians did not visibly assume the asceticism and preference for devoting themselves to God (may He be glorified and exalted!) that Galen describes; but a century later, they had become so widespread that they surpassed philosophers in good deeds.*

Ibn Juljul records that Galen was one of the Greek sages who lived under the empire of the Caesars following the establishment of Rome. His birth and upbringing were at Pergamum.

In the following story Ibn al-Dāyah* provides information about Galen's domicile and residence:

My master, the prince Ibrāhīm ibn al-Mahdī, asked the physician Jibrīl ibn Bukhtīshūʿ,* 'Whereabouts in Roman territory was the house of Galen?' He stated that in Galen's own age it had been located in central Roman territory, but 'at the present time it lies on the edge,' and he noted that in Galen's day the eastern Roman border in the Euphrates region was at the settlement called Nighyā in the district of al-Anbār.* 'The border between the Persians and the Romans to the North was Armenia and in the West, Egypt, except that the Romans had control of Egypt and Armenia at certain times.'*

I—Ibn al-Dāyah—took Jibrīl's statement about Roman control of Armenia 'at certain times' as something that should be rejected, and denied the Romans had conquered Armenia apart from the zone called Armenianus in their tongue, 'for it is a fact that the Romans refer to the people of this country as far as this boundary as Armenians.' Ibrāhīm ibn al-Mahdī testified to this and gave irrefutable proof in the form of an Armenian rug,* the finest workmanship I had seen from Armenia. On it were pictures of young women playing a variety of Roman games in a garden. It was embroidered in Latin mentioning the name of a Roman king. And so I gave in to Jibrīl.

The conversation returned to the topic of Galen and Jibrīl said, 'The name of the town where he was born and had his residence was Smyrna.* His house was in the neighbourhood of Qurrah, which was only two parasangs from Smyrna.'

Jibrīl continued, 'When Hārūn al-Rashīd was encamped under Qurrah, I asked him if I might go to the house of my supreme Master that I might eat and drink in it, to accord me supremacy over my peers. He laughed and said, "I fear the Roman army will come out and seize you!" I countered, "It is impossible that the Romans could come so near your camp." So he gave the order for five hundred men to be assigned to me so that I could reach the area. When I said, "Commander of the Faithful, fifty would suffice," he laughed, saying, "Assign him one thousand horsemen. For I know he hates the thought of having to provide them with food and drink!" So I said, "There is no actual need for me to visit Galen's house!" At that he laughed even more: "You will certainly carry this out, with a thousand horsemen at your side!" '

Jibrīl went on. 'I was the most sorrowful and the unhappiest of men as I left, for I had not even prepared for myself the wherewithal to cater for ten. The situation was not resolved until all the bread, carcasses, and salt were delivered to me. It was enough for all those with me and a lot was left! And so I did indeed get to stay in that place and eat there. The young men of the army went and raided the wine and meat stores of the Romans. They ate the meat sliced on bread and drank wine with it. I withdrew at the end of the day. Ibrāhīm ibn al-Mahdī asked me, "Was anything obvious in the layout of Galen's house to show that he enjoyed honour?" "Yes, the layout was extensive. I saw it had ranges to the east, west, and south, but I did not see one pointing north. This was the way Roman philosophers built their houses, and we know the Persian nobility used the same style. And I am planning the same, for any house that lacks sunshine is prone to disease. Galen was in the service of the Roman kings because of his wisdom and the Roman kings were men of ambition in everything they did. If you were to compare Galen's dwelling with other Roman dwellings, you would appreciate its large area and the size of its buildings. Although I only saw it in a dilapidated condition, from the fact that I found ranges with intact roofs I could tell he was a man who enjoyed high honour." '

Ibrāhīm ibn al-Mahdī fell silent so I—Ibn al-Dāyah—said, 'My dear Jibrīl, the kings of the Romans were indeed men of ambition, as you have described, but the ambition they showed in their gifts and presents was no different from the ambition they showed in their personal honours. This failing affects servant and served. If you consider the plot occupied by the palace of the king of Rome and the plot owned by Galen, then consider the palace of the Commander of the Faithful and your own residence, the ratio of Galen's residence to that of the king of Rome will resemble the ratio between your own and the residence of the Commander of the Faithful!'

Jibrīl was sometimes impressed by my often penetrating claims and praised me in front of Ibrāhīm ibn al-Mahdī, but at other times he was annoyed and flew into a temper. Thus he said, 'What has your mention of "ratio" got to do with me?' I replied that, by mentioning 'ratio' I deliberately used a technical term from the discourse of the Roman sages. 'You are their leading student and I intended to make things relevant to you by speaking the language of your masters. The point of my remark that the ratio of Galen's estate to the king of Rome's estate resembles the ratio of your estate to the Commander of the Faithful's estate is this: if Galen's estate is a half or a third or a fourth or a fifth or some such proportion, does its proportion relative to the king of Rome's estate resemble that of your estate in relation to the Commander of the Faithful's estate, or is it less? Suppose that the estate of the Commander of the Faithful were one parasang by one parasang and the proportion of yours was a tenth of a parasang by a tenth of a parasang, and that the estate of the king of Rome was a tenth of a parasang by a tenth of a parasang and Galen's estate was a tenth of a tenth of a parasang by a tenth of a tenth of a parasang. In that case the proportion of Galen's estate in relation to the king of Rome's estate would resemble the relationship between the scale of your house and that of the Commander of the Faithful exactly.' 'But Galen's estate is not like that!' he said. 'It's hugely smaller in scale than my estate is in relation to the estate of the Commander of the Faithful!'

Jibrīl continues:

I and my father Bukhtīshūʿ ibn Jūrjis and my grandfather Jūrjis ibn Jibrīl led lives that were not like the caliphs' and yet their lives were like those led by caliphs and crown-princes and brothers and uncles of caliphs and their notable clients and commanders, and in terms of liberality every single one of them benefited from the generosity of the caliphs' hearts. But every one of the companions of the king of Rome lived in a state of poverty and enjoyed little influence. So how could I, Jibrīl, possibly be like Galen who enjoyed no privileges because his father was a husbandman and a keeper of parks and gardens? How could it be that someone with that standard of living might be like me? Two generations of my own family served the caliphs, who showered them with gifts, and other inferior men besides. The caliphs have bestowed gifts upon me and raised me up from the rank of physician to their own company and companionship. And if

I said that the Commander of the Faithful has no brother or relative or commander or governor who does not treat me right without being disposed to do so out of love for me—and whether he is so disposed or grateful to me for some treatment I gave him or for a brilliant occasion where I was in attendance or for a flattering story I told at court to his advantage, well, all of these people lavish their benefactions and gifts upon me. But if the size of my house is 10 percent of the house of the Commander of the Faithful and Galen's is 1 percent of the house of the king of Rome—he was still much more distinguished than me!

In his *Examination of the best physician*, Galen includes the following narrative:

I learned the demonstrative method from the time of my youth, and more importantly I rejected pleasures at the start of my study of medicine, despising and repudiating the worldly trappings people fight over, so that I excused myself the trouble of getting up early to attend at men's doors in order to accompany them as they rode out from their residences and waiting around by the doors of kings so that we might escort them to their palaces and stay by their side. Nor did I waste my time or upset myself wandering around after men, which people call 'greeting'. Instead I devoted my time entirely to practising medicine and to reflecting and thinking about it. I generally stayed up at nights to explore the treasures left us by the ancients. If anyone can say he acted as I did and possesses the natural intelligence and quick understanding that allow familiarity with this great science, he ought to be trusted even before people encounter his diagnoses and conduct with patients, and he should be judged superior to those who lack the talents we have specified and the practices we have listed.

I once attended one of the public meetings where people assemble to find out more about a physician's knowledge. I demonstrated many examples of dissection including one where I took an animal and split open its belly to pull out the intestines. I called on the physicians who were present to replace them and suture the belly in the correct manner. Not one of them ventured to do this. We took care of it ourselves and thereby revealed our skill, experience, and dexterity. It became clear to anyone of intelligence in the audience that a man who was to be in charge of the wounded must possess the skills I did. When the high priest made me responsible for their affairs, he was absolutely delighted because none of the many men in my care died, barring two individuals, whereas sixteen had died in the care of the physician before me. Later a second high priest put me in charge and he was even luckier as a result of the appointment because not one of those under me died although they had many very serious injuries.*

Al-Mubashshir ibn Fātik writes that:

Galen travelled to Athens, Rome, Alexandria, and other places in his quest for knowledge. He was taught by the physician Arminus* after having initially learned geometry, philology, grammar, and other subjects from his father and others. He also studied medicine with a woman called Cleopatra.* He acquired many drugs from her, especially those relevant to treatments for women. He travelled to Cyprus to see calcite being mined. He also visited the island of Chios* to view the manufacture of *terra sigillata*.* He handled and verified these products for himself. He also journeyed to Egypt and stayed there for a while examining Egyptian medicaments, particularly opium in the territory of Asyūt, a province of Upper Egypt. He left and was making for Syria en route to his home town but fell ill on the way and died at al-Faramā. This is a city by the Green Sea (the Mediterranean) in the last of the provinces of Egypt.

The cause of Galen's death was allegedly diarrhoea. The story goes that when the illness became protracted he treated it by every means but without effect, and his pupils remarked that 'the Sage does not know how to treat his own illness' and so were negligent in looking after him. It was summer time. He had a jar fetched containing water. He produced something and tossed it in the jar, left it for an hour, then broke the jar open, and lo and behold the liquid had frozen. From it he made a drug which he drank and used as an enema but to no effect. 'Do you know why I did this?' he asked his pupils. 'So that you don't think I am incapable of curing myself! For this is a "chronic" illness, which means a disease for which no medicine exists, in other words, death.' I—Ibn Abī Usaybiʿah—regard this tale about Galen as a fiction.

Another physician describes how to freeze water outside the proper season for it. He asserts that 'You take a *ratl* of good quality Yemeni alum,* crush it finely, and place it in a brand-new crock, introduce six *ratls* of clean water and leave the whole thing in an oven, lined with clay, until one-third and only one-third remains. Once it has thickened, lift it out into a container with the top well sealed. When you want to use it, take a new ice-tray with pure water, stir in ten *mithqāls* of the alum solution, let it stand for one hour and it will become ice.'

Al-Mubashshir ibn Fātik writes:

Galen's father took considerable care over him and spent a great deal of money doing so. He paid tutors handsome salaries to bring them from faraway towns. From boyhood onward Galen showed a passion for knowledge based on logic, pursuing it with real eagerness, effort, and aptitude. As a result of his eagerness for knowledge, when he left his teacher's side and was on the way home he would go over what he had been taught so he could reach the same level. The boys who were with him in class kept cursing him: 'You there, you've got to learn to take some time to laugh with us and come out to play!' Often he did not answer them because he was so engrossed in his studies but sometimes he asked, 'What motivates you to laugh and play?'—'We like it!'—'Well, there is a reason why I am motivated to renounce such things and that is my loathing of your behaviour and my love of my own!' People used to marvel at him, saying, 'In addition to your father's enormous wealth and magnificent reputation he has been blessed with a son who thirsts for knowledge!' His father was a geometer who retained an interest in agriculture. His grandfather was the leading carpenter, and his father's grandfather was a surveyor.

In *Avoiding distress** Galen says that a significant quantity of his books and effects was destroyed by fire in the great imperial storehouses of the city of Rome. He records that among the losses were numerous books of his which existed only in a single copy.

Al-Mubashshir ibn Fātik writes that:

Among Galen's losses in the fire and among Galen's most precious possessions were books of his in white brocade bound with black wool-silk, on which he had spent a fortune.

Al-Mubashshir also records that:

Galen was light-brown in complexion, with fine features, broad shoulders, large hands, long fingers, and beautiful hair. He loved songs and music and reading books. He walked with an average gait, led a cheerful life, talked a lot, was hardly ever silent, and loved disparaging fellow-professionals. He was well travelled, had a pleasant odour, and wore clean clothes. He was fond of riding and walking. Galen enjoyed access to kings and leaders but was never enrolled in the service of any of the kings, though they continued to honour him. When they needed Galen to treat a difficult disease, they bestowed upon him numerous gifts of gold and other things when they recovered. He records this in many of his books. If a king asked him to stay in his service, he left his town for another so he would not lose his way by serving him.

In his *Stories of the philosophers and the sages*, Hunayn ibn Is'hāq records many sayings and stories that illustrate his culture, manners, and wisdom, including two on the topic of love:

> Passionate love is finding someone attractive—as well as wanting them.
>
> No one fully deserves the name of lover unless his imagination, reasoning, memory, heart, and liver are besotted with his beloved whenever they are apart so that he refuses food and drink because his liver is preoccupied and cannot sleep because his brain is busy with imagining, remembering, and thinking about him, and thus each abode of his soul is distracted.

Hunayn records that on the stone of Galen's signet ring was inscribed: If a man hides his illness, it will be completely impossible to find a cure for it.

In *The choicest sayings and best maxims* al-Mubashshir ibn Fātik records the following sayings, among others:

> There is more hope for a sick man with an appetite than for a healthy man without one.
>
> In the past men used to meet and drink and argue with one another about the effects of drinks upon temperaments and of music upon the faculty of passion. But when they meet today they compete about who has the largest cup to drink from.

I—Ibn Abī Usaybiʿah—have copied a Galenic saying that occurs elsewhere:

Let me present you with an allegory of the four humours. Yellow bile, which is in fact red bile, is like a woman who has a sharp tongue but is good and pious. She causes harm by her tirades and quick temper but soon calms down without disaster. Blood is like a rabid dog. When he enters your house, catch him and put him out or kill him. Phlegm, when it moves in the body, is like a king entering your house. Though you fear his villainy and tyrannical behaviour, you cannot cross him or harm him but have to remain friendly when you usher him out. Black bile in the body is like a spiteful man whose motives are unsuspected but then pounces with a sudden attack. Then there is nothing odious he will leave undone, and he will not calm down without a great deal of trouble.

Galen composed a very large number of books.* In the following, I—Ibn Abī Usaybiʿah—list titles translated into Arabic by Hunayn ibn Is'hāq and others known to me to be in general circulation. I also refer to Galen's aims in writing them.

The *Pinax*,* which means the Catalogue. His purpose in this work is to give details of the books he wrote, the aim of each of them, the motive for writing it, the identity of the addressee, and his own age. It is in two sections. The first records his books on medicine, the second books on logic, philosophy, rhetoric, and grammar.

On bones. This consists of a single section. Hunayn says: 'Galen entitled this book "for beginners" and not "to beginners" because there is a difference in his mind between the phrases. If a book's title is "to beginners", he indicates by this phrase that he is adapting his teaching to suit the abilities of beginners, and presupposes the existence of a course for advanced students on the topic beyond this course. If he entitles a book "for beginners", he indicates that it contains a complete course on this subject but that the course is in fact "for beginners".'

The great book of the pulse. Galen composed this book in sixteen sections.

Readily available drugs. These are the drugs that are termed 'locally available'. The book consists of two sections. Hunayn says that, 'An additional section on this subject was added to it and attributed to Galen but in fact it is by Philagrius.' He also says that 'a great deal of nonsense' has been inserted 'into the book including novel and surprising prescriptions and drugs that Galen had never seen or heard of.'

Medical terms. His aim is to clarify the senses in which physicians use the terms they employ. He wrote it in five sections. We have found translated into Arabic only the first section.

On proof. He composed this book in fifteen sections. His aim is to show that the method used to demonstrate a proposition proceeds by necessary steps. Hunayn says that, 'To the present day none of my contemporaries has managed to locate a complete copy in Greek despite the fact that Jibrīl had made a great effort and I too combed the lands of Iraq, greater Syria, and the whole of Palestine and Egypt till I reached Alexandria. I found no trace except for about half of it at Damascus, but the sections were not consecutive or complete. Jibrīl also found sections from it and not all of them were the same as the ones found by me. As for me, I had no desire to translate any of it without the prospect of being able to read the entire thing, both because of its damaged and defective condition and because my soul longed to discover a complete version. Later I translated what I had into Syriac.'

Generation of the fetus born at seven months.

Hunayn ibn Is'hāq himself says:*

We have also come across a number of books with Galen's name in the title but not by him. Rather, some of them are snippets of Galenic phrases arranged by other people to make books. In other cases they are by other authors and someone put Galen's name in the title. He did this either out of a desire to multiply the number of Galenic books in his possession, or because of a lack of sense that is a constant attribute of fools so that when they come across a book with a number of sections and find the first section begins with a certain person's name, they assume all the other sections are by the same person. Two of these titles are:

Treatise entitled 'The physician, by Galen'. Galen in fact mentions it at the start of his Catalogue, where he informs us that it was falsely attributed and inauthentic.

On medicine according to the views of Homer. It consists of two sections. The phraseology and language of both of them are very similar to Galen's. However, the argument is weak and at the end the views are improbable and quite unlike the Galenic system.

Hunayn made this list of Galen's books when he was 48 years old. Since he lived till 70, he must have come across many additional works of Galen's, as we have found works ascribed to Galen in translations by Hunayn or others of which there is no record in Hunayn's work.* Some examples of these are:

Commentary on Hippocrates' pains in women. It consists of a single section.

Treatise on sleep, insomnia, and emaciation.

Treatise on prohibiting burial within twenty-four hours.

Secret remedies. He gives us hints and suggestions of these remedies in his works. It is in a single section. Hunayn ibn Is'hāq says, 'In this book Galen intended to describe all the secret and special remedies he had assembled over his lifetime and had successfully tried out on many occasions but had concealed and kept back from all but a select number of practitioners who possessed the right qualities and the soundest judgement. Someone made a version of it but made a mess. I applied myself and made an Arabic version.'

In conclusion, I—Ibn Abī Usaybiʿah—note that Galen certainly wrote many other books that translators did not find or that were lost in the course of time, in particular those mentioned in the second section of Galen's catalogue of his works called the *Pinax*. Should anyone wish to refer to their titles or to Galen's aim in each of them, he should consult this book.

CHAPTER 6

THE ALEXANDRIAN PHYSICIANS, THEIR CHRISTIAN AND OTHER CONTEMPORARIES

This short chapter on the enigmatic John the Grammarian and the Alexandria School bridges the gap between Galen in second-century Rome and the earliest days of Islam in the seventh century. Though Ibn Abī Usaybiʿah has rejected John's chronology of Galen's coevality with Jesus in the previous chapter, John is nevertheless important as he provides a human link from Galen, on whom he writes commentaries, to the advent of Islam, when, in legend at least, he is still alive.

The chapter also describes the, possibly idealized, seven-stage Alexandrian Summaries of the Sixteen Books of Galen used at Alexandria for teaching. Greek in origin but only extant in Arabic, this curriculum maps closely with the list of Galen's books given by Hunayn. Building on this and Chapter 4, Ibn Abī Usaybiʿah shows here how medicine became an academic subject, taught through books and a syllabus.

According to an account by Ibn Butlān, the Alexandrians who summarized Galen's Sixteen Books and composed commentaries upon them were seven in number: Stephanus, Gessius, Theodosius, Archelaus, Nicolaus, Palladius, and John the Grammarian, all of whom followed the way of the Messiah.

I—Ibn Abī Usaybiʿah—say, these Alexandrians read only the Sixteen Books of Galen at the academy where the art of medicine was taught in Alexandria.* Each day, the masters would agree on a passage, and that passage would be read and explicated. Subsequently, they made summaries and abridgements, and later, each composed his own commentary. Of these commentaries, I have found that of Gessius* to be superior to those of his fellows. One of these Alexandrians, namely John the Grammarian, was granted such a long life that he was still living at the time of the advent of Islam.

The following account from the *Fihrist* of Ibn al-Nadīm, the Baghdadi bookseller, says:

John the Grammarian at the beginning of his career acted as bishop over several Egyptian churches, adhering in belief to the Jacobite sect.* But then he turned against the belief in the Trinity professed by Christians, and as a result a conclave of bishops was convened to examine him. So ably did he argue his case, however, that his accusers were quite won over: they sought to conciliate him, asking him in respectful terms to renounce the position he had adopted and to desist from proclaiming it. In vain: he stood his ground, and was removed from his office.

When Egypt was conquered by the army of ʿAmr ibn al-ʿĀs, the victorious general* called on John the Grammarian to pay his respects and acknowledge his eminence.

ʿUbayd Allāh relates in *The merits of physicians*:

John the Grammarian acquired his skill in philosophy as a result of something that happened to him when he was young and earning his living as a wherryman. He had a great love of knowledge even then, and one day, when his passengers chanced to be a group from the Museum in Alexandria, he overheard them discussing the theories that they had been investigating. He was captivated at once, began to look into the domain of learning, but reflected, 'Here I am, over 40 years old, with no skill of any kind. All I know how to do is pilot this wherry. How can I possibly hope to succeed at any aspect of learning?' As these thoughts were running through his mind, what should he see but an ant that was trying to carry a date pit up a slope. It lost its grip repeatedly but persisted, and by degrees got the pit up the slope. The struggle continued all day as John watched in fascination. He said to himself, 'If a tiny creature can attain its goal through persistent effort, surely I can do as well, or better.' He sold his wherry, entered the Museum, and embarked on the study of grammar, language, and logic, in all of which he proved to be outstandingly brilliant. He was dubbed 'the Grammarian' because he had begun with the subject of grammar, and, indeed, he was particularly renowned for his accomplishments in that field.*

I have read in a history of the Christians that John the Grammarian was present at the Fourth Council held in a city named Chalcedon.

The Council comprised 630 bishops, who sat in judgement on Eutyches* (that is, John the Grammarian) and his colleagues. Although he was pronounced anathema,* he was not banished, in contrast to others who had been so condemned, for he was skilled in the art of medicine, and his colleagues needed his ability. Accordingly, he remained in Constantinople and lived there unmolested until the emperor Marcian died.

Eutyches, or John the Grammarian, also known by the Greek name 'Philoponus' (that is, 'lover of toil'),* was thus allowed to go on living in Constantinople. Marcian's successor was the emperor Astīriyūs,* who became gravely ill two years after Eutyches had been pronounced anathema. The physician was consulted, examined and treated the emperor, and in due course he recovered. 'Ask for what you will,' said the grateful emperor. 'Imperial Majesty,' replied Eutyches, 'there is indeed something that you can do for me. The bishop of Dorylaeum* and I have become enemies. He has treated me unjustly, having encouraged the patriarch of Constantinople, Archbishop Flavian, to convene a "synod" which wrongfully pronounced me anathema. I should be most grateful, Sire, if you would convene another council to review the matter.' 'I shall be happy to do so,' replied the emperor.

The emperor was as good as his word. He wrote to Dioscorus, the patriarch of Alexandria, and to John, the patriarch of Antioch, ordering them to come to Constantinople. Dioscorus obeyed, with thirteen bishops in his train, but John did not appear.* The emperor therefore instructed Dioscorus to examine Eutyches' case and withdraw the pronunciation of anathema. 'If you withdraw it,' said the emperor to him menacingly, 'I shall reward you abundantly, and if you do not I shall have you put to death painfully.' The patriarch, having decided that being rewarded was preferable, convened a conclave with his thirteen bishops and, taking a lenient view of the matter, they declared the pronunciation of anathema withdrawn. At that, the bishop of Dorylaeum and his colleagues left Constantinople.* Eutyches came to be known as John the Grammarian. He remained a staunch Jacobite to the day of his death, who never accepted Melkite doctrine.*

Among the works of John the Grammarian are commentaries on many of the works of Galen.

The following account is from the *The useful book on the method of medical learning* by Ibn Ridwān:

The reason why, out of all Galen's works, the Alexandrians used only the Sixteen Books for teaching purposes, was that any student who applied himself to those books, if he was genuinely talented, truly interested in the subject, and ambitious, would be so inspired by the evidence of Galen's mastery of the field of medicine as he found it in them, that he would study the rest of Galen's works of his own accord.

There were seven successive levels in the Alexandrians' employment of the Sixteen Books.

For the first level, they would use the books as an introduction to the art of medicine. Following attainment of that level, the novice would be ready to perform simple medical procedures.

At this level, only four of the books were used—*The sects*, *The small art*, *The small book of the pulse*, and *To Gaucon*. These provided an adequate instruction in the art of medicine for the student. And for the expert physician, they were a repository of information that he could consult to refresh his knowledge.

At the second level, again, four of Galen's Sixteen Books were used—*The elements*, *Mixtures*, *Natural faculties*, and *The small book of anatomy*. These works contain information about the natural factors by which the body subsists. Every keen student who reads them will undoubtedly want to study all other works relating to the nature of the human body.

At the third level, only a single work was used: *Causes and symptoms*, consisting of six treatises. Galen had originally composed those treatises as separate works, but the Alexandrians combined them into a single book. It explains diseases, their causes, and the symptoms characteristic of various disorders.

At the fourth level, two works were used. The first of these was the *Classification of the diseases of the internal parts*, in six volumes, which explains the various disorders affecting the parts inside the body. Such disorders are not perceptible to the sight, they can be detected only by symptoms, each of which must be assessed by the physician. Galen never composed a work on disorders affecting external parts of the body, for the simple reason that such disorders are plainly visible. The fledgling physician can thus diagnose them through observation in the presence of his masters. The second book was *The great book of the pulse*. This is a highly useful means of identifying disorders and discovering whether the patient's body has sufficient strength to withstand his particular disorder.

At the fifth level, three works were used. The first of these was *Fevers*. The second was *Crises*. From it the student learned about the times of diseases, and was thus enabled to administer treatment that was appropriate at every time. He also learned to recognize, for every disorder, the tendency of the patient's condition, that is, whether he was likely to recover or not. The third work was *Critical days*. From this, the student acquired knowledge of the times of the crisis and its causes and symptoms.

At the sixth level, a single work was used, *Method of healing*, in fourteen volumes, which set forth the rules governing the treatment of every disorder in accordance with the Dogmatic doctrine.* Every student who immersed himself in this work would inevitably go on to read *Simple drugs* and Galen's works on compound drugs.

At the seventh level, again, only a single work was used, *Regimen for healthy people*, in six volumes, which explains how the health of every human body can be maintained. Every student of this work will inevitably go on to read *Foods*, *Good and bad juices*, *The thinning diet*, and *Requirements for exercise*, a subject also addressed in Galen's work *Exercise with a small ball*.

It is clear, then, that while the Alexandrians relied exclusively on the Sixteen Books in teaching the art of medicine, this approach was well calculated to motivate learners, stimulate their interest in the art of medicine, and impel them to study all of Galen's works.

I—Ibn Abī Usaybiʿah—say, we are also indebted to the Alexandrians for the numerous summaries of philosophical and medical works that they prepared, including in particular their summaries of Galen's works and their commentaries on the works of Hippocrates.

CHAPTER 7

ARAB AND OTHER PHYSICIANS OF THE EARLIEST ISLAMIC PERIOD

Ibn Abī Usaybiʿah now moves to his own world, chronicling the Arabs and physicians of the early Islamic period. The chapter starts with a contemporary of the Prophet Muhammad, the semi-legendary al-Hārith ibn Kaladah, accounts of whom were elaborated over time to the point where it is impossible to assess the historical figure. He is said to have studied medicine at Gondeshapur in Sasanid Persia, but, unhistorially, to have held learned discussions with the Persian ruler Khusraw Anūshirwān (d. 579) and to have been consulted during the final illnesses of the last two of the Orthodox caliphs nearly a hundred years later. The chapter ends with Zaynab, the only woman who has an entry and about whom frustratingly little is known. Several poems are included in this chapter and we start to see the wide role they play in daily social interactions, here providing clues to an illness and a catalyst for release from prison, as well as a moving elegy from another of the very few women mentioned.

1. AL-HĀRITH IBN KALADAH (*d.* 634–5)

AL-HĀRITH came originally from al-Tāʾif, but he travelled to various countries, and he learned the art of medicine in Persia and from the people of Gondeshapur, where he practised for a time; he was familiar with diseases and remedies. Partly there in Persia, and partly in Yemen, he learned the art of playing the lute as well.

While he was in Persia he practised as a physician. The people there relied on him, and he became known far and wide among them as a man of great skill and learning. Once, when he had successfully treated a prominent dignitary, the grateful patient gave him money and a slave-girl whom al-Hārith named Sumayyah. Some time thereafter, al-Hārith became homesick for his native land and returned to al-Tāʾif, where he became renowned among the Arabs for his medical skill, with patients flocking in large numbers to consult him.

There Sumayyah, his *mawlā*,* became the mother of Ziyād ibn Abīhi,* whom the caliph, Muʿāwiyah, recognized as being the son of his own father, Abū Sufyān. Abū Sufyān was said to have bedded Sumayyah illicitly in al-Ṭāʾif, as a result of which she became pregnant with Ziyād. She had also borne two sons previously, both of whom were known by the patronymic 'ibn al-Hārith' (son of al-Hārith).

Al-Hārith was born in the *Jāhiliyyah*, the 'time of ignorance' before the advent of Islam, and became a Muslim. The Messenger of God, may God bless him and keep him, used to advise anyone who was ill to seek treatment from him. He lived during the times of the Messenger of God, may God bless him and keep him.

Ibn Juljul calls al-Hārith 'the physician of the Arabs'. He recounts an anecdote from a man who tells how, when he once fell ill in Mecca, the Messenger of God, may God bless him and keep him, came to visit him and said 'Send for al-Hārith, for he is a man who knows medicine.' Al-Hārith came to see the sick man, and after examining him, he said, 'There is nothing much wrong with him. I recommend some boiled soup with a little date paste and fenugreek in it.' The patient drank the soup and soon recovered.

Al-Hārith knew of many types of treatment and was familiar with the customary practices and requirements of the Arabs in treating diseases. He could discourse eloquently on the art of medicine and other matters, as may be seen from an account of a visit that he undertook to the court of Khusraw Anūshirwān. The monarch bade him enter, and al-Hārith went in and stood erect before him.

'Who are you?' asked Khusraw.

'I am al-Hārith,' he replied.

'What is your trade?' the king asked.

'Medicine,' he said.

'Are you an Arab?' said Khusraw.

'I am,' said al-Hārith, 'Arab to the marrow, from the very depths of Arabia.'

'What have the Arabs to do with doctors,' said the king, 'ignorant and feeble-minded as they are, and with the filthy food they eat?'

'Majesty,' replied al-Hārith, 'if they are as you say, all the more reason why they need someone to enlighten their ignorance, straighten whatever is awry with them, tend their bodies, and balance their constitutions. A rational man knows these things for himself and can identify the site of his ailment, but for the most part he avoids illness altogether by leading a well-regulated life.'

'How do they know what it is that you are explaining to them?' Khusraw asked, 'for if they knew what reason is, they would not have become such a byword for ignorance.'

'Soothe a child, and you can treat him,' said al-Hārith, 'charm a snake, and you can handle it.'

He then added, 'Your majesty, reason is something that is apportioned by God, exalted is He, among His servants, as He apportions people's livelihoods. Every individual obtains his particular share, or more if God grants it, and thus it is that one man may be wealthy and another destitute, one foolish and another wise, one feeble and another determined. All is in the hand of Him who is mighty and omniscient.'

Khusraw marvelled at what al-Hārith had said. 'Do you consider, then,' he enquired, 'that the character, mores and behaviour of the Arabs are admirable in certain respects?'

'Majesty,' answered al-Hārith, 'they are generous, with daring hearts; they are eloquent in their matchless tongue; they have true lineages and noble reputations. Words fly from their mouths like arrows shot from a bow, softer than the air in springtime, sweeter than the beneficent water of the spring Salsabīl* in Paradise. They are sharers of food in the parching drought, smiters of heads in the battle. Their might is unassailable: anyone under their protection is safe from harm, their women cannot be ravished, and the noblest among them cannot be humbled. They do not acknowledge any man as superior to them, except perhaps a mighty king, one unrivalled and incomparable, having no peer, whether subject or monarch.'

After listening to al-Hārith's eloquence, Khusraw sat up straight, looking benevolent, and said to his courtiers, 'There is one who sounds worthy of being taken seriously. He praises his people and describes their virtues, and his words have the ring of truth. He speaks like a man of reason, one who has been taught wisdom by experience of life.' Khusraw then invited al-Hārith to take a seat, and when he was seated the king asked him, 'How much do you really know about the art of medicine?'

'Enough, in all conscience,' replied al-Hārith.

'What is the root of medicine?' Khusraw asked.

'*Al-azm*,' said al-Hārith.

'And what is *al-azm*?' said the king.

'Keeping the lips closed,' answered al-Hārith, 'and being gentle with the hands.'

'Well said!' Khusraw exclaimed. 'Now, what is the most serious of illnesses?'

'Shovelling in more food on top of food already taken will be the ruin of any creature; it is enough to kill wild beasts in the heart of the steppe,' said al-Hārith.

'Very true,' said Khusraw. 'What do you think of the practice of cupping?'

'When the moon is waning,' said al-Hārith, 'on a bright, cloudless day; you should be in a good mood and your blood-vessels in repose as a result of some unexpected happiness or avoidance of care.'

'What do you have to say about entering the bath?' asked the king.

'Never enter the bath when you have just eaten,' answered al-Hārith, 'and moreover, never copulate with your women while intoxicated, or rise from your bed at night while naked, or sit down to eat while angry. Avoid stress, for you will be more relaxed; eat sparingly, and you will sleep better.'

'What of wine?' asked the king.

'The finer the flavour, the more enjoyable,' said al-Hārith, 'the purer, the more wholesome, and the sweeter, the more stimulating to desire. Do not drink it neat, for it will leave you with a headache and bring on an array of illnesses.'

'What is your opinion of fruit?' asked Khusraw.

'All fruits are good when they are properly ripe,' said al-Hārith, 'but do not eat them once they are past their peak of perfection and have begun to spoil. The pomegranate and the lemon are the best fruits. The rose and the violet are the best aromatic plants, and endive and lettuce are the best vegetables.'

'What do you have to say about drinking water?' said the king.

'Water,' said al-Hārith, 'is life itself. It sustains the body. A reasonable amount of water is beneficial, and to drink it when thirsty is good; on the other hand, to drink water upon rising from sleep is harmful. The more wholesome it is, the better. The clearest and purest water is that which comes from great rivers, cold and limpid, unmixed with water from marshes or hills, flowing down over pebbles large and small as they grow.'

'What of its taste?' asked Khusraw.

'It has no definable taste,' was the answer, 'only it is redolent of life.'

'What of its colour?' said the king.

'Its colour is not identifiable to the eye,' replied al-Hārith, 'for it imitates the colour of whatever is holding it.'

The king then asked him, 'How many natural elements make up the body?'

'The body comprises four humours,' said al-Hārith: 'black bile, which is cold and dry, yellow bile, which is hot and dry, blood, which is warm and moist, and phlegm, which is cold and moist. Four humours are just right for equilibrium and support.'

The king said, 'Now summarize "hot" and "cold" for me in a few words.'

'Sweetness is hot,' returned al-Hārith, 'and acidity is cold; whatever is pungent is hot, and whatever is bitter is in balance, for the essence of bitterness is that it is a blend of hot and cold.'

'What is the best treatment for wind?' said the king.

'Gentle enemas,' returned the physician, 'and warm, mild oils.'

'What is your idea of a diet?' Khusraw enquired.

'Moderation in all things,' responded the physician, 'for overeating constricts the vital spirit and blocks up the passages through which it moves.'

'What are your views on women and intercourse with them?' said the king.

'Frequent intimacy is harmful,' al-Hārith replied, 'and beware of it with an elderly woman like a worn-out waterskin, for she will drain your strength and make you ill. Her ejaculate is mortal poison, her breath sudden death; she will take everything from you but give you nothing. A young woman, on the other hand, ah! her ejaculate is sweet and fresh, her embrace coquetry and flirtatiousness, her mouth cool, her saliva sweet, her breath fragrant, her passage tight; she will enhance your strength and make you more active.'

The king then said, 'To what women does the heart most incline, and who amongst them most gladdens the eye?'

'With luck,' al-Hārith returned, 'you will meet one who will be like a cherished egg, softer than butter, sweeter than honey, more delightful than Paradise and the abode of eternity. She will be more fragrant than the jasmine or the rose. Being near her will be a pleasure, and for the two of you to be alone together will be happiness.'

Khusraw's shoulders shook with laughter. 'What are the best times for copulating with women?' he asked.

'Towards the end of the night, for the belly is empty, the breathing is regular, the heart is inclined to desire and the womb is warm then,' replied al-Hārith, 'but you may prefer to take her in daylight, when your eyes will admire the beauty of her face, your mouth gather the fruits of her loveliness, your ear take pleasure in the sweetness of her voice, and all your limbs be at ease because of her.'

'Well spoken, O Arab!' exclaimed Khusraw 'It is clear that you have been given knowledge and granted an abundant share of understanding and intellect,' and he ordered that al-Hārith should be generously rewarded and all that he had said written down.

We are told that al-Hārith once saw some people out in the sun and advised them to go into the shade. 'The sun,' he said, 'will wear out your clothing, take away its fragrance, cause the colour to fade, and may cause some disease that is now dormant to flare up.'

The Commander of the Faithful, ʿAlī ibn Abī Tālib, may God be pleased with him, is reported to have said, 'He who would live forever (but no one lives forever) would do well to take wholesome food, to eat sparingly, to drink when thirsty, to drink little water, to rest after his midday meal, to go out walking after his evening meal, and not to go to bed until he has paid a visit to the latrine. Entering the bath soon after eating will infallibly make a man fall ill. Going to the bath once in summer is better than going there ten times in winter. To eat dried salt meat at night will speed you to the grave. Sexual relations with an old woman will destroy the life of any man living.'

Apparently when al-Hārith lay dying, people came to his bedside and begged him to give them some words for their guidance when he would no longer be with them. 'Choose none but a young woman as a wife,' he said, 'eat only the meat of young animals; eat none but ripe fruit; when you are ill, avoid treatment until the illness has become too much for you to bear; use the depilatory paste, called *nūrah** every month, as it gets rid of phlegm, eliminates bile, and causes flesh to grow; whenever one of you has had his midday meal, he should have a nap immediately afterwards, and after his evening meal, he should go for a stroll forty paces in length.'

Another saying that has been attributed to al-Hārith runs as follows: 'Avoid taking medicine if you possibly can. Drink it if you must, but it is likely to do you as much harm as good.'

Ibn Juljul relates an anecdote that he heard from al-Hasan ibn al-Husayn, who had had it from Saʿīd ibn al-Umawī, who in turn had had it from his uncle, Muhammad ibn Saʿīd, who had heard it from ʿAbd al-Malik ibn ʿUmayr:*

Once there were two brothers,' he said, 'of the tribe of Thaqīf, who had the greatest affection for one another; all history can show no comparable example of such fraternal devotion. The elder brother, being about to set out on a journey, asked the younger to look after his wife in his absence, and the younger brother promised to do so. One day, however, by the merest chance, his glance happened to light upon her, and forthwith he was hopelessly smitten with love for her, and pined away with desire. The elder brother, when in due course he returned, sought the advice of various physicians, but none of them could say what ailed the younger brother. Finally he consulted al-Hārith. 'I see a pair of clouded eyes,' said the physician, 'but it is not clear what is causing the trouble. I should like to try something: let us see what will happen if he has a drink of date-wine.' The young man drank the date-wine, and when he had been affected by it, he recited the following lines:

O, have pity, O have pity,
just a little, for what I shall be!
Bring me, you two,* to the tents
in al-Khayf so that I can visit them:
A certain gazelle I saw today
in the abodes of the Banū Kunnah,
With a smooth cheek, raised in the tribe,
and with a sweet voice.

'Now we shall try giving him a little more date-wine,' said al-Hārith, so they gave him some more, and when it had done its work on him, the patient recited,

Neighbour friends, greetings!
and halt here so that you can speak

And perform a need for me
 and convey greetings and blessings.
A rain-cloud came from the sea,
 water-loaded,* with a wailing sound.
She is not my brother's wife but she asserts
 that I am a brother-in-law to her.

At this, the elder brother divorced his wife on the spot and said, 'She is yours, my brother.' But the younger brother answered, 'By God, I shall not marry her,' and he went to the end of his days without marrying her.

2. AL-NADR, SON OF AL-HĀRITH* (*d.* 624)

His mother was a maternal aunt of the Messenger of God, may God bless him and keep him. Like his father, he travelled widely, meeting with prominent men and scholars in Mecca and elsewhere and associating with learned Christians and Jews. He was a keen student of learning himself, and acquired an extensive knowledge of the sciences of antiquity, philosophy, and other lore. In addition, he learned all that his father had to teach him about medicine and other subjects.

Al-Nadr was a bitter adversary of the Messenger of God, may God bless him and keep him, speaking against him frequently, seeking to denigrate him in the eyes of the Meccans and attempting to hinder his mission. Little did he know, unhappy man, that prophethood is the greatest calling, that divine providence is most sublime, and that what God has decreed must inevitably come to pass. Al-Nadr imagined that with his knowledge, his accomplishments, and his philosophy he could resist prophethood, but how far below the Pleiades is the planet Earth!

At Badr, where the Muslims encountered the polytheists of Quraysh, the Quraysh numbered between 900 and 1,000 men, whereas there were a mere 313 of the Muslims. But God aided Islam and granted victory to His Messenger, peace be upon him, while the ranks of the unbelievers were shattered. The flower of Quraysh were killed, and a number of them were captured. Of the latter, some arranged for their release through the payment of ransom, while others were put to death at the orders of the Messenger of God, may God bless him and keep him, and these included al-Nadr who was executed after the battle was over and the Muslims had left Badr. He was mourned in the following lines by his sister, Qutaylah:

Rider! One supposes you will reach al-Uthayl
    on the morning of the fifth night, if you are lucky.
Convey there, to a dead man, a greeting
    unceasingly brought by swift camels,
From me to him, with tears that are shed
    profusely and other tears that choke.
Let al-Nadr hear if I call him,
    if a dead man can hear or speak.
The swords of the Umayyads came down on him—
    Ah, the bonds of kinship that were torn up there!
As a captive he was led to his death, exhausted,
    walking like a hobbled beast, a prisoner in bonds.
O Muhammad, offspring of a noble dam
    among her people, and whose sire was a sire of highbred stock,
It would not have harmed you if you had pardoned him:
    a man sometimes pardons even when enraged.
Al-Nadr was the nearest relative you took on account of his error
    and the one most deserving to be freed, if there was to be
        any freeing.
If you had accepted a ransom I would have ransomed him
    with the most precious ransom that anyone could spend.

It is said that the Prophet, may God bless him and keep him, said, 'If I had heard those lines before killing him, I would not have killed him.' This elegy by Qutaylah is reputed to be the noblest, purest, most restrained, and most dignified poetry ever uttered by a woman lamenting a murdered kinsman.

3. IBN UTHĀL (active 670)

He was a prominent Christian practitioner who was one of the most distinguished of the physicians of Damascus. When the caliph, Muʿāwiyah ruled in Damascus, he showed Ibn Uthāl particular favour, visiting him frequently and conversing with him in friendly fashion at all hours of the night and day. Muʿāwiyah's reason for cultivating Ibn Uthāl was that the physician possessed extensive knowledge of simple and compound medicines, including highly toxic ones. Many prominent Muslims and persons of rank died of poisoning in Muʿāwiyah's time.

When Muʿāwiyah was seeking to have his son Yazīd proclaimed as heir to the caliphate, he said to the people of Syria, 'The Commander of the Faithful has grown old and wrinkled, his bones have become fragile; he has not much time left. He would like to designate his successor: whom will you have?' The people answered, "ʿAbd al-Rahmān.'* Muʿāwiyah said nothing, keeping his intentions in the matter to himself, but he schemed to have the Christian physician Ibn Uthāl brought into the presence of ʿAbd al-Rahmān. The physician gave him a dose of poison, and he died.

Word of the murder was brought to ʿAbd al-Rahmān's nephew, Khālid, while he was in Mecca. Khālid had no great regard for his late uncle, for he was a Hāshimite* by persuasion, like his father, whereas ʿAbd al-Rahmān had fought on the side of Muʿāwiyah. But a prominent man went to Khālid and said, 'Will you look on as your uncle's bones are desecrated in Syria by Ibn Uthāl, while you spend your time strutting about here in Mecca, full of your own importance and letting your cloak trail behind you as you go?' Khālid was infuriated at this. He summoned a *mawlā* of his, a sturdy, audacious fellow named Nāfiʿ, and informed him of the situation. 'This Ibn Uthāl must be killed,' he said.

They travelled to Damascus together, and there they found that their quarry, Ibn Uthāl, was in the habit of visiting Muʿāwiyah, staying until late in the evening. Accordingly, they lay in wait for him in the Damascus mosque, each with his back against a column. 'I shall strike him myself; just guard my back,' said Khālid to his man. When Ibn Uthāl came abreast of him, Khālid sprang upon him and stabbed him to death. The people who had been accompanying Ibn Uthāl came running up, but upon finding themselves confronted with Nāfiʿ, they fled, and Khālid and Nāfiʿ made their escape.

When Muʿāwiyah was told what had happened, he said, 'This is the doing of Khālid.' After a search, Khālid was taken and brought to Muʿāwiyah. 'May God not reward you, O visitor,' he said, 'you have murdered my physician.' 'I have killed the one who was obeying an order,' replied Khālid, 'but not the one who gave the order.' 'God's curse upon you,' exclaimed the caliph. 'I swear by God, had Ibn Uthāl testified even once that God is one and Muhammad His Messenger, I would have you put to death.'* Khālid, did not have to endure anything more than a term of imprisonment.* However, Muʿāwiyah compelled Khālid's clan to pay 12,000 dirhams as bloodwit for Ibn Uthāl. Of that sum, 6,000 dirhams went into the Treasury, while Muʿāwiyah kept the other 6,000 for himself.

After Muʿāwiyah had had Khālid imprisoned, he composed the following lines in his cell:

> If my steps are close together,
> as someone walks who is hobbled, in detention,
> How can I then walk around in the beds
> of the wadis, dragging my loin-cloth on my traces?
> No more of this. But do you see
> a fire, lit at Dhū Murār?
>
> It is not lit because of the cold
> for someone warming himself, nor for fumigation.
> How is it that the length of your night
> is not diminished by the length of the day?
> Will Time grow shorter,
> or does a prisoner grow weary of prison?

When Muʿāwiyah heard these verses, he had Khālid released at once. Khālid returned to Mecca.

Another account relates:

Muʿāwiyah was afraid that popular sentiment was inclining in favour of ʿAbd al-Rahmān. He therefore dropped a hint to his physician, who gave ʿAbd al-Rahmān a drink of poisoned honey which scorched him inside. Muʿāwiyah thereupon remarked, 'There is no stroke of good fortune to compare with one that disposes of someone you dislike.'

In the year 38/658, ʿAlī ibn Abī Tālib, peace be upon him, dispatched al-Ashtar to Egypt as governor after the previous governor had been murdered. When Muʿāwiyah got wind of this, he sent a message secretly to a local official there: 'If you will kill al-Ashtar, for twenty years you shall keep your land-tax for yourself.' When al-Ashtar arrived, the official received him hospitably and enquired what he most preferred to drink. 'Honey,' was the answer. 'I have some honey from Barqah,' said the official. He poisoned a cup of it and served it to al-Ashtar, who subsequently died. When ʿAlī learned of his death, he said, 'Fallen on hands and mouth.'*

4. HAKAM AL-DIMASHQĪ (*d.* 825)

This physician followed in his father's footsteps, and had an extensive knowledge of the treatment of disease and therapeutic procedures, and was skilled in the art of compounding ingenious medications. He lived in Damascus and was granted a long life.

Ibn al-Dāyah tells us:

I have heard from his son, ʿĪsā, that Hakam died in Damascus in 210/825. Asked how old his father had been, he replied that Hakam had been 105 years old when he died, and that he had still had all his wits about him and had forgotten none of his learning. It was remarked that, 'His life spanned half of history.'*

The same author says:

I have also heard from ʿĪsā that once, while he and his father were out riding in the city of Damascus, they passed by the establishment of a cupper, with a great throng of people crowding around. 'Someone in the crowd noticed us,' he said, 'and called out, "Make way there, make way, this is the famous physician Hakam and his son ʿĪsā." The crowd made way for us, and we saw a man whom the cupper had bled by opening his basilic vein.* Now the basilic vein overlies an artery, and between making too wide an incision and not being sufficiently careful about determining the location of the vein, the cupper had nicked the artery, and was now unable to staunch the bleeding. We tried our usual methods, applying rags, cobweb, and soft animal hair, to no avail. My father asked me if I could suggest any other recourse, but I replied that I knew of none. At that, he called for a pistachio nut. Cracking it in two, he discarded the kernel, took one of the two half-shells, and placed it on the site of the incision. Taking a strip of coarse linen torn from the hem of someone's robe, he bound the arm so tightly with it, pressing the half-shell upon the incision, that the patient cried out, "Help!" After having tightened the bandage even further, my father had the man carried to the Baradā River, where he ordered him to immerse his arm in the water. Making a smooth place on the riverbank, he told the patient to lie there, and he gave instructions that he should be fed soft-boiled egg yolks. Finally, he left one of his students to keep watch, telling him to be sure not to allow the patient to take the incised arm out of the water, except at prayer times, unless there appeared to be a serious risk of his dying of cold, in which case he might remove it for a short time, but he must immerse it again afterwards. Thus matters remained until the end of the day. My father then sent the man home, instructing him not to cover the site of the incision or loosen the bandage for a full five days. The patient complied, but when my father went to see him on the third day, he found that both the forearm and the upper arm were greatly swollen. Accordingly, he loosened the bandage somewhat, saying to the man, "Some swelling is easier to bear than death." On the fifth day, he removed the bandage, and we found the pistachio shell stuck to the flesh. My father said to the man, "You have escaped death thanks to this shell. Do not remove it, but leave it strictly alone until it falls off of itself; otherwise, you are a dead man." '

'On the seventh day' (ʿĪsā's account continues), 'the shell fell off, revealing a thick scab the shape and size of a pistachio nut. My father told the patient not to fiddle with it, scratch the skin around it, or pick any of it off with his fingers. It gradually flaked away until, more than forty days later, the site of the incision became visible, and the man had completely recovered.'

5. ʿĪSĀ SON OF HAKAM AL-DIMASHQĪ
(*d.* after 839)

ʿĪsā, who was known as Masīh,* was the author of the *Greater compendium* which bears his name and has made him famous.

Ibn al-Dāyah tells us:

I have heard from ʿĪsā that when Ghadīd, the *umm walad** of Hārūn al-Rashīd, suffered an attack of colic, she sent for him and also for two astrologers, al-Abahh and al-Tabarī.* She first asked ʿĪsā what treatment he considered appropriate. 'This is such a severe colic,' he replied, 'that your life is in danger; it must be treated with an enema at once.' She then turned to the two astrologers and asked them to select a suitable time for the procedure. Al-Abahh said, 'Your illness is not of such a kind that you can afford to delay until some astrologer deems the time propitious. My view is that you should undergo treatment immediately, lest your condition become worse. That is what ʿĪsā advises, and I concur.'

ʿĪsā's account continues: Then she asked al-Tabarī for his opinion. He replied, 'Today, the moon is in conjunction with Saturn; tomorrow, it will be in conjunction with Jupiter. In my opinion, you should put off the treatment until then.'* Al-Abahh broke in to say, 'I fear that by the time the moon has closed with Jupiter, this colic will have progressed to such an extent that no treatment will be of any avail.' Both Ghadīd and her daughter, Umm Muhammad, thought that that was an unlucky thing to say and Ghadīd announced that she would follow al-Tabarī's advice.

Before the moon had caught up with Jupiter, however, Ghadīd had died. When the moon was in conjunction with Jupiter, al-Abahh sent a message to Umm Muhammad, saying, 'This is the time selected by al-Tabarī for the treatment: where is our patient?' This message enraged Umm Muhammad even further, and she remained hostile to al-Abahh to the day of her death.

Ibn al-Dāyah recounts:

Once, in Damascus, ʿĪsā and I fell into a discussion about onions. He launched into a diatribe against the things, listing various reasons why they were to be regarded as unwholesome. I should mention here that ʿĪsā and Salmawayh* lived like monks and never had a good word to say about foods that promoted sexual activity, holding that it made bodies die and souls perish. I did not think it appropriate to discuss the onion's ability to enhance sexual performance; I simply commented that in the course of my present trip from Samarra to Damascus I had personally experienced one beneficial property of onions. 'What was that?' he asked, and I told him that I had tasted the water of a certain well along the way and found it to be brackish, but that after having eaten some raw onions, I had tasted the water again and found it noticeably less brackish. ʿĪsā was seldom heard to laugh, but he burst out laughing at that. Then a concerned look came over his face, and he said, 'I am distressed to see a man of your intelligence go so wrong. You encountered the very least attractive feature of onions, their very worst defect, and mistook it for a virtue. Is it not true that when some disturbance befalls the brain, the senses—smell, taste, hearing, and sight—are adversely affected?' I acknowledged that that was indeed the case. 'You see,' he said, 'one of the properties of onions is that they induce a disturbance in the brain, and that is what happened to you. You perceived the water to be less brackish only because the onions you had eaten had perturbed the functioning of your brain.'

Ibn al-Dāyah's narrative continues:

When ʿĪsā accompanied me as far as al-Rāhib, his final words to me were, 'My father died at the age of 105, his face unwrinkled and its brightness undimmed, thanks to some practices of his. I shall now tell you what they were, and recommend that you follow them: never to eat dried salt meat, and to wash the hands and feet with water as cold as you can bear upon emerging from the bath. If you do these things, you will find them beneficial.' I have followed his advice, except that I have sometimes sucked on a small piece of dried salt meat, perhaps once in the course of a year or longer.

6. TAYĀDHŪQ (*d. c.*709)

An eminent physician, and the author of a number of pointed aphorisms and pithy sayings about the art of medicine. God granted him a long life: he saw the advent of the dynasty of the Umayyads, who held his medical skill in high esteem. He became closely associated with al-Hajjāj, whom the caliph ʿAbd al-Malik had appointed governor, and served him faithfully. Al-Hajjāj relied on Tayādhūq, had confidence in his treatment, and consequently paid him lavishly and provided abundantly for him.

Tayādhūq provided al-Hajjāj with much medical advice, including, 'Marry none but a young woman; eat only meat from a young animal, well cooked; when you eat a meal during the day, there is no harm in having a nap, but when you eat a meal at night, do not go to bed without having gone for a walk, even if only fifty paces.'

On one occasion, this advice drew the following comment from a bystander: 'If the matter is as you say, how is it that Hippocrates is dead, Galen is dead, and all the others are dead? Not one of them is left; how can that be?' Tayādhūq replied, 'A fair question, young man. Now, pay attention: they managed their lives well in so far as it was in their power to do so, but they were overcome by things over which they had no control.' By that he meant death and external accidents such as heat, cold, falling, drowning, injury, grief, and the like.

Further advice given by Tayādhūq to al-Hajjāj included the following admonitions: 'Do not eat until you are hungry; do not engage in sexual relations if you feel reluctant; never hold in your urine; and always remember that the bath is a good servant but a bad master.'

Tayādhūq also told al-Hajjāj, 'Four things that are injurious to life and have been the death of many a man are entering the bath too soon after eating, sexual relations on a full stomach, eating dried salt meat, and drinking cold water while the stomach is empty. A fifth, sexual congress with an old woman, is not far behind.'

One day, al-Hajjāj was suffering from a bad headache. He summoned Tayādhūq, who advised him to bathe his feet in hot water and then to rub them with oil. One of al-Hajjāj's eunuchs who was present at the time remarked, 'By God, I have never seen a physician so clearly lacking in knowledge of the art of medicine. Here is the emir with a headache, and the remedy you prescribe is all about his feet!' 'And why not?' Tayādhūq retorted, 'consider your own case.' 'How do you mean?' the eunuch asked. 'Why,' said Tayādhūq, 'here your testicles have been removed, and see! the hairs of your beard have all fallen out.' Al-Hajjāj and all the company laughed heartily at this.

Tayādhūq lived to be very old. He finally died in Wāsit about the year 90/709.

7. ZAYNAB, THE PHYSICIAN OF THE BANŪ AWD (eighth century)

A woman who was known as 'the physician of the Banū Awd.* She was a skilled medical practitioner, and was particularly renowned among the Arabs for her expertise in treating sore eyes and wounds.

An author gives an account he had heard transmitted through reliable sources:

I consulted a woman of the Banū Awd and asked her to treat an inflammation of the eyes from which I was suffering. She applied some salve, and then said, 'Lie down, so that the salve can spread over your eyes.' I complied, and as I did so I quoted a line of poetry:

> Will the vagaries of Fate take me away before I have visited
> the physician of the Banū Awd, Zaynab, in spite of the distance?

At this, the woman burst out laughing, and then said, 'Do you know who it is to whom that verse refers?' 'No,' I said. 'None other than myself, by God,' she said, 'I am the Zaynab whom the poet mentions; I am the woman doctor of the Banū Awd. Do you know who wrote the verse?' 'No,' I said again. 'It was your uncle!' she said.

CHAPTER 8

SYRIAC PHYSICIANS OF THE EARLY ABBASID PERIOD

This longer chapter covers the Christian physicians called Suryāniyyūn *in Arabic. The adjective* suryānī *does not mean 'Syrian', but rather (speaker of) 'Syriac',* a form of Aramaic that had become the language of the eastern Christians in the Middle East. Many of them served as court physicians to the Abbasid caliphs, including the six members of the distinguished and very successful Bukhtīshūʿ family with whom the chapter starts. An astonishing list of wealth is given for the earnings of Jibrīl ibn Bukhtīshūʿ, personal physician to Hārūn al-Rashīd and an equally astonishing account of the extravagant luxury in which his son Bukhtīshūʿ ibn Jibrīl (physician to the caliph al-Mutawakkil) lived. While Christians were clearly generally accepted and well treated at the Muslim court, their potentially delicate position is illustrated by the story of Hunayn's temporary downfall at the hands of other Christian physicians—related in the first of four autobiographies included in the book.*

I PROPOSE to begin with accounts of Jūrjis, his son Bukhtīshūʿ, and other distinguished members of that family* in chronological order, proceeding subsequently to accounts of other eminent physicians of that period.

1. JŪRJIS (*d.* 769)

Jūrjis was experienced in the art of medicine and was knowledgeable about remedies and methods of treatment. He enjoyed the favour of the caliph al-Mansūr, who employed him as his physician, and in consequence acquired considerable prestige and substantial wealth.

We read in the account of Pethion the Translator that the caliph al-Mansūr first sent for Jūrjis when he fell ill in 148/765, the year in which he built Baghdad, the City of Peace.

As soon as Jūrjis reached the caliph's court, al-Mansūr ordered Jūrjis brought before him. Upon entering, Jūrjis greeted the caliph in Persian and in Arabic. Al-Mansūr was greatly impressed at his visitor's distinguished demeanour and cultivated speech, and bade him be seated. He then asked the physician about various matters, Jūrjis responding with composure. Finally al-Mansūr said, 'It seems to me that in you I have found the man I have been looking for,' and he proceeded to describe his illness and its initial symptoms. 'I shall treat you in a way you will like,' replied Jūrjis, whereupon the caliph ordered the servants to bring a robe of honour for the physician and said to his chamberlain, 'Have one of our best suites of apartments placed at his disposal, and see to it that he is treated as an honoured guest.'

The next morning, Jūrjis went to see al-Mansūr. He felt his pulse, examined a phial of his urine, and persuaded him that a lighter diet would be beneficial. With such conservative treatment, the caliph was soon his old self again. Delighted at this outcome, he ordered that the physician's every wish should be granted. A few days later, however, al-Mansūr said to the chamberlain, 'That man looks discontented to me. You haven't been depriving him of something that he usually drinks, have you?' 'Well,' said the chamberlain, 'we're not allowing him to bring wine into the palace.' At this, the caliph cursed him roundly and said, 'You yourself shall go and obtain for him as much wine as he wants.' So the chamberlain went out to Qutrabbul (a village near Baghdad, known for its wine) and brought back as much good wine as he could carry for Jūrjis.

Pethion's account continues:

In the year 151/768, Jūrjis went to see the caliph on Christmas Day. 'What shall I eat today?' enquired al-Mansūr. 'Whatever the Commander of the Faithful chooses,' replied Jūrjis, and made as though to take his leave, but when he had reached the door, the caliph called him back. 'Who are your servants here?' he asked. 'My pupils,' answered Jūrjis. 'I have heard that you have no wife,' said al-Mansūr. 'I do have a wife,' Jūrjis said, 'but she is old and frail, and cannot leave home to be with me here,' and he left the caliph's presence and went to the church. The caliph then ordered his eunuch to select three attractive Roman slave-girls and to deliver them to Jūrjis, along with the sum of 3,000 dinars.

The eunuch did as he had been ordered to do, and when Jūrjis returned home, his pupil informed him what had happened and showed him the slave-girls. Jūrjis was not pleased. 'Pupil of Satan! Why did you let them into my house? Go at once and take them back to their owner.' Mounting his mule, he rode with his pupil and the slave-girls to the caliph's palace and handed the girls over to the eunuch. When al-Mansūr heard of the matter, he sent for Jūrjis and asked him why he had returned the slave-girls. 'Such persons cannot stay in the same house with me,' answered Jūrjis, 'because we Christians marry one woman only, and as long as she lives, we take no other wife.' Al-Mansūr was filled with admiration, and immediately gave orders that Jūrjis should be allowed admittance to the quarters of his wives and concubines and that he should serve as their physician. This incident enhanced his prestige even further in the caliph's eyes.

Pethion continues:

In the year 152/769, Jūrjis fell gravely ill. The caliph sent servants daily to obtain news of him, and when his condition worsened, al-Mansūr ordered him carried in a litter to the public reception salon of the palace. There the caliph walked out to greet him and ask how he was. 'O Commander of the Faithful, may God prolong your life,' cried Jūrjis, his tears flowing copiously, 'if you would but permit me to return to my native land to see my wife and children and, if I die, to be buried there with my ancestors.' 'Fear God and accept Islam, Jūrjis,' said the caliph; 'for I can guarantee that you will see Paradise.' 'No,' replied Jūrjis, 'I shall die in the religion of my ancestors. I want to be with them wherever they are, be that Paradise or Hell.' The caliph laughed at this. 'I have been in the best of health ever since I set eyes on you,' he said. 'I used to fall ill all the time, but no longer.' The caliph ordered that Jūrjis should be given 10,000 dinars and allowed to return to his native land, with a servant to accompany him. 'If he should die on the way,' said al-Mansūr to the servant, 'have his body carried to his home, so that he can be buried there as he desires to be.' In the event, however, Jūrjis was still alive when they reached Gondeshapur, his ancestral home.

2. BUKHTĪSHŪʿ SON OF JŪRJIS
(*d.* probably before 801)

Bukhtīshūʿ means 'servant of the Messiah', for in the Syriac language *bukht* means 'servant' and *Yashūʿ* is Jesus,* peace be upon him. Bukhtīshūʿ

ibn Jūrjis knew as much as his father did about the art of medicine and was no less skilful. The time of his greatest renown was during the reign of Hārūn al-Rashīd, whom he served as personal physician.

We read in Pethion's *History* that when the caliph al-Hādī fell ill, he sent an emissary to Gondeshapur to bring Bukhtīshūʿ to Baghdad, but the caliph died before he had arrived.

The chamberlain had said to him, 'We have heard reports of a competent physician, named ʿAbdīshūʿ. The caliph immediately ordered the physician brought to him. When he arrived, the chamberlain took him to see al-Hādī. 'Have you seen my phial of urine?' the caliph asked him. He replied, 'I have, O Commander of the Faithful, and I am here to prepare a remedy for you. Nine hours from now you will be well and rid of your illness.' After leaving the caliph's presence, he said to the other physicians, 'Do not be concerned. You shall all return to your homes later today.' The caliph had given ʿAbdīshūʿ 10,000 dinars to enable him to purchase the necessary medicines, but he had the money carried to his own home instead. He did have some medicinal materials brought to the palace, however. Calling the physicians together in a room not far from where the caliph lay, he said to them, 'Pound this stuff vigorously so that he can hear you; that will keep him easy in mind. By the end of the day, there will be nothing to keep you here any longer.' Every hour, the caliph would call him in and enquire about the remedy, and he would answer, 'It is almost ready, you can hear my servants pounding the ingredients,' whereupon the caliph would say no more. After nine hours the caliph died, and the physicians were allowed to go home. This incident took place in 786.

Pethion continues:

In the year 174/790, when Hārūn al-Rashīd was afflicted with a migraine, he complained to Yahyā ibn Khālid [ibn Barmak], 'These physicians are no good at all.' 'When your brother al-Hādī fell ill,' said Yahyā, 'your father sent to Gondeshapur for a man by the name of Bukhtīshūʿ.'

It was not long before Bukhtīshūʿ reached the court. Upon being presented to Hārūn al-Rashīd, he greeted him in both Arabic and Persian. Hārūn al-Rashīd, ordered one of the servants to bring a phial of urine from the stables. No sooner had Bukhtīshūʿ set eyes on it than he exclaimed, 'O Commander of the Faithful, this is no human urine.' 'Wrong! This is urine from one of the ladies of the harem.'

'O venerable shaykh,' said Bukhtīshūᶜ, 'as I stand here, I declare this is no urine ever excreted by any human being. If the matter is as you say, it would appear that the lady in question has been turned into an animal.' 'What makes you so sure that it is not human urine?' asked the caliph. 'Its consistency, colour, and odour are not those of human urine,' replied Bukhtīshūᶜ. 'Who taught you your art?' enquired Hārūn al-Rashīd. 'My father, Jūrjis,' said the physician. 'It is true, his father's name was Jūrjis,' chorused the others, 'and he had no peer in his time.' The caliph then turned to Bukhtīshūᶜ and asked him, 'What sort of diet would you recommend for the patient who produced this urine?' 'Some of your best barley,' said the physician. This sally caused Hārūn al-Rashīd to roar with laughter. He then ordered that Bukhtīshūᶜ should be given a fine, costly robe of honour and a well-filled purse. 'Bukhtīshūᶜ,' he said, 'shall henceforth be chief among my physicians; the others shall be under his orders and shall obey him in all matters.'

3. JIBRĪL SON OF BUKHTĪSHŪᶜ (*d.* 827–8)

Jibrīl ibn Bukhtīshūᶜ was renowned for his outstanding skill in medical practice. He was a high-minded person, and was very successful, enjoying as he did the favour of caliphs who held him in high regard and treated him with the utmost generosity. Thanks to them, he became wealthier than any other physician.

Pethion informs us that one of Hārūn al-Rashīd's concubines, having extended one arm, found herself unaccountably unable to lower it again. The physicians treated her with embrocations and oils, but to no avail. 'I know of a highly competent doctor,' said the vizier, Jaᶜfar,* to Hārūn al-Rashīd, 'he is Jibrīl, Bukhtīshūᶜ's son.' Accordingly, Jaᶜfar had him summoned, and when he arrived, Hārūn al-Rashīd asked him 'What do you know about medicine?' 'I can cool what is unnaturally hot, warm what is unnaturally cold, moisten what is unnaturally dry, and dry what is unnaturally moist,' said the physician. The caliph laughed. 'Well, that's what the art of medicine is all about, isn't it?' he said, and then he explained the strange case of the slave-girl. 'O Commander of the Faithful,' said Jibrīl, 'will you bring her out here for all to see, so that I can do what I want to do, and I beseech you, give me time, without becoming angry at once.'

Hārūn al-Rashīd had the girl brought out, and as soon as Jibrīl saw her he ran towards her, stooped, and seized the hem of her robe, as though intending to lift it revealingly. The girl, in her alarm for her modesty, found that her limbs functioned after all: she lowered her arms, grasped the cloth with both hands, and held it down. 'There, O Commander of the Faithful,' said Jibrīl, 'she is cured.' 'Hold both your arms out, the left and the right,' said Hārūn al-Rashīd to the girl, and she did so, much to the amazement of all the company, including the caliph, who awarded Jibrīl 500,000 dirhams then and there, and made him chief of all the court physicians.

Jibrīl was asked about the cause of the girl's illness. 'During coitus,' he explained, 'a thin humour flooded this girl's limbs, owing to the movement and the diffusion of heat. When the movement of sexual intercourse ceased suddenly, the remnant of that humour congealed inside all her nerves, and there was nothing that could release it but further movement. My idea was to cause its heat to diffuse and thus release that remnant of humour.'

Pethion continues:

Jibrīl went from strength to strength, to such an extent that Hārūn al-Rashīd got into the habit of saying to his companions, 'Anyone who wants a favour from me, speak to Jibrīl about it; I do anything he asks me to do, and grant him whatever he requests.' Even military commanders used to come to Jibrīl when they needed anything. His situation was unassailable, for Hārūn al-Rashīd never had a day's illness for fifteen years after Jibrīl had entered his service, and the caliph attributed his continued good health to his physician.

Ibn al-Dāyah relates the following account:

Sulaymān, a eunuch from Khorasan who was a *mawlā** of Hārūn al-Rashīd's, told me that one time when the court was in al-Hirah, he was in attendance on the caliph as he ate his noon meal one day, when in came a platter with a splendid fat fish on it, which was placed before Hārūn al-Rashīd, along with some stuffing for it. Hārūn al-Rashīd made as though to take some of it, but Jibrīl stopped him, making a sign to the steward to indicate that he should take the dish away. The caliph was well aware of the byplay. When the table was cleared and Hārūn al-Rashīd washed his hands, Jibrīl left the room.

Sulaymān continued, 'Hārūn al-Rashīd ordered me to follow him, keeping out of sight, to see what he was up to, and to report back.

I obeyed, but I think Jibrīl had spotted me, for he acted very cautious. He went to a room and called for food; when it was brought, it included the fish. He then called for three silver goblets. Into one of these he put a morsel of the fish and over it poured some wine, without adding any water, and said, "This is for Jibrīl." He put another morsel of the fish in another of the goblets, poured snow-cooled water over it, and said, "This is for the Commander of the Faithful if he does not eat anything else with his fish." Lastly, he put another morsel of the fish into the third goblet, added several pieces of meat of various kinds, sweets, pickled dishes, chicken, and vegetables, and poured snow-cooled water over all, saying, "This is the food of the Commander of the Faithful if he eats other things with his fish." He then handed the three goblets to the steward and said "Keep these until the Commander of the Faithful awakens from his siesta."

'Jibrīl then attacked the fish himself and ate until he could eat no more. From time to time, feeling thirsty, he would call for a cup of wine, which he would drink neat. Finally he went to sleep. When Hārūn al-Rashīd awoke, he sent for me and asked me what I had found out. When I told him what I had seen, he had the three goblets brought to him. In the goblet into which the wine had been poured, the morsel of fish had disintegrated completely; there was not a scrap of it left. In the second goblet, the one which had had snow-cooled water poured into it, the morsel of fish had swollen to more than double its former size, while the contents of the third goblet, the one with the meat and fish, had developed a putrid odour. Hārūn al-Rashīd immediately ordered me to take 5,000 dīnārs to Jibrīl. "Who could blame me for regarding that man with affection, considering the way he looks after me?" he said. Accordingly, I delivered the money to Jibrīl.'

The following account is also from a work by Ibn al-Dāyah:

I have heard the following story from the eunuch Faraj, in his own words: 'My patron, brother of the caliph al-Maʾmūn, was the governor of Basra. When Jibrīl was having his house built on the esplanade, he asked my patron to give him five hundred teak logs.* Now a teak log costs 13 dinars, and to my patron, that seemed like a pretty costly gift, so he refused.'

The eunuch's account continues:

I said to my patron, the governor, 'I suspect Jibrīl is planning to do you a bad turn, out of resentment.' 'Jibrīl is no threat to me,' he replied, 'for I never ask him to prescribe medicine for me or to treat me.' But subsequently, my patron decided to visit his brother, the caliph. When he was seated in al-Maʾmūn's presence, Jibrīl said to the caliph, 'You do not look well, Sire,' and he went up to him and felt his pulse. Then he said, 'The Commander of the Faithful should take a drink of oxymel* and postpone his noon meal until we know what is going on.' The physician took his pulse over and over again. Finally, some of his servants came in with a loaf of bread and some squash, beans, and such like. Jibrīl said to the caliph, 'I do not think the Commander of the Faithful should eat meat today. He should eat only foods like these.' The caliph ate them, and then took his siesta. When he awoke, Jibrīl said to him, 'O Commander of the Faithful, the aroma of date wine causes increased heat. It would be best to withdraw,' and al-Maʾmūn withdrew. All the expense my patron had incurred had gone for naught! Then Jibrīl remarked to me, 'You know, the saving on teak logs hardly outweighs the cost of a visit to the caliph.'*

Another account by Ibn al-Dāyah runs as follows:

I had this story from Jūrjis ibn Mīkhāʾīl. His uncle, Jibrīl ibn Bukhtīshūʿ, had told him that on 1 Muharram 187/30 December 802, he had been concerned because the caliph, Hārūn al-Rashīd, had no appetite. He could find nothing from an examination of the caliph's urine or from his pulse to indicate that he was ill, so he said, 'O Commander of the Faithful, you are in excellent health, praised be God, so I am at a loss to explain why you have not finished your breakfast.' This was two days and a night before he had the vizier, Jaʿfar executed.

He ordered Jaʿfar, who was also fasting, to attend him at supper. Hārūn al-Rashīd ate almost nothing, and Jaʿfar said to him, 'O Commander of the Faithful, you should try to eat something.' 'I prefer to spend the night without too much in my stomach,' said the caliph, 'so that I feel hungry when I awake and take my morning meal with the ladies of the harem.'

Early Friday morning, the caliph went out on horseback, accompanied by Jaʿfar, for a breath of fresh air. As I watched, I saw that he had put his arm into Jaʿfar's sleeve all the way to his body, drawn him close, embraced him, and kissed his eyes. They trotted off for a distance of more than 1,000 cubits, hand in hand all the way, and then returned. 'By my life,' said Hārūn al-Rashīd to Jaʿfar, 'you have made this a day of happiness. Now I am going to be occupied with my household.'

Then, turning to me, he said, 'Jibrīl, I am going to have something to eat. You stay here.' Accordingly, I went with Jaʿfar, who had some food brought, and we ate together. He then sent for the blind singer, Abū Zakkār, to perform for the two of us alone.

So we passed the time until after the night-prayer. Then Masrūr the Elder* and his second-in-command, Harthamah,* accompanied by a substantial body of men-at-arms, burst in upon us. Harthamah grasped Jaʿfar by the arm and said to him, 'On your feet, filthy swine!'

Jibrīl's account continues:

No one said anything to me, so I went home, comprehending nothing. But before I had been there half an hour, a messenger from Hārūn al-Rashīd arrived and said I was to go and see the caliph at once. When I entered, there was Hārūn al-Rashīd with Jaʿfar's head in a large basin in front of him.* 'Good evening, Jibrīl,' he said. 'Weren't you asking me this morning why I was not hungry?' 'Yes, O Commander of the Faithful,' I said. 'It was because I was thinking about that which you see before you,' he explained, 'but tonight, Jibrīl, tonight I am as hungry as a camel. Bring on supper! You will be astonished at how much I can eat now. I was eating a bite at a time so as not to overburden my stomach and make myself ill.' He then called for food, fell to at once, and ate heartily.

I—Ibn Abī Usaybiʿah—say: the length of Jibrīl ibn Bukhtīshūʿ's service as Hārūn al-Rashīd's physician was twenty-three years.

An account book was found in his son's cabinet containing records written by Jibrīl's secretary with corrections in Jibrīl's own handwriting, showing how much he had earned during his service. His regular salary was 10,000 silver dirhams monthly, which makes 120,000 dirhams a year, and thus in the course of his twenty-three years he made a total of 2,760,000 dirhams.

For his accommodation, he received 5,000 dirhams monthly, making 60,000 dirhams a year. Over his twenty-three years, then, he received 1,380,000 dirhams.

In addition, he received a special allowance of 50,000 dirhams in the month of Muharram each year, making a total of 1,150,000 dirhams over twenty-three years.

He also received robes worth 50,000 dirhams every year, making a total of 1,150,000 dirhams over twenty-three years. The following is a list of his robes: gold brocade, twenty; silk brocade, twenty; Mansūrī silk,* ten; ordinary silk, ten; Yemeni embroidery, three; embroidery from Nisibis, three; cloaks, three. The linings were of sable, fox, marten, ermine, and squirrel. In addition, he was paid 50,000 silver dirhams every year at the beginning of the Christian fast, totalling 1,150,000 dirhams over twenty-three years.

Furthermore, every year on Palm Sunday he received a robe of gold or silk brocade, embroidery, or similar worth 10,000 dirhams, making 230,000 dirhams over twenty-three years.

He also received 50,000 silver dirhams on the ʿĪd al-Fitr* every year, totalling 1,150,000 dirhams over twenty-three years, in addition to robes reportedly worth 10,000 dirhams, making a total of 230,000 dirhams.

He bled Hārūn al-Rashīd twice yearly, and was paid 50,000 silver dirhams, thereby earning 100,000 dirhams in the year, and 2,300,000 dirhams over twenty-three years.

Similarly, he gave Hārūn al-Rashīd a dose of physic twice a year, receiving 50,000 dirhams each time, thereby earning 100,000 dirhams a year, and 2,300,000 dirhams over twenty-three years.

In addition, from Hārūn al-Rashīd's entourage he received (over and above clothing, perfumes, and horses worth 100,000 dirhams) 400,000 dirhams yearly. This makes a total of 9,200,000 dirhams over twenty-three years.

His revenue from estates in Gondeshapur, al-Sūs, Basra, and the Sawād came to 800,000 silver dirhams net of contributions due, equalling 18,400,000 dirhams over twenty-three years.

His revenue from surplus contributions due was 700,000 silver dirhams per year, making 16,100,000 dirhams over twenty-three years.

From the Barmakids,* he received 2,400,000 silver dirhams yearly, as follows: Yahyā ibn Khālid, 600,000 dirhams; Jaʿfar ibn Yahyā, the vizier, 1,200,000 dirhams; and al-Fadl ibn Yahyā, 600,000 dirhams. This makes a total of 31,200,000 dirhams over twenty-three years.

All the above figures relate to the time of Jibrīl's employment in the service of Hārūn al-Rashīd, which was twenty-three years, and in the service of the Barmakids, which was thirteen years. They do not include substantial gifts that are not shown in the account book; these amount to a total of 88,800,000 silver dirhams, comprising one of 85,000,000 dirhams, one of 3,400,000 dirhams, and one of 400,000 dirhams.

The following is a record of expenditures from these revenues, and from the unlisted gifts, for various purposes, according to the account book: in gold, 900,000 dinars, and in silver, 90,600,000 dirhams.*

A detailed breakdown of the objects of those expenditures is as follows: personal items, approximately 2,200,000 dirhams yearly, making a total for the above-mentioned period of 27,600,000 dirhams; houses, gardens, parks, slaves, horses, and camels, 70,000,000 dirhams; tools, wages, workshop facilities, and so forth, 1,000,000 dirhams; estates that Jibrīl purchased as his exclusive property, 12,000,000 dirhams; gems and items of jewellery valued at 500,000 dinars, 50,000,000 dirhams; outlays for charity, donations, good works and alms, and the cost of his share of amounts paid to avoid confiscations during the above-mentioned years, 3,000,000 dirhams; sums written off as a result of defalcation on the part of persons entrusted with investments, 3,000,000 dirhams.

After all this, when Jibrīl was at the point of death, he made a will bequeathing 7,000,000 dinars to his son, Bukhtīshūʿ, and naming al-Maʾmūn as executor. The caliph scrupulously saw to it that Bukhtīshūʿ received every dirham of his legacy.

A number of aphorisms are attributed to Jibrīl ibn Bukhtīshūʿ, including the following:

'There are four things that will destroy life: shovelling in more food before the previous food has been digested; drinking on an empty stomach; marriage with an elderly woman; and luxuriating in the bath.'

4. BUKHTĪSHŪʿ SON OF JIBRĪL (*d.* 870)

Bukhtīshūʿ ibn Jibrīl was an honourable man, an individual of excellent character, and he enjoyed a highly successful career as well, achieving greater eminence and prestige and acquiring greater wealth than any other physician of his time; his wardrobe and carpets were as fine as any that the caliph al-Mutawakkil had to show.

Pethion gives the following account:

During the caliphate of al-Wāthiq, the vizier and chief judge hated Bukhtīshūʿ ibn Jibrīl, for his virtuous character, piety, honour, generosity, and general decency made them envious. Accordingly, they took advantage of every opportunity to besmirch his reputation with the result that in 230/844, al-Wāthiq turned against Bukhtīshūʿ, confiscated his property and much of his wealth, and had him exiled to Gondeshapur [from Baghdad].

During the caliphate of al-Mutawakkil, however, Bukhtīshūʿ ibn Jibrīl's fortunes improved: he became a man of enormous importance, prestige, reputation, prowess, and wealth. His apparel, perfumes, carpets, estates, opulent lifestyle, and conspicuous expenditure beggared description, with the result that the caliph became envious and had him arrested. However, when al-Mustaʿīn became caliph, he restored Bukhtīshūʿ to his old position and treated him most favourably.

Subsequently, when al-Wāthiq's son became caliph taking the name al-Muhtadī, he followed the example of al-Mutawakkil in favouring physicians and granting them advancement and privilege, and Bukhtīshūʿ ibn Jibrīl enjoyed particular prestige. He told the caliph about the exactions to which he had been subjected during the reign of al-Mutawakkil, and al-Muhtadī ordered that the physician should be given access to the vaults, and that everything which he recognized as belonging to him should be restored to him, with no verification and nothing deducted. Consequently, he recovered everything that he had lost; the caliph allowed him to take it all, and granted him full protection.

Another man relates the following anecdote, told to him by his grandfather:

I called on Bukhtīshūʿ ibn Jibrīl (the old man said), on a very hot day. I found him seated in a sitting-room cooled by a punkah,* with two windows for ventilation and a blacked-out window between them. In the centre was a pavilion with an embroidered cover the material of which was impregnated with rose water, camphor, and sandalwood oil. Bukhtīshūʿ, to my astonishment, was wearing a heavy robe and wrapped up in a shawl. When I joined him in the pavilion, however, I found the air icy cold. He laughed, and ordered his servants to bring a robe and shawl for me. 'Boy,' he said to one of them, 'uncover the sides of the pavilion,' and there were a number of open doors, and beyond them, banks of snow, with servants fanning it. That was the source of the cold that I had felt. Bukhtīshūʿ then called for supper, and a most elegant table was brought in, with dainties of every kind upon it. Then came roast chickens, done to a turn; the cook entered and gave each of them a shake, whereupon it fell to pieces. Seeing how impressed I was with their colour and the tenderness of the meat, he explained that the chickens were fed nothing but almonds and cottonseed, and were given pomegranate juice to drink.

I called on him again one bitterly cold day in the dead of winter. I found him wearing a padded robe and gown, sitting in a sort of pergola inside his house, in the midst of a most beautiful garden. The pergola was covered with a fine display of sable fur, surmounted by a cover of dyed silk, felt from the west, and Yemeni leather mats. Before Bukhtīshūʿ stood a gilded brazier, made of silver and pierced with holes, in which a servant dressed in a resplendent tunic of gold and silver brocade was keeping a fire of aloes-wood burning. I joined him in the pergola, and found that it was very warm inside. Bukhtīshūʿ laughed and ordered the servants to bring me a tunic of gold and silver brocade as well. He then told them to uncover the sides of the pergola, and I beheld adjoining compartments with wooden lattices over iron ones, and in them braziers in which fires of tamarisk* charcoal were burning, with servants pumping bellows like so many blacksmiths. Bukhtīshūʿ then ordered supper, which proved to be excellent and impeccable as usual. Some very pale chickens were served. I did not like the look of them, for I feared that they were underdone, but the cook came in and shook them, and they fell to pieces. I asked him about them, and he said that the chickens were fed on shelled almonds and given milk to drink.

The following account is also related by the same source:

One day, al-Mutawakkil said to Bukhtīshūʿ ibn Jibrīl, 'Invite me to dinner.' 'Your wish is my command, Sire,' said the physician. 'I want it to be for tomorrow,' said the caliph. 'It will be my pleasure, Sire,' Bukhtīshūʿ replied.

It was summer, and the weather was very hot. 'Everything seems to be in order,' said Bukhtīshūʿ to his friends and associates, 'except that we don't have enough canvas for punkahs.' Accordingly, he sent out emissaries to buy all the canvas they could find. They brought back the canvas, and they also brought with them all the upholsterers and craftsmen they could find. They cut enough canvas to make punkahs for the entire house—courtyards, chambers, salons, apartments, and latrines—so that wherever the caliph might go, he would find one. Then Bukhtīshūʿ thought of the smell of new canvas, which disappears only after the fabric has been in use for a time. He ordered his servants to buy as many melons as they could in the town, and had his staff and slave-boys sit and rub the canvas with the melons all night long, so that in the morning the smell had vanished.

Al-Mutawakkil arrived and saw how many punkahs there were and how fine they were. 'How have you managed to get rid of the smell?' he asked, whereupon Bukhtīshūʿ told him the story of the melons, which astonished him. The caliph, his cousins, and his secretary, al-Fath ibn al-Khaqan, ate at one table, while the princes and chamberlains were seated at two gigantic leather mats, the like of which had never been seen in the possession of any of Bukhtīshūʿ's peers.

Equal baskets of food were distributed to the domestics, servants, intendants, equerries, valets, salt-boys, and other staff, every man his own basket. 'In this way,' Bukhtīshūʿ said, 'I have escaped opprobrium; for had I seated them at tables, one might have been content and another dissatisfied, one might have said he had eaten enough, while another might have said he had not had sufficient. By giving each one his own basket, I have ensured that there was enough for all.' The caliph looked at the banquet and pronounced it most splendid.

Al-Mutawakkil then said that he wished to take a nap, and asked Bukhtīshūʿ to put him in a well-lighted room that was free of flies, thinking thereby to cause him embarrassment. But Bukhtīshūʿ had had urns of treacle placed on the roofs of the house, so that the flies would gather there, and there was not a single fly in the lower apartments. When al-Mutawakkil lay down for his nap, he began to smell extraordinarily sweet odours that he was unable to identify. The caliph was so puzzled that he told his secretary, al-Fath, to find out where it came from. He found outside, in every possible nook and cranny, small, narrow, window-like apertures stuffed full of aromatic plants, fruits, perfumes, and scents, with mandrake* in them, and melons with the flesh scooped out and replaced by mint and Yemeni basil sprinkled with rose water, essence of saffron, camphor, aged wine, and turmeric. Servants had been stationed at these apertures, and by each servant was a brazier containing incense, which was burning and giving off its aromatic smoke. The inner side of the chamber was painted with a coat of white lead pierced with tiny, imperceptible holes, and it was through these that the delightful odours and perfumes were wafted in.

When al-Fath returned and told al-Mutawakkil what he had seen, the caliph was greatly astonished, and was consumed with envy of Bukhtīshūʿ ibn Jibrīl because of his generosity and the impeccable hospitality he had provided and thereafter felt a brooding hatred towards the physician. A few days later, Bukhtīshūʿ was out of favour: the caliph dismissed him and confiscated an incalculable quantity of wealth from him. His wardrobe was found to include four thousand pairs of pantaloons made of fine *sīnīzī* linen,* each of them with a waistband of Armenian silk.

One day the caliph was sitting on a divan in the middle of his private apartment, and Bukhtīshūʿ ibn Jibrīl sat down beside him, as he usually did. Bukhtīshūʿ was wearing an outer robe of Roman silk brocade, which was split a little up the front. As the caliph talked with him, he took to playing with the split, which he followed up as far as the edge of the waistband of Bukhtīshūʿ's pantaloons. They had reached a point in their conversation where al-Mutawakkil had occasion to ask Bukhtīshūʿ, 'How can you tell when a confused person is in need of firm treatment and stability?' 'When he has followed the split in his physician's robe up as far as the waistband of his pantaloons,' answered Bukhtīshūʿ, 'then we adopt firm treatment.' The caliph fell over and lay on his back laughing, and awarded Bukhtīshūʿ splendid robes of honour and a substantial purse on the spot.

Al-Bīrūnī relates* that when Zubaydah,* the wife of Hārūn al-Rashīd, was near death from a severe case of hiccups, her hiccupping could be heard from outside the chamber where she lay. Bukhtīshūʿ's father had ordered the servants to carry some large jars up to the roof overlooking the courtyard, line them up on the edge all the way around, and fill them with water, with a servant behind each jar. When he clapped his hands, they pushed the jars over the edge, making such a crash that Zubaydah started up in alarm—and found that her hiccups had vanished.

After Bukhtīshūʿ's death, his son and his three daughters suffered the exactions of viziers and other officials, who took so much of their wealth that in the end they scattered and went their diverse ways. Bukhtīshūʿ died on a Sunday with eight days remaining in the month of Safar in 256/29 January 870.

5. JIBRĪL IBN ʿUBAYD ALLĀH (*d.* 1006)

Jibrīl ibn ʿUbayd Allāh was a deeply learned man who was proficient in the art of medicine, both as a skilful practitioner and as a master of medical theory. He is the author of a number of admirable works on the subject. The art of medicine had been practised for several generations in his family, and each of his predecessors had been outstanding in his own time and had been celebrated as the marvel of the age.

The following account is taken from a biography by his son:

My grandfather, he says, was an official in the financial service for the caliph al-Muqtadir at the palace. He soon died, leaving two young children, including Jibrīl ibn ʿUbayd Allāh, my father. On the night of my grandfather's death, al-Muqtadir dispatched eighty chamberlains to carry off all the chattels, furniture, and plate that they could find. After he had been interred, his wife fled to ʿUkbarā. Subsequently, she married a physician. When she died, not long after her marriage, her husband took possession of everything and ordered the boy out.

Young Jibrīl made his way to Baghdad with almost nothing but the clothes he stood up in. There he was taken in by a physician who was a close associate of al-Muqtadir and one of his personal physicians; he provided Jibrīl with a place to stay and taught him the rudiments of medicine. There were maternal uncles of his living in the Roman quarter, and sometimes he would go to visit them. They were not sympathetic towards him, criticizing his interest in learning and the art of medicine; they would ridicule him, saying, 'This young fellow wants to be like his grandfather Bukhtīshūʿ and his great-grandfather Jibrīl; he is not content to be like his uncles.' Jibrīl, however, paid no attention to their jeers.

One day, an emissary arrived from Kirmān, bringing Muʿizz al-Dawlah* a striped ass, a man seven *shibr* tall, and another man who was only two *shibr* in height. As it happened, he stayed not far from the shop where my father Jibrīl carried on his practice. The emissary used to go there frequently, to sit and chat with him in friendly fashion. One day, he sent for my father to ask whether he should have himself bled. My father advised him to have the operation done, performed it himself, and then attended his patient assiduously for two days. Afterwards, the emissary sent him, in the fashion of the Daylamites,* a tray holding bandages, a metal basin, a ewer, and all the necessary implements. He then summoned my father and asked him to visit his household to see what medical care they might require. The establishment included a slave-girl of whom the emissary was very fond. She was suffering from a chronic loss of blood. When my father saw her, he devised a plan featuring a medicine that he made her drink. In less than forty days she was whole and well, much to the emissary's satisfaction; some time later, he summoned my father and presented him with 1,000 dirhams, a gown of siglaton* fabric, a Nubian robe, and a turban of gold and silver brocade. The slave-girl also gave him 1,000 dirhams and two of every kind of apparel. In addition, he was given an equipage of a mule and litter with a black slave to run behind. Thus outfitted, he made a more magnificent showing than any of his maternal uncles. When the uncles saw him, they would salute him and receive him graciously, whereupon he would say to them, 'It is the clothes you are honouring, not me!'

The reports of Jibrīl's medical accomplishments caused the physician's renown to spread. The emir ʿAdud al-Dawlah* (whose career as ruler of Shīrāz was beginning just about that time) was informed of his skill and sent for him. Subsequently, ʿAdud al-Dawlah entered Baghdad, bringing Jibrīl with him as one of his personal attendants. There, Jibrīl overhauled the hospital, drawing two salaries, one by virtue of his post as physician, which was 300 shujāʿī dirhams,* and the other by virtue of his post at the hospital, which was also 300 shujāʿī dirhams.

On one occasion, the vizier, may God be pleased with him, was afflicted with a severe stomach complaint, and wrote to ʿAdud al-Dawlah asking him to dispatch a physician. The vizier was well known for his accomplishments, and accordingly ʿAdud al-Dawlah ordered that all the physicians should determine which of them was most suitable to be sent to attend the vizier. The assembled physicians unanimously advised him to send Jibrīl ibn ʿUbayd Allāh for they were envious of his preferment and welcomed an opportunity of having him removed from their midst. This verdict met with ʿAdud al-Dawlah's approval and he sent him off to Rayy.

The vizier received him most hospitably and selected a knowledgeable man of the city to debate with him. This man asked Jibrīl various questions about the pulse. Jibrīl proceeded to expound aspects of the matter beyond the scope of the questions, going into detailed explanations, raising difficult points and resolving them so that there was not a man in the room who could withhold his respect and admiration.

The vizier bestowed splendid robes of honour upon him that day, and asked him to write a treatise exclusively devoted to the ailments affecting each part of the body, from head to foot. The result of this commission was Jibrīl's *Lesser compendium*, which discusses nothing but the disorders of every part of the body, just as the vizier had requested. He was greatly impressed with it and gave Jibrīl a present worth 1,000 dinars. After that, Jibrīl never tired of saying, 'I wrote two hundred pages, and I received 1,000 dinars for them.'

The same author continues:

Jibrīl ibn ʿUbayd Allāh returned to Baghdad after the death of ʿAdud al-Dawlah and lived there for many years, devoting himself exclusively to writing. During this time he completed his *Greater compendium* and a treatise entitled *On the refutation of the Jews*, containing a wide variety of information, evidence of the soundness of the doctrine of the coming of the Messiah and that he has already come, showing that continuing to wait for him is an error, and the validity of the Eucharist and the bread and wine.

Jibrīl stayed in Mayyāfāriqīn for another three years and died there on Friday 8 Rajab in 396/10 April 1006, when he was 85 years of age, and was buried in an oratory just outside the city.

He is the author of eight works, including *The reason why wine is used in the Eucharist although it is a prohibited substance*.

6. ʿUBAYD ALLĀH SON OF JIBRĪL
(*d.* after 1058)

ʿUbayd Allāh was thoroughly versed in the art of medicine. As such, he ranked among the most distinguished physicians and was an unrivalled master of his profession. In addition to his profound understanding of medicine, on which he wrote numerous works, he possessed an extensive knowledge of the Christians and their various sects. He was a contemporary of Ibn Butlān, with whom he associated frequently; they were close friends.

ʿUbayd Allāh is the author of the following works among others:

The merits of physicians,* a work in which he recounts some of the memorable events in the lives of physicians and some of their noteworthy achievements. He composed it in 423/1032.

The medical garden, composed for a learned scholar.

Memorandum for the sedentary and provision for the traveller.

7. AL-TAYFŪRĪ (active 785)

Al-Tayfūrī was a keen-witted and eloquent man, despite a peculiar way of speaking that he owed to his place of origin, for he was a native of one of the villages of the Kaskar district, in the Sawād.

The following account is taken from a work by Ibn al-Dāyah:

I have heard first hand from al-Tayfūrī as follows:

When al-Hādī became caliph, he determined to have his son, Jaʿfar ibn Mūsā, receive the oath of allegiance as his successor. Al-Hādī held a ceremonial session and the tribes all took the oath of allegiance to Jaʿfar, swearing to uphold him and proclaim al-Hādī's brother, Hārūn al-Rashīd, deposed. Then it was the turn of the caliph's entourage and the Arab shaykhs. Last of all were the captains, most of whom took the oath before midday.

There was one exception, however. Harthamah was summoned to attend the ceremony and was ordered to take the oath. 'O Commander of the Faithful,' he said, 'to whom am I to swear fealty? My right hand is occupied with a declaration of fealty to you, the Commander of the Faithful, while my left is occupied with a declaration of fealty to Hārūn. Which hand am I to use to make my declaration?' 'You will declare Hārūn deposed and swear fealty to Jaʿfar,' retorted the caliph, but Harthamah said, 'O Commander of the Faithful, I am a man who is greatly in need of your advice. By God, even if I thought you would have me burned alive for speaking the truth to you, that would not prevent me from doing so. This declaration of fealty is nothing more nor less than an oath. I have sworn fealty to Hārūn in the same words that you are now asking me to utter to swear fealty to Jaʿfar. If I declare Hārūn al-Rashīd deposed today, perhaps I will declare Jaʿfar ibn Mūsā deposed tomorrow, and so with all those who had sworn fealty to Hārūn in that way and have now betrayed him.'

Al-Hādī said angrily, 'Be off with you, God curse you!' He ordered him escorted out of the ʿĪsābādh palace and stripped of his command. He then remained silent for half an hour or so, then raised his head and said to his servant, 'Catch up with the filthy swine and bring him back to the Commander of the Faithful.'

The servant caught up with Harthamah, brought him back to the palace, and as soon as he entered, al-Hādī said to him, 'The most eminent Arab shaykhs and the captains have pledged their fealty, and only you refuse to take the oath?' Harthamah replied, 'O Commander of the Faithful, the matter is as I have told you before: one cannot proclaim Hārūn deposed today, yet remain loyal to Jaʿfar tomorrow.'

Al-Tayfūrī's account continues as follows:

Al-Hādī turned to his assembled throng and said, 'You have disgraced yourselves. By God, Harthamah has spoken the truth and kept his faith, whereas you have been false to yours.' So saying, the caliph ordered that Harthamah was to receive a purse of 50,000 dirhams, and that the district where the servant had taken him was to be granted to him in leasehold, and to this day that district is known as Harthamah's Camp. The people dispersed in a state of great alarm, such as might be felt by any man of importance who had been the cause of embarrassment to the caliph.

Al-Hādī died only a few nights later, and Hārūn al-Rashīd succeeded to the caliphate. By God, he treated Jaʿfar's mother most generously, while Jaʿfar ibn Mūsā himself received more benefits than ever, including marriage to Hārūn al-Rashīd's own daughter, Umm Muhammad.

8. ZAKARIYYĀ SON OF AL-TAYFŪRĪ
(active 835)

The following account is taken from a work by Ibn al-Dāyah, in his own words:

Zakariyyā once said to me: I was with the emir al-Afshīn* in his camp during his campaign against Bābak.* He had decided to have a list of all the merchants in the camp* drawn up. The list was prepared and when the reader came to the section on apothecaries, al-Afshīn said, 'Zakariyyā, I consider that it is more important to have an accurate record of apothecaries than of other tradesmen. I want you to look into the practice of every one of them, to determine which of them are honest professionals and which are charlatans.'

'May God preserve the emir,' I replied, 'there was an alchemist who was an associate of the caliph al-Maʾmūn, who said, "The art of alchemy is beset by a plague, and the name of that plague is apothecaries. Any apothecary, when asked for something or other, regardless of whether he actually has the item or not, will invariably say that he has it, saying, 'Here you are, sir, this is what you asked for.' If the Commander of the Faithful wishes to test the truth of this, he might make up a fictitious name and send out a number of his servants to try to buy some of it from apothecaries." "I know just the name," said al-Maʾmūn, "I shall send them out for some *saqtīthā*—that is the name of an estate not far from Baghdad." He dispatched a number of emissaries with instructions to ask the city's apothecaries for some *saqtīthā*, and every one of those apothecaries said that he had some in stock, handed over something from his shop, and took the emissary's money. Various things were brought back to the caliph, a handful of seeds, a fragment of stone, or a bit of fur. Al-Maʾmūn was so pleased with the alchemist's frankness and honesty that he granted him some land. His descendants hold that land to this day, and live from the revenues that it yields. Now, if the emir should think fit, he might put these apothecaries to the test in the same way as al-Maʾmūn did.'

At this, the emir al-Afshīn had a volume of registers brought to him and selected twenty places out of it. He then sent men out to try to buy drugs having those names from the apothecaries in camp. Some of them said they were not familiar with any such drugs, but others said they had them, took the emissary's money, and gave him something out of their stock. Al-Afshīn then had all the apothecaries brought before him, and when they were all assembled, he had licences issued to those who had said that they were not familiar with the names of the drugs his emissaries had asked for. Those apothecaries were allowed to remain in the camp, while all the others were ordered to leave. Not one of them was permitted to stay, and a crier was sent around to announce that any of them who was found in the camp might lawfully be killed. Finally, al-Afshīn wrote to the caliph al-Muʿtasim, asking him to send out a number of honest apothecaries and physicians. This pleased the caliph, and he sent al-Afshīn some apothecaries and physicians as he had requested.

9. SALMAWAYH (*d. c.*841)

Salmawayh was the personal physician of the caliph al-Muʿtasim. The caliph treated him with such generosity that words are inadequate to describe it. When decrees from al-Muʿtasim relating to appointments and other matters came in to the relevant administrative departments, they would be in Salmawayh's handwriting. Salmawayh was a sincere Christian. He performed many good works, led an exemplary life, and was a man of intelligence and good judgement.

The following account comes from Is'hāq ibn Hunayn, who states that he had it from his father as follows:

Salmawayh was more knowledgeable about the art of medicine than any of his contemporaries. Al-Muʿtasim was wont to address him as 'My father'. When the physician was ill, the caliph went to visit him, and said to him in tears, 'Give me some advice that will be useful to me once you have left me.' 'You have treated me with the utmost kindness, Sire,' replied Salmawayh. 'Beware of that mountebank Yūhannā ibn Māsawayh. When you consult him about some complaint, he is likely to prescribe a number of remedies; if so, take the one that contains the fewest ingredients.' When Salmawayh died, al-Muʿtasim refused to take any food on the day of his death. He ordered that the physician's bier should be carried to the caliphal palace, and there the full Christian rites performed, complete with prayers, candles, and incense. The caliph was present throughout, demonstrating to the highest degree his esteem for Salmawayh. He mourned him sadly.

Al-Muʿtasim was blessed with a good digestion. Salmawayh had bled him regularly twice a year, administering a laxative afterwards, and had put him on a diet from time to time. Yūhannā ibn Māsawayh, however, wishing to show the caliph something to which he was not accustomed, administered the laxative before bleeding him, explaining, 'I am afraid that your yellow bile will become agitated.' But when his patron took the laxative, his blood became fevered and his temperature rose; he lost flesh, fell prey to a growing array of disorders, wasted away, and finally died, twenty months after Salmawayh.

The following account is taken from Ibn al-Dāyah:*

On one occasion, Salmawayh and I were talking about Yūhannā ibn Māsawayh. I went on at length about him, saying that I knew how learned he was. 'Yūhannā is a catastrophe for anyone who consults him,' retorted Salmawayh, 'and for anyone who relies on his treatment, his extensive book-knowledge, his elegant explanations, and misleading descriptions of his patients' morbid conditions. The beginning of medicine,' he lectured me, 'is a knowledge of the scope of the disorder, so that a judicious treatment can be applied, and Yūhannā is the most ignorant of all God's creation both as regards the assessment of an illness and as regards the determination of an appropriate treatment. If his patient is feverish, he treats him with cold medicines and excessively cooling foods in an attempt to drive down his temperature, with the result that the patient's stomach and body become chilled and must be treated with warming medicines and foods. Yūhannā then proceeds in the same way as he did before, taking excessive measures to counteract the chill, and consequently the patient becomes ill from overheating. In short, his patient is always ill from being either chilled or overheated, and that weakens the body. People consult physicians in order to preserve their health while they are healthy and to help restore Nature when they are ill, and Yūhannā cannot serve either of those functions, owing to his inability to gauge disease and remedy alike correctly. Such a person cannot call himself a physician.'

The following account is from another author, quoting his father:

My grandfather and Salmawayh the physician were good friends. My grandfather told me that he had called on Salmawayh at home once when he was in the bath. Salmawayh emerged, bundled up in clothing, with sweat running down his brow. A servant came to meet him, carrying a small table laden with a roast chicken, some green stuff in a cream-coloured porcelain bowl, three loaves of flatbread, some tamarisk fruits, and some vinegar in a plate. He ate it all, and then called for a quantity of wine amounting to about two dirhams, which he mixed and drank. Finally he washed his hands with water, and then he proceeded to change his clothes, which he perfumed with incense. When he had finished doing all this, he began to speak, but my grandfather said, 'I want you to tell me what it is that you have just done.' 'I have been treating myself for consumption for thirty years,' Salmawayh replied, 'and in the course of those years I have eaten only what you saw just now: a roast chicken, endive steamed and roasted in almond oil, and precisely this much bread. Upon emerging from my bath, I need to take action to neutralize the heat, as otherwise it would act upon my body and deprive it of some of its humours. Accordingly, I use food to divert it. The heat acts on the food instead, leaving me free for other matters.'

10. MĀSAWAYH (*d.* probably early ninth century)

The following account is taken from Pethion:

Māsawayh worked at the hospital in Gondeshapur, where he pounded medical ingredients. He did not know how to read so much as a single letter of any language. He could, however, diagnose and treat various disorders, and he could distinguish effective medicines from ineffective ones. The physician Jibrīl ibn Bukhtīshūʿ took him under his wing and treated him generously.

Another account relates:

Māsawayh was a trainee at the Gondeshapur hospital for thirty years. When he heard of Jibrīl ibn Bukhtīshūʿ's appointment as personal physician to the caliph Hārūn al-Rashīd, he said, 'Here Jibrīl has risen to the stars, while we here at the hospital are never going to get any further.' Jibrīl was responsible for the hospital, and when he heard of the younger man's remark he had Māsawayh dismissed and turned out and his salary discontinued. Finding himself without resources, Māsawayh went to Baghdad, intending to grovel and offer his apologies to Jibrīl. He haunted the physician's door, but was never invited to enter; when Jibrīl rode by, Māsawayh would utter blessings upon him and implore his pity, but the great man ignored him.

Finally, confronted with the prospect of starvation, he went to the Christian quarter known as the Roman quarter, on the eastern side of the city, and asked the priest to give him sanctuary in the church. 'It may be that some generous soul will enable me to return to my native place,' he said, 'for Jibrīl bears me a grudge and will not speak to me.' 'You spent thirty years at the hospital,' said the priest, 'surely you know something about the art of medicine?' 'I do, by God,' replied Māsawayh, 'I can dose a patient with physic, cure a sore eye, bind up wounds, or whatever is required.' At this, the priest went and brought out a chest of medical supplies to enable Māsawayh to treat patients, and set him down at the entrance to the women's quarters in the palace of the vizier of Hārūn al-Rashīd. There he sat, occasionally earning a trifle, until by degrees he found he was making a very fair living.

It so happened that a eunuch in the employ of the vizier came to be afflicted with a painful eye condition. Jibrīl summoned oculists to treat him but to no avail: his pain grew worse and worse, so that he was unable to sleep. One morning, as he emerged from the palace, sleepless and suffering, half out of his mind with pain and in a foul temper, he caught sight of Māsawayh and shouted at him, 'What are you doing here, O shaykh? If you are any good at your trade, make me better; otherwise, be off with you!' Māsawayh accompanied him into the palace; and there he everted the man's eyelid and applied collyrium* to it, dripped one kind of medication on to his head, and made him sniff another. The patient became calm and fell asleep. The next morning, he sent Māsawayh a bowl with a loaf of fine bread, a kid, a chicken, sweetmeats, and a number of dinars and dirhams in it, together with a message: 'This is to be your daily allowance. The dirhams and dinars are to be your monthly stipend from me.' Māsawayh was so overcome that tears came into his eyes.

Only a few days later, it was the vizier himself who was suffering from an inflamed eye. Jibrīl summoned oculists to treat him, and they dosed him with a number of remedies, but without effect. Then one night the eunuch smuggled Māsawayh into the palace. Māsawayh applied collyrium to the vizier's eye for a third of the night, then gave him a purgative. After this treatment, the vizier felt better. When Jibrīl came in the next day, the vizier said to him, 'Do you know, Jibrīl there is a man here named Māsawayh who is the most skilful and knowledgeable oculist you can imagine.' 'You don't mean the fellow who has been plying his trade outside the gate there?' asked Jibrīl. 'The very man,' said the vizier. 'I know him,' sneered Jibrīl, 'I used to give him menial tasks to do, but he proved useless, so I gave him the sack. If you agree, may I suggest that you summon him while I am here.'

Jibrīl imagined that when Māsawayh came in he would remain standing and behave respectfully in his presence, but he entered, greeted the company, and sat down facing Jibrīl, quite at his ease. 'Turned physician, then, have we, Māsawayh?' said Jibrīl. 'I have never ceased to be a physician,' replied Māsawayh calmly, 'for I have served at the hospital for thirty years. And you presume to speak to me in such terms!' Jibrīl, alarmed at the possibility that the younger man might go on to say more, made haste to take his leave, greatly disconcerted.

The vizier then gave orders that Māsawayh was to be paid a monthly salary of 600 dirhams and fodder for two horses, and that accommodation for five slave-boys should be placed at his disposal. The vizier also ordered him to bring his family from Gondeshapur, giving him an ample sum of money to cover the cost of the move. Accordingly, he brought his household to Baghdad. His son Yūhannā was very young at the time.

Shortly thereafter, it was the turn of Hārūn al-Rashīd to suffer from a painful sore eye and the vizier had Māsawayh sent for. The caliph bade him approach, and Māsawayh looked at his eyes. 'What we want just now is a cupper,' he said. After al-Rashīd's legs had been cupped, Māsawayh dripped medication into his eyes, and two days later the caliph was better. He ordered that the physician should be paid a stipend of 2,000 dirhams a month, together with 20,000 dirhams a year for incidental expenses, fodder for his mounts, and accommodation. Māsawayh joined Jibrīl and the other physicians as a member of the caliph's medical staff. At that time he was on a footing of equality with Jibrīl; Jibrīl, however, was wealthier.

The following account regarding Māsawayh and his son, Yūhannā, is from Ibn al-Dāyah, quoting Jibrīl ibn Bukhtīshūʿ:

Hārūn al-Rashīd had assigned me responsibility for a hospital (Jibrīl said), and I decided to bring in Dahishtak, the superintendent of the Gondeshapur hospital, to take charge of my hospital. However, he demurred on the grounds that he would not be assured of drawing a regular salary. The Catholicos,* Timothy I,* urged me to excuse Dahishtak and in the end I did so.

Dahishtak then said to me, 'Jibrīl, I should like to offer you a present in return for your kindness in excusing me. He explained that it was one of his own men. 'He used to pound medical ingredients for us when he was young,' he said. 'No one knows who his father was, and he seems to have no family. He has lived at the hospital for forty years; he must be 50 years old by this time, perhaps even older. He cannot read so much as a letter of any language, but he knows every ailment there is, and he also knows what medicines should be used to treat any given disorder. I am making you a present of him. Assign him to whichever of your pupils you like, then make that pupil superintendent of your hospital; you will find that it will be better run than would be the case if you had appointed me superintendent.' I replied that I should be delighted to accept his present.

Dahishtak then returned home and sent his man to me. His name was Māsawayh. At that time Dāwūd ibn Sarābīyūn was personal physician to Ghadīd, *umm walad** of Hārūn al-Rashīd, and the house where Māsawayh lived was not far from Dāwūd's. Dāwūd was a light-hearted, frivolous sort of fellow, and Māsawayh was subject to the weaknesses of the lower orders, so he revelled in the pleasures to which Dāwūd introduced him. It was not long before he came to me dressed quite differently, all in white. I asked him how he was doing, and he informed me that he was in love with a slave-girl who belonged to Dāwūd ibn Sarābīyūn, a Slav girl, he said, named Risālah. Would

I buy her for him? I did buy the girl, for 800 dirhams, and gave her to Māsawayh, and she subsequently bore him two sons, Yūhannā and his brother, Mīkhāʾīl.*

Nor was that the end of the matter: I, Jibrīl, went on to treat his children as though they had been my own flesh and blood. I furthered their advancement and gave them precedence over the sons of noble practitioners and learned scholars. I saw to it that Yūhannā was elevated to a position of responsibility, making him superintendent of the hospital and chief among my pupils. And now he repays me by giving vent to his own self-importance and talking arrant rubbish! It was precisely because of this sort of pernicious nonsense from jumped-up persons of the lower orders that the Persians strictly prohibited people from engaging in occupations other than those that had been practised by their fathers and grandfathers.

11. YŪHANNĀ SON OF MĀSAWAYH (*d.* 857)

Yūhannā was a skilful, sagacious, learned physician, experienced in the art of medicine as well as a good stylist and the author of well-known treatises. He commanded the respect and enjoyed the favour of caliphs and kings.

The caliph al-Wāthiq was avidly and tenaciously attached to Yūhannā. One day Yūhannā was drinking in his company when a cup-bearer brought him a drink that was not pure and sweet, as was the practice of cupbearers when they were insufficiently rewarded. When Yūhannā drank the first cup he said: 'O Commander of the Faithful, the taste of this drink is unlike that of any others.'

The Commander of the Faithful was angry with the cupbearers, and ordered 100,000 dirhams to be given to Yūhannā at once, within the hour. In the afternoon, he asked if the money had been handed over. The chief steward replied 'Not yet,' and so the caliph doubled the sum. After the evening prayer he enquired and again was told that the money was not yet delivered. So he ordered 300,000 dirhams to be sent. The steward said to the treasurer: 'Send out Yūhannā's money this minute or nothing will be left in the treasury,' and the money was dispatched at once.

Ibn al-Dāyah says:

The teaching session of Yūhannā attracted the largest audience of any I have seen in the city of Baghdad, whether conducted by a physician or theologian or philosopher, for every type of educated person assembled there. Yūhannā had a great capacity for being funny, which was part of the reason for the large gatherings. His impatience and quick temper surpassed even those of Jibrīl, his sharpness expressing itself in droll statements. His teaching was especially enjoyable in the sessions when he examined phials of urine.

I, along with a colleague known as Mule's Balls, took upon ourselves to memorize his witticisms. I made a pretence of studying the books on logic under his guidance, while the other pretended to be his pupil studying Galen's books on medicine.

Ibn al-Dāyah also relates:

A man complained to him about scabies that afflicted him. So Yūhannā directed him to draw blood from the median cubital vein of the right arm, but was informed that he had already done so. Then Yūhannā advised him to bleed the median cubital vein of the left arm, but the man had also done that. So he ordered him to drink a decoction, but he said 'I already did'. Then Yūhannā ordered him to drink a purgative stomach medicine, and again the man informed him that he had done it. Then he prescribed drinking cheese-water for a week and churned cow's milk for two weeks, but yet again the man informed him that he had done it.

Yūhannā then said to him: 'There is one thing left, not mentioned by Hippocrates or Galen, but which we have observed can work, so try it, for I hope that it might bring about your recovery, God willing. Buy two sheets of paper and cut both of them into small pieces. Then write on each piece of paper "May God have mercy upon those who pray for the health of the afflicted". Take half of them to the eastern mosque in Baghdad and the other half to the western mosque, and distribute them at the Friday gatherings. I hope God will answer their prayers, since no medical treatment has helped you.'

Ibn al-Dāyah adds:

The Christians censured him for taking concubines. They said to him: 'Either follow our code of conduct and confine yourself to one woman and remain our deacon, or remove yourself from the office of deacon and take any number of concubines you like.' To which he replied: 'It is true that in one passage we are commanded not to take two women or two garments.* But who made the Catholicos, that motherfucker,* more entitled to take twenty garments than the wretched Yūhannā to take four concubines? Go tell your Catholicos that if he transgresses the rules of his religion, then we shall as well.'

Ibn al-Dāyah also relates:

The house of his father-in-law, al-Tayfūrī, was in the Christian quarter in the eastern part of Baghdad, next to the house of Yūhannā, who owned a peacock which used to perch on the wall between his house and al-Tayfūrī's. Al-Tayfūrī had a son called Dāniyāl who became a monk. He arrived in Baghdad one night during the month of August, a time of great heat and intense sultriness. Whenever the heat of the night oppressed the peacock, it screeched. The screeching woke Dāniyāl, who was clad in a monk's woollen robe, and he tried several times to drive the peacock off, but it did no good. So Dāniyāl raised his iron bar, hit the peacock on the head, and killed it. Yūhannā went out riding and came back to happen upon his peacock dead at the door of his house. He started swearing vengeance on its killer, when Dāniyāl came out and said to him: 'Do not vilify the killer, for I am the one who killed it, and I owe you several peacocks in its stead.' Yūhannā replied, in my presence: 'I don't like a monk with a huge cock and balls as big as a camel's hump!', although he said that using more foul language. Dāniyāl answered him: 'Likewise, I don't like a deacon who has several women, and whose chief woman's name is Qarātīs—for it is a Byzantine name, not Arabic, and for Christians the name Qarātīs means a woman who makes a cuckold of her husband, and a woman cannot do that unless she commits adultery.' Yūhannā was ashamed and entered his house defeated.

Ibn al-Dāyah relates as well:

Yūhannā had accompanied the caliph al-Wāthiq to a stone bench alongside the Tigris River. Al-Wāthiq had a fishing rod that he cast into the Tigris in order to catch fish, but he caught nothing. He turned to Yūhannā, who was on his right, and said: 'Get away from my side, you ill-omened man.' Yūhannā replied: 'O Commander of the Faithful, do not say absurd things. Yūhannā is from Khuzestan, and his mother was Risālah the slave, bought for 800 dirhams. Good fortune brought him up to be the fellow-drinker of caliphs, their participant in evening conversations, and their close companion—so much so that the riches of the world overwhelmed him, beyond all his expectations. It is absurd to say that such a person is ill-omened. However, if the Commander of the Faithful wishes, I can tell him who really is an unlucky person.' Al-Wāthiq said 'And who is such a person?' to which Yūhannā replied: 'He who is the descendant of four caliphs, to whom God entrusted the caliphate but who forsook its palaces and gardens to sit on a small stone bench alongside the Tigris River, unprotected from the gales that might drown him, in order to be like the poorest and lowest in the world, namely the fisherman.'

Ibn al-Dāyah continues:

Yūhannā had a son called Māsawayh ibn Yūhannā. This Māsawayh resembled his father in his appearance, his speech, and his movements, only he was simple and took a long time to understand anything, and then would forget it a moment later. Yūhannā used to make a show of loving his son as a precaution against the criticism of his father-in-law, al-Tayfūrī, and his children. But in fact he hated the son even more than he hated Sahl the Beardless, who had attacked his honour by claiming that he 'had put it in his mother's vagina'.*

Ibn al-Dāyah then continued:

Once in the house of a man, Yūhannā said: 'I have been afflicted with a long face, a high cranium, a broad forehead, and blue eyes, but I have been blessed with intelligence and a memory for everything that takes place within my hearing. My wife, the daughter of al-Tayfūrī, was the most beautiful woman I had seen or heard of, although she was stupid and simple-minded. Our son received all of our bad qualities and was not endowed with any of our good qualities, and if it had not been for the authorities, I would have dissected this son of mine while living, just as Galen dissected apes and humans, so that I might learn the causes of his stupidity. I would have relieved people of his character, and I would have provided the people in the world with the knowledge of the structure of his body and the pathways of his arteries and veins and nerves, describing it as a science.'

Ibn al-Dāyah continued:

In my opinion, this story was told to al-Tayfūrī and his children deliberately to sow evil and strife, and to laugh at what would occur, and indeed things turned out as imagined.

Yūhannā's son, Māsawayh, fell ill a few nights later. A messenger from the caliph al-Muʿtasim arrived from Damascus, to request Yūhannā's immediate presence. After seeing his son, Yūhannā recommended blood-letting for him, while al-Tayfūrī and his sons were of the opposite opinion. Yūhannā performed the blood-letting and left for Syria on the following day. On the third day after his father's departure, Māsawayh died. At the funeral, al-Tayfūrī and his sons swore that Yūhannā had intended to kill his son, and they offered as proof what Yūhannā had said previously in the man's house.

A man criticized Yūhannā in front of the caliph al-Mutawakkil. Yūhannā answered him saying 'If your ignorance was replaced with intelligence, and then it was distributed amongst one hundred black-beetles, each one of them would be more intelligent than Aristotle.'

It is also recorded:

The caliph al-Maʾmūn went down to the Budandūn river, his brother, al-Muʿtasim accompanying him. They put their feet in to cool them, for the water was very cold, pure, and pleasing. Al-Maʾmūn said to al-Muʿtasim: 'Right now I would like to eat Azadh dates* and drink this cold water with them.' Immediately he heard the sound of post-horses and someone calling 'This is the Iraqi courier system.'* A large silver tray of fresh dates was produced, and al-Maʾmūn was astonished that his wish had been fulfilled. Afterwards, he took his leave and had a short rest, but then rose from that with a fever. Blood-letting was performed, but on his neck there appeared a swelling of a type he had had before, which the physician treated by letting it suppurate, open, and then heal. His brother, al-Muʿtasim, said to Yūhannā, who was the physician, 'What a peculiar situation! For you are an outstanding and unparalleled physician, and yet this swelling repeatedly returns to the caliph. Can you not destroy it by cutting it out so that it does not return? By God, if this affliction returns, then I will have you beheaded!'

Yūhannā was taken aback by this speech and left. He later related the event to a man he trusted who said: 'Did you not detect what al-Muʿtasim's purpose was? He directed you to kill al-Maʾmūn so that the swelling would not return. It could be only that, for he knows that a physician cannot keep diseases from attacking people. Indeed, he told you not to let him live by saying that the disease must not return.'

Then Yūhannā pretended to be ill and ordered a student of his to attend al-Maʾmūn in his place. The student returned each day and informed him of al-Maʾmūn's condition. Then Yūhannā ordered the student to open the swelling, but the student exclaimed 'May God protect you! It has not reddened and has not reached the stage of suppuration.' Yūhannā said to him: 'Go on and open it, as I've told you, and do not consult again with me.' And so the student did as directed, but al-Maʾmūn died, may God have mercy on him.

I—Ibn Abī Usaybiʿah—say that Yūhannā acted in that way because he was deficient in virtue, in religion, and fidelity. He was not a Muslim, but he did not even hold to his own religion, as is evident from the stories given above. No sensible person should give credit to or rely upon someone who does not have a religion which he follows and to which he is bound.

Yūhannā died in Samarra on Monday, 25th of Jumādā II 243/19 October 857 during the reign of al-Mutawakkil.

Some of the sayings of Yūhannā:

When asked about what was good without any accompanying evil, he replied: 'Drinking a small amount of a wine that is pure.'

When asked about what is evil without any good, he said, 'Having intercourse with an old woman.'

'Eating apples restores the soul.'

'You should have food that is new and wine that is old.'

Forty-six books were written by Yūhannā including:

The book of fevers, in branch-diagram format.

On phlebotomy and cupping.

The preparation and administration of laxatives, with details of each drug and its benefits.

A book explaining why physicians should avoid treating pregnant women during certain months of their pregnancy.

On taking the pulse.

On the toothbrush and toothpowders.

12. HUNAYN IBN IS'HĀQ (*d.* 877)

Hunayn was a fluent, eloquent speaker and an accomplished poet. He lived for a time in Basra, where he studied Arabic grammar. Subsequently, he removed to Baghdad and acquired an unrivalled mastery of the art of medicine.

The following account is taken from Ibn al-Dāyah:

Hunayn was diligent and keenly interested in learning the art of medicine, but his initial experience was unfortunate. I used to see Hunayn reading under Yūhannā's guidance, and constantly asking questions, much to Yūhannā's annoyance. Another factor that did not endear him to his teacher was the fact that Hunayn came from a family of money-changers in al-Hīrah. The people of Gondeshapur, in particular its physicians, looked down on the people of al-Hīrah and did not encourage tradesmen's sons to enter their profession.

It so happened one day that Hunayn asked Yūhannā a question about a passage that he had read. Yūhannā lost his temper. 'The people of al-Hīrah are not fit to learn the art of medicine!' he snapped. 'You would do better to go and borrow 50 dirhams. For one dirham you can buy some little baskets; for three more, you can buy some orpiment.* Spend the rest on copper coins* of Kufa and al-Qādisiyyah, coat them with orpiment, put them into the baskets, and then sit by the roadside crying "Fine coins for alms and gratuities!" You will make a better living selling such coins than you ever will by practising this profession. Leave my house!' Hunayn slunk out in tears, humiliated and downcast. We saw no more of him for two years.

It so happened, however, that the caliph Hārūn al-Rashīd had a Roman slave-girl named Kharshā whom he had appointed as storewoman. She had a sister, or it may have been a niece, who sometimes brought the caliph an item of clothing from Kharshā's wardrobes. Al-Rashīd noticed eventually that he had not seen the sister for some time, and asked what had become of her, whereupon Kharshā informed him that she had given the girl in marriage to a kinsman. 'What!' said the caliph angrily, 'without first obtaining my permission and purchasing her? For she is my property,' and he ordered a retainer to ascertain the identity of the husband and to chastise him. The man soon ran the husband to earth, and, without even saying a word, had him seized and castrated. However, the girl had already conceived, and in due course gave birth to a son. Kharshā adopted the boy and brought him up in the Roman fashion, with the result that he read and spoke perfect Greek. He was known familiarly as Ibn al-Khasīy, that is, 'son of the eunuch'.

Once when he fell ill, I—Ibn al-Dāyah—went to visit him. There at his home, what should I see but a man with such a head of hair that it partially covered his face, reciting some Greek poetry. I addressed the hirsute man tentatively: 'Hunayn?' and he acknowledged that it was indeed he. 'Yūhannā, the son of that whore Risālah,' he went on, 'said that no ʿIbādī* was capable of learning the art of medicine. May I renounce the Christian religion if I undertake the study of medicine before I have achieved a more comprehensive mastery of the Greek language than anyone else in this age! No one knows about this, and I must ask you to keep my identity to yourself.' After that, it was a good three years, perhaps closer to four, before I saw him again.

I had gone to call on Jibrīl ibn Bukhtīshūʿ, and there in his house was none other than Hunayn, translating parts of one of Galen's works on anatomy. Jibrīl was addressing him with the utmost respect as 'Rabban ('teacher') Hunayn'. Seeing my astonishment, Jibrīl said to me, 'I tell you, if God preserves him, he will outshine not only Sergius* but other translators as well.'

When I left, I found Hunayn outside the front door, where he had been waiting for me. After greeting me, he said, 'I once asked you not to divulge what I had been up to, but now I should like you to make the matter known.' 'I shall be making Yūhannā feel like a fool when I tell him how I heard Jibrīl praise you to the skies,' I said. 'To make him feel even more of a fool, show him this,' said Hunayn, taking from his sleeve a copy of the translation he had done for Jibrīl, 'without telling him that it was I who translated it, and then, once you see that he thinks it really excellent, tell him.' I hastened to show the manuscript to Yūhannā that very day, without so much as going home first.

Having read the selections, Yūhannā expressed the utmost astonishment. 'Is it possible for someone in our present age to be inspired by Christ, do you think? I am sure this could have been produced only with the help of the Holy Ghost.' 'As a matter of fact,' I said, 'it is the work of Hunayn ibn Is'hāq, whom you ordered out of your house and advised to sell doctored copper coins.' 'Impossible!' he gasped. At length, however, he realized that I must be speaking the truth, and he begged me to use my good offices to effect a reconciliation. I was able to bring this about, and from then on Yūhannā treated Hunayn with the utmost respect and generosity.

The following account quotes the physician Hunayn himself:*

During the reign of al-Mutawakkil, messengers from the caliph's palace came to my house one night looking for me and saying that the caliph wanted me. They, including Zarāfah,* the majordomo himself, dragged me out of bed and set off with me at a run, finally showing me into the caliph's presence with the words, 'This is Hunayn, Sire.' 'Pay Zarāfah what I said I would give him,' said al-Mutawakkil to his attendants, whereupon the servant was given a purse of 30,000 dirhams. The caliph then turned to me and said, 'I am hungry,' so I suggested a menu. After he had finished his meal, I asked them why I had been summoned as I had been, and was informed that a singer had performed before the caliph, and that the caliph had asked who the composer of the song was and had been told that it was Hunayn ibn Balūʿ al-ʿIbādī.* Al-Mutawakkil had then ordered Zarāfah to have him brought to the palace. 'Unfortunately, I do not know him, O Commander of the Faithful,' said Zarāfah. 'I must have him,' cried the caliph. 'Bring him here, and you shall have 30,000 dirhams.' So he brought me instead, but by the time I got there, al-Mutawakkil had forgotten about the matter, owing to the date-wine that he had imbibed.

I—Ibn Abī Usaybiʿah—now continue. Hunayn was born in 194/809 and died on 6 Safar 264/18 November 877, when he was 70. He is said to have died of chronic diarrhoea.

I have also been able to ascertain an account from none other than Hunayn himself, for I have come upon an epistle written by him in which he describes the trials and misfortunes that befell him at the hands of the celebrated physicians of his time who were hostile to him.

The following, then, is part of Hunayn's own account, in his own handwriting, regarding his return to favour following their trickery:

These Christian physicians, whom I have known all their lives and who learned their art under my tutelage, most of them, are the very ones who want to see my blood shed, even though I am indispensable to them. 'Hunayn has no skill in matters of illness and has never diagnosed a disease. He wants to be like us, that's all, and to be known as Hunayn the Physician rather than Hunayn the Translator,' they have been known to say. They add, 'Whenever Hunayn calls on a patient, regardless of whether the people of the house belong to the cream of society or the common folk, they snigger when he comes in, and they laugh behind his back when he leaves.'

Whenever I heard this kind of thing, I was so distressed that I thought of killing myself from sheer exasperation and indignation. I had no way of striking back, for one man alone cannot resist a large number of enemies who are in league against him. All the while, however, I realized that they were doing what they did out of nothing but envy. Despite all this, I did not complain to anyone about my situation, bad though it was. On the contrary, I would praise my detractors on official occasions, in the presence of eminent persons.

Here I propose to tell the story of the final pit that they dug for me, my last and most recent trial. The physician Bukhtīshūʿ ibn Jibrīl* had concocted a scheme against me that he brought off successfully and thus did me harm, as he had sought to do.

He began by obtaining an icon bearing an image of Our Lady Mary holding Our Lord Jesus Christ in her lap, with angels all about them. It was very beautifully painted and most lifelike, and Bukhtīshūʿ had paid a high price for it. He had it sent to al-Mutawakkil, taking care to ensure that he was there in person to take it from the hands of the servant who brought it in and present it to the caliph himself.

Bukhtīshūʿ then proceeded to kiss the image repeatedly. 'Why are you doing that?' asked the caliph. 'Do all you Christians do that?' 'Yes, we do, O Commander of the Faithful,' said Bukhtīshūʿ, 'and with greater fervour than this, for I am acting with exceptional restraint, being in your presence. But despite the preference of our Christian community, I know of one man who is in your service, enjoying the benefits and drawing a salary. Although a Christian, he despises and spits on her. He is a miscreant and a heretic; his Christianity is merely a cloak; in reality, he denies the existence of God and does not believe in the message of the prophets.' 'Who is this person?' asked al-Mutawakkil. 'Hunayn, the Translator,' replied Bukhtīshūʿ. 'I shall have him sent for,' said the caliph, 'and if the matter proves to be thus, I shall make an example of him; I shall have him tortured repeatedly and thrown into the Mutbaq prison [in Baghdad].'

'I should be greatly obliged if the Commander of the Faithful would wait an hour or so before having him brought in,' said Bukhtīshūʿ ibn Jibrīl, and he took his leave.

Immediately upon leaving the palace, he came straight to my house. 'Hunayn' he said, 'I must tell you that the Commander of the Faithful has been given an icon that he finds most impressive. The caliph has been saying how fine it is, and if we praise it in his hearing, he will urge us to share his enthusiasm for it on every possible occasion. He has already been on at me about it. "Look at this painting, it is your Lord and His mother," he said, "isn't it exquisite?" "Well, Sire," I said to him, "it is a painting such as may be seen in baths and churches and other places that are decorated. It is of no importance to us." "If you are telling the truth," he said, "let me see you spit on it."

'So I spat on it, and then took my leave, while he roared with laughter. As you can imagine, I did that only to encourage him to throw the icon away and not keep badgering us about it and throwing us into prominence. Someone might become resentful, and that would make matters worse. So if by any chance he sends for you and asks you the same sorts of questions, it would be best for you to do what I did. I have already been to see those of our colleagues who attend his receptions and urged them to do the same.' I swallowed his instructions and he was successful in making a complete fool of me.

Not more than an hour after he had left, a runner from the caliph arrived and escorted me to the palace. I saw the icon as soon as I entered, set before al-Mutawakkil. 'Hunayn,' he said, 'would you not say that this picture is beautiful and impressive?' 'Indeed it is.' I agreed politely. 'What do you think of it?' he persisted. 'Why, Sire,' I said, 'it is a painting, such as may commonly be seen in baths or churches.' 'But,' he said, 'is it not a picture of your Lord and His mother?' 'God forbid,' I said, 'that God should be portrayed or that there should be a likeness of Him.' 'In a word, then, it has no power, either for good or for ill?' asked the caliph. 'If that really is the case,' he said, 'spit on it.' So I did spit on it, and no sooner had I done so than he had me arrested.

The caliph then had the Catholicos, Theodosius, sent for. As soon as he entered and saw the icon that was set before the caliph, even before having formally greeted al-Mutawakkil, he fell upon it, embraced it, and kissed it repeatedly, his eyes brimming with tears. Some of the servants went to restrain him, but the caliph motioned them back. At length, he got to his feet, and treated al-Mutawakkil to an elaborate formal salutation. The caliph returned his greeting and bade the Catholicos be seated, whereupon he took a seat, cradling the icon on his knees.

'What is this?' said the caliph, 'who are you to take something that was set before me and hold it on your knees without my permission?' 'O Commander of the Faithful,' replied the Catholicos, 'I have a better claim to this object that was set before you, even though the rights of the Commander of the Faithful, may God prolong his life, are paramount. My religion does not allow me to leave an image of Christ and Our Lady on the floor in a place where its significance is not understood. An image like this should be hung in a place where due honour is shown it, where lamps with the finest oils are kept burning before it.'

'You may go on holding it for the time being,' said al-Mutawakkil. 'A boon, Sire,' said the Catholicos. 'Would the Commander of the Faithful give me this icon so that I may ensure that the icon is shown the reverence that is its due?' 'The image is yours,' said the caliph, 'but what I want you to tell me is, what should be done with a man who spat upon it, in your view?' 'If the man was a Muslim,' replied the Catholicos, 'no blame could be attached to him, for he would not be aware of the significance of this icon. If the offender was a Christian, acting in full awareness of what he was doing, he would in effect have spat on Mary, the mother of Our Lord, and upon Our Lord Jesus Christ Himself.'

'What punishment would you consider appropriate for a person who had done such a thing?' asked al-Mutawakkil. 'O Commander of the Faithful,' replied the Catholicos, 'I have no authority to apply lash or rod, nor do I have a narrow cell in which to confine an offender, but I should declare him anathema, forbid him to enter a church or to receive the Eucharist, and forbid all Christians to associate with or speak to him. He would remain an object of abhorrence to us all until such time as he repented and mended his ways.

'Once he had changed for the better, we should apply the words of our Scripture: "Except ye forgive sinners, your sins shall not be forgiven you." The anathema would be withdrawn, and we and he should be as we had originally been.' At these words, the caliph told the Catholicos to take the icon and ordered that he should be given a bag of dirhams to spend on it. The Catholicos then left, while al-Mutawakkil remained lost in thought for a time, wondering at the prelate's behaviour and his veneration for such an object of adoration. 'A strange matter altogether,' he remarked at length.

Then the caliph ordered me to be brought in. Calling for a whip and ropes, he ordered his men to strip me and tie me up, and I was given a hundred lashes then and there. He then ordered me held in close confinement and subjected to torture, while a gang of ruffians was sent to confiscate all my goods and chattels. In addition, al-Mutawakkil ordered my houses demolished down to groundwater level. Meanwhile, I remained imprisoned in the palace for six months under dreadful conditions, so that any who caught sight of me could not but pity me. Furthermore, every few days he would send a number of toughs to beat me and torture me anew.

I was subjected to this treatment until, after I had been in detention for three months and five days, the caliph was laid low by a grave illness, was unable even to move, and was given up for lost. He himself despaired of his life. However, the physicians who were my enemies never left his side for a moment, plying him with remedies, and all the while urging him to do something about me. 'What would you have me do with him?' 'Best to relieve the world of him altogether, O Commander of the Faithful,' they said. There were few who were supporting me and many who were working against me, and I despaired of my life.

After all this incessant badgering, the caliph finally agreed to have me put to death. 'I shall have him executed tomorrow morning and thus relieve you of him,' he told his physicians. They were delighted, and went home well pleased with themselves, while a flunky from the palace came to inform me of the decision. Forthwith I prayed to God mighty and glorious, asking Him to favour me once more with the grace that I had previously known at His hands, for I was sorely distressed. I had done nothing to deserve such a fate, but rather was a victim of those who had conspired against me.

The next thing I knew, I was being shaken and a voice in my ear was saying, 'Arise, praise and glorify God, and take heart, for God has delivered you from your enemies and placed the health of the Commander of the Faithful in your hands,' and I praised and glorified God until the dawn.

In due course the keeper appeared and opened the door of my cell. Hardly had he taken his seat when a slave-boy entered with a barber in tow. My hair was trimmed, and then I was taken to the bath, where I was washed clean and anointed with perfume, for so our master, the

Commander of the Faithful, had ordered. When I emerged from the bath, I was arrayed in fine clothes and then taken to the keeper's lodge, where I waited until all the physicians had assembled before the caliph, each in his assigned place. Finally the caliph called out, 'Bring Hunayn in!' The physicians were quite sure that they were about to see me executed. I was ushered into the salon.

Al-Mutawakkil bade me be seated immediately in front of him. Then he spoke: 'I have forgiven you for the sin you committed, acceding to the request of one who has spoken for you. Praise God, then, for your life. Now take my pulse and prescribe whatever treatment you think fit.' I took his pulse and advised him to take purging cassia fruit pulp* and manna,* together with the other ingredients in the usual formula for the use of that medication, for he was suffering from constipation. My enemies, the other physicians, urged him not to touch it. 'Silence,' roared the caliph. 'I have been told to take whatever he prescribes for me,' and he ordered the medicine prepared immediately. It was brought in, and he swallowed it then and there.

Then al-Mutawakkil spoke to me. 'Hunayn,' he said, 'I must beg your pardon for what I did to you. A very powerful person has interceded with me on your behalf.' 'I pardon you willingly, Sire,' I replied, 'and how could I do otherwise, since you have spared my life?' The caliph then turned to the assembled physicians. 'I want you all to listen carefully to what I am about to say.

'When you all left yesterday evening,' he went on, 'you were aware that I intended to have Hunayn put to death this morning. Afterwards, I lay awake in severe pain for half the night, but finally fell asleep and had a dream. In my dream, I was sitting in a narrow place. You physicians were there, but you were a long way off, with my household staff and retinue. I was saying to you, "You wretched crew, why are you looking at me like that? What are we doing here? Is this a fitting place for such a person as myself?" but you answered not a word. In the midst of all this, a great, terrifying light suddenly shone upon me in that place. As I sat there petrified with fear, I became aware that somehow there was a man standing beside me, a man with a face glorious in its beauty, and with another man, dressed in fine clothing, standing behind him. He greeted me, and I returned his greeting. Then he asked, "Do you know who I am?" "No," I answered, whereupon he said, "I am Jesus Christ." This caused me to quake with alarm. "Who is that person with you?" I asked. "Hunayn ibn Is'hāq," Christ replied. "Please excuse me," I said, "but I am unable to stand to shake hands* with you." "You are to excuse Hunayn," he said, "and forgive him for his sin, for that is what God has done. You are also to take whatever he advises you to take. Do that, and you will recover from your illness."

'I then awoke,' the caliph went on, 'overcome with distress at the suffering I had caused to be inflicted upon Hunayn and reflecting on the powerful intercessor who had approached me on his behalf. It is now incumbent upon me to ensure that justice is done. I am dismissing the lot of you: Hunayn shall henceforth be my attending physician, in accordance with Christ's bidding. Furthermore, those of you who demanded that I have him put to death shall be liable for the blood-price. Every one of you who was here last night is to bring me 10,000 dirhams, and any of you who fails to pay that amount will be beheaded.'

By the time the day was drawing to a close, the medicine had sent al-Mutawakkil to the lavatory three times, with the result that he was greatly relieved and feeling better. 'Hunayn,' he said, 'you shall have whatever you wish, for I have the utmost regard for you, and you shall enjoy much greater prestige than you formerly did. I shall compensate you many times over for what you have lost, make your enemies subject to your authority, and raise you in rank above all other practitioners of your art.' He then ordered that three houses belonging to him should be fitted up for me, such houses as I had never lived in, nor ever known any other physician to live in, in all my life. They were valuable houses, worth thousands of dinars. The caliph had them formally conveyed into my possession, with documents signed in the presence of witnesses, for he wished to ensure that they should be my property and the property of my descendants, and that no one should have any grounds for challenging my ownership of them. Once this was done, he had the houses fitted out with everything I might need.

He then arranged for me to be carried thither with five of his own finest mules, splendidly equipped, and he also gave me three Roman servants. Furthermore, he ordered that I was to be paid 15,000 dirhams a month, besides a lump-sum payment covering all the time I had been in prison, which made a very handsome total indeed. In addition to all this, the caliph's household staff, the ladies of the harem, the remainder of his retinue, and his family all contributed incalculable quantities of coin, robes of honour, and estates in fee. I became al-Mutawakkil's chief physician, with all the others, friends and enemies both, subject to my authority.

I found that as often as not, the first person to come to my door in the morning to seek my influence in obtaining a favour from the Commander of the Faithful, or to consult me about some illness that had baffled him, would be one of those enemies at whose hands I had known the hardships that I have described. And, by the truth of my Lord, never did I seek to avenge myself upon a single one of them for what he had done to me. I became an object of wonder for my willingness to oblige these physicians. Moreover, I would translate books for them, as meticulously as always, without charging them anything, so anxious was I to show my good will towards them, whereas formerly it had been usual for me to be paid a purse of silver dirhams equal in weight to the translated work.

I—Ibn Abī Usaybiʿah—here insert the following comment. I have found large numbers of these works, and have purchased a good many of them. They are written in *muwallad* Kufic script* in the handwriting of Hunayn's secretary. The letters are written very large, with broad strokes, and the lines are widely spaced. The paper is very heavy, being three or four times as thick as the paper manufactured nowadays, while the sheets are trimmed to about one third the size of a sheet of Baghdadi paper.* Hunayn had his works published in that fashion to make them bulkier and increase their weight, inasmuch as he was paid weight for weight in silver dirhams. It is thus clear that he used that particular type of paper deliberately. Small wonder, then, that the manuscripts have lasted so well for so long.

Resuming Hunayn's narrative:

I have set forth the foregoing account only to help the intelligent reader understand that trial and tribulation may befall anyone, be he wise or foolish, old or young. This being the case, a reasonable man should never despair of God's providence or the prospect of deliverance from his predicament. Rather, he should cling fast to his trust in his Creator. Praised be God, who granted me a second life and enabled me to overcome the wicked machinations of my enemies. Praised be He now and forevermore!

Here ends the account of Hunayn ibn Is'hāq in his own words, which I have reproduced faithfully.

Hunayn ibn Is'hāq is the author of 111 works on a wide variety of subjects including:

On the eye, a work of three treatises and 209 question-and-answers, written for Hunayn's two sons, Dāwūd and Is'hāq.

Treatise consisting of a definitive list of works not mentioned by Galen in the Catalogue of his books, but which in the author's judgement are undoubtedly by Galen; he reasons that they must be later than *On my own books*.*

Letter addressed to al-Tayfūrī on rose pastilles.

On infants born at eight months, in question-and-answer format. Hunayn composed this work for an *umm walad** of al-Mutawakkil's.

On tickling.

Letter addressed to Salmawayh in response to his request for a translation of Galen's *On habits*.

Stories of the philosophers and the sages and of the culture of the teachers of old.*

On rainbows.

On his trials and tribulations.*

13. IS'HĀQ SON OF HUNAYN (*d.* 910)

Is'hāq ibn Hunayn rivalled his father in respect of his skill as a translator, his knowledge of languages, and his fluent style. However, he translated only very few medical works.

Is'hāq served the same caliphs and officials as his father had served, including in particular the vizier, al-Qāsim ibn ʿUbayd Allāh, to whom he was devoted, and who rewarded him with advancement and entrusted him with confidential matters. Al-Qāsim once learned that Is'hāq had treated himself with a laxative. The vizier composed the following teasing verses and sent them to Is'hāq:

Let me know how you were last night
and how you felt
And how often the she-camel took you
to the Empty Abode.*

To which Is'hāq replied,

> I was fine, happy,
> and relaxed in body and spirit.
> As for travelling, the she-camel,
> and the Empty Quarter,
> My respect for you made me forget,
> O goal of my hopes!

Is'hāq possessed a fund of entertaining stories and anecdotes including the following:

I once had a patient who came to me complaining of pain in the belly. I gave him an electuary.* 'Take it first thing in the morning,' I said to him, 'and then come to see me in the evening and tell me how you are feeling.' That evening, his slave-boy appeared bearing a message from his master: 'Dear Sir, I took the medicine, and (as I value you) have had ten bowel movements, some reddish and resembling ropy saliva, and some greenish, the colour of chard, reminiscent of a salad. Afterwards, I suffered from colic in my head, with dizziness in my navel. Would you very kindly take Nature to task for this in such a way as you may deem appropriate, if God, exalted is He, so wills.'

After reading this extraordinary missive, I said to myself, 'Well, there is only one way to answer such a crackpot as this,' and I wrote back to him, 'I have received your message. I shall go to see Nature as you ask, and will send you the answer once we have met. Yours faithfully.'

Towards the end of his life, Is'hāq ibn Hunayn suffered from paralysis, which ultimately proved fatal to him. He died in Baghdad in the month of Rabiʿ II 298/December 910, during the reign of al-Muqtadir.

Is'hāq ibn Hunayn is the author of fifteen works including *On the origins of the art of medicine*,* in which the names of a number of sages and physicians are mentioned.

CHAPTER 9

PHYSICIANS WHO TRANSLATED WORKS ON MEDICINE AND OTHER SUBJECTS FROM GREEK INTO ARABIC, AND THEIR PATRONS

This text of this short chapter has not been included in this edition. In it are listed thirty-eight translators and eleven patrons 'other than caliphs'. None have more than four or five lines of description.

The translators appear to be listed randomly and several, such as Hunayn ('His works are superb examples of the translator's art') and his son, Is'hāq ('conversant with all the languages that his father knew, and comparable to him as a translator'), are given biographies in Chs. 8 and 10. Pethion the Translator, though cited extensively in Ch. 8, is described with one sentence, 'I have found his translations barbarously ungrammatical; his grasp of the Arabic language was most imperfect.' Many are Christian, translating from Greek to Syriac as well as Arabic; some are also physicians.

The eleven patrons, where they can be identified, are naturally men of status and often officials of the caliph, though a physician and bishop also appear. They are also usually scholars or men of learning. Muhammad ibn Mūsā (see also Ch. 10, nos. 1 and 2) typifies this. He was part of the caliph al-Mutawakkil's court, a scholar and patron of the arts who commissioned many translations, from Hunayn in particular. In the brief descriptions, Ibn Abī Usaybiʿah focuses on the patronage alone, writing, for example, 'He was generous to translators, who grew fat on his bounty,' and that another 'spent nearly 2,000 dinars a month on translators, scholars and copyists and many books were translated under his patronage, including Greek works.' Further background information on the patrons themselves is generally lacking.

Occasionally, here and elsewhere, the study of Arabic, and a person's mastery or sound knowledge of the language is referred to. It should be remembered that native speakers would have grown up speaking 'Middle Arabic', the unwritten dialects or non-standard Arabic of the Middle Ages, rather than the Classical Arabic used for academic study and writing. It would therefore have been necessary for any scholar or physician, as well as the educated elite, including many Persians, to acquire proficiency in the more formal, scholarly, written language as part of their education.

It is also worth noting that Arabic was not as predominant in the Middle East as it is today. Greek was still used for a considerable time in Syria and Palestine, Coptic was only slowly replaced by Arabic among Egyptian Christians, while Syriac (a form of Aramaic) was spoken among Syrian Christians. In Iraq the rural population still spoke another form of Aramaic (Nabatī, 'Nabataean') for several centuries, while the Turkish Seljuq rulers (from the eleventh century) used Persian.

CHAPTER 10

IRAQI PHYSICIANS AND THE PHYSICIANS OF AL-JAZĪRAH AND DIYĀR BAKR

This and the following chapters are based on geography, rather than chronology, though the entries within are organized roughly chronologically. Chapter 10 contains entries on physicians active in Iraq, which in pre-modern times meant the land between the Euphrates and the Tigris north of Baghdad. Not many of them are household names, but they include famous scholars such as the early philosopher Yaʿqūb al-Kindī and Ibn Butlān, a Christian. Also in this chapter, some of the harshness of medieval life is brought into focus. In the entry on Thābit ibn Sinān, we are given a relatively graphic account of the imprisonment and torture of Ibn Muqlah, while in Ibn Butlān's entry, one of the great famines is described and in the entry on Ibn Safiyyah, the caliph al-Mustanjid is imprisoned in an over-heated bathhouse by one of his emirs and left to die.

1. YAʿQŪB AL-KINDĪ (*d. c.*870)

YAʿQŪB AL-KINDĪ was the philosopher par excellence of the Arabs and he was descended from Arabian chieftains. He could trace his lineage back twenty-three generations. His father had been Governor of Kufa for the caliphs al-Mahdī and Hārūn al-Rashīd, and his great-great-great-great-grandfather had been a companion of the Prophet—God bless him and keep him—before which he had been the chieftain of all of the tribe of Kindah.*

Ibn Juljul says:

Yaʿqūb al-Kindī was a Basran* of noble stock whose grandfather had been a provincial governor for the Hashemites.* He settled in Basra where his estates were, then moved to Baghdad where he was educated. He was learned in medicine, philosophy, arithmetic, logic, musical composition, geometry, the properties of numbers, and astronomy, and prior to him there had been no philosopher in the Islamic period. In his compositions he followed the example of Aristotle and he composed many works. Yaʿqūb attended the ruling classes and treated them with courtesy. He translated a great many books of philosophy, clarifying their difficulties, summarizing their complex parts and simplifying their obscurities.

The son of Ibn al-Dāyah, says in his *Book of fortunate outcome*:

During the time of the caliph al-Mutawakkil, Muhammad and Ahmad, the sons of Mūsā ibn Shākir,* used to plot against all those who had a reputation for advanced learning. They had already caused the engineer, Sind ibn ʿAlī, to be sent to Baghdad* after having estranged him from al-Mutawakkil and had also plotted against Yaʿqūb al-Kindī so that al-Mutawakkil had him flogged. They sent people to Yaʿqūb's house to confiscate all his books and placed them in a repository that was given the name 'Kindiyyah'. They had been able to do this because of al-Mutawakkil's passion for automata.*

The caliph approached them with regard to excavating the canal known as the Jaʿfarī canal* [to the north of Samarra]. The Banū Mūsā (sons of Mūsā ibn Shākir) delegated this matter to al-Farghānī who had built the new Nilometer in Egypt. Al-Farghānī's knowledge, however, was greater than his good fortune since he could never complete a work. He made an error in the mouth of the canal and caused it to be dug deeper than the rest so that water which filled the mouth would not fill the remainder of the canal. Muhammad and Ahmad, the two Banū Mūsā, protected him, but al-Mutawakkil demanded that they be brought before him and had the engineer, Sind ibn ʿAlī, summoned from Baghdad. When Muhammad and Ahmad realized that Sind had come they felt sure they were doomed and feared for their lives. Al-Mutawakkil summoned Sind and said to him, 'Those two miscreants have left no foul words unsaid concerning you, and they have squandered a great deal of my money on this canal. Go there and examine it and inform me whether it has a defect, for I have promised myself that if what I have been told is true, I will crucify them on its banks.' All of this was seen and heard by Muhammad and Ahmad.

Then Sind went out with the two of them and Muhammad said to him, 'O Sind, the power of the freeman dispels his grudges, and we flee to you for the sake of our lives, which are our most valuable possessions.

We do not deny that we have done wrong, but confession effaces the commission,* so save us as you see fit.' He said to them, 'I swear by God that you know well the enmity that I have for Yaʿqūb al-Kindī, but what is right is right. I swear I will not speak in your favour until you return his books to him.' Then Muhammad ibn Mūsā had Yaʿqūb's books taken to him and obtained his signature that this had been done.

When Yaʿqūb al-Kindī's note arrived saying that he had received them to the last, Sind said, 'I am obliged to you both for returning the man's books to him, and I am further obliged to inform you of that which you failed to recognize; the fault in the canal will remain hidden for four months due to the rising of the Tigris. Now the astrologers agree that the Commander of the Faithful will not live that long. I will tell him immediately that there was no error in the canal on your part so as to save your souls, and if the astrologers are correct then the three of us will have escaped. But if they are wrong and he lives longer until the Tigris subsides and the water disperses then all three of us are doomed.' Muhammad and Ahmad were grateful and mightily relieved to hear these words.

Then Sind visited al-Mutawakkil and said to him, 'They did not err.' And the Tigris rose and the water flowed into the canal and the matter was concealed. Al-Mutawakkil was killed two months later* and Muhammad and Ahmad were saved.

It is said that someone recited the following verses to Yaʿqūb al-Kindī:

> Four parts of you sweeten four parts of me,
> and I don't know which one aroused my grief:
> Is it your face in my sight, or your taste* in my mouth,
> your speech in my ears, or your love in my heart?

At which al-Kindī said, 'By God, he has classified the matter philosophically!'

I—Ibn Abī Usaybiʿah—say, among the sayings of al-Kindī is this:

> Let the physician be mindful of God, exalted is He, and let him not take risks, for there is nothing that will compensate for the loss of human life.

Among al-Kindī's words are these from his testament to his son:

My son, a father is a lord and a brother is a snare and a paternal uncle is an affliction and a maternal uncle is a calamity and a son is a distress and relatives are scorpions. Saying no averts a blow, saying yes dispels blessings.*

Yaʿqūb al-Kindī wrote 283 books and epistles on a wide range of subjects including:

On First Philosophy, excluding physics or metaphysics.

On the fact that philosophy may only be attained through knowledge of the mathematical sciences.

On the nature of the infinite, and in what way it is termed infinite.

An explanation of the impossibility of the infiniteness of the matter of the world but rather its potential for infinity.

On ornithomancy and divination with regard to number.*

On the sphericity of the world and all that is in it.

On the arrangement of musical notes which indicate the natures of the heavenly bodies. It resembles the treatise on (musical) composition.

An essay apologizing for his own death before reaching the term of a natural life, which is 120 years.

On the geometry of palaces.

On the division and construction of triangles and squares.

On the drugs which cure painful flatulence.

On making food without using foodstuffs.

An essay demonstrating the utility of medicine when the art of astronomy is used in conjunction with its indicators.

On speech impediments.

On prognosis by deduction via the heavenly bodies, arranged by questions.

On the great phenomenon that was observed in the year 222 of the Hijrah (836).*

On the nature of time, the nature of eternity, moments in time and the present.

On making panes of glass.

On the manufacture of whistling kettles.

On the chemistry of perfumes.

On removing stains from clothes and suchlike.

2. THĀBIT IBN QURRAH (*d.* 901)

Thābit ibn Qurrah was one of the Sabians* who lived at Harrān. He was a money-changer there. Then Muhammad ibn Mūsā (of the Banū Mūsā) took him as a companion when he left Anatolia because he thought him to be eloquent. It is said that Thābit studied with Muhammad and was educated in his house and that Muhammad felt obliged towards him so he presented him to the caliph al-Muʿtadid. Thābit was the first of the Sabians to attain a high rank in Baghdad and at the caliph's court.

In his day no other was the equal of Thābit ibn Qurrah in the art of medicine nor in any other branch of philosophy. He authored many works known for their good quality, and in addition, a great many of his descendants and relatives rivalled him in their fine output and skill in the sciences. Thābit also made some fine observations of the sun which he undertook in Baghdad and collected in a book in which he explained his ideas regarding the solar year.

He was a good translator into Arabic, used beautiful expressions, and had superb knowledge of Syriac and other languages.

His grandson, Thābit ibn Sinān says:

When the caliph al-Muwaffaq became displeased with his son, al-Muʿtadid, he kept him under arrest in the house of the vizier, who approached Thābit ibn Qurrah so that the physician might visit al-Muʿtadid and keep him company. Thābit remained constantly with al-Muʿtadid who enjoyed his company immensely. Thābit used to visit him three times every day to converse with him, console him, and teach him about the lives of the philosophers, and about geometry, astronomy, and the like. Al-Muʿtadid was struck with Thābit and his situation became pleasant because of him. When he emerged from house arrest he said to his most trusted servant, 'Badr,* after you, which man has benefited me most?' Badr said: 'Who would it be, master?' He replied: 'Thābit ibn Qurrah.'

When al-Muʿtadid became caliph he endowed Thābit with great estates, and he often used to let him sit with him in the presence of the elite and the commoners alike; and Badr and the vizier would remain standing while Thābit would be sitting next to the caliph.

Thābit was once walking with al-Muʿtadid in a garden in the caliph's palace intended for exercise, when al-Muʿtadid leant on the arm of Thābit and they walked together. Suddenly, al-Muʿtadid forcefully wrested his arm from Thābit's arm at which Thābit took fright as al-Muʿtadid was very fearsome. Then al-Muʿtadid said, 'O Abū l-Hasan (in private he would use Thābit's *kunyah* or nickname and in public his given name), I absent-mindedly put my arm on your arm and leant upon it and this is not right, for the learned should be uppermost and none should be higher than them.'

With regard to Thābit's wonderful medical treatments, his grandson relates:

One of my ancestors related to me that one day when my grandfather Thābit ibn Qurrah was passing by a shop on his way to the palace of the caliph he heard a shrieking and wailing. He said: 'The butcher who had this shop has died!?' Amazed at his words, the people said, 'Yes master, by God, suddenly last night!' Thābit said: 'He has not died. Take me to him.' They took him to the house and he asked the women to stop slapping themselves and shrieking and asked them instead to prepare some food suitable for a convalescent. Then he indicated to one of the servants to strike the butcher on the ankles with a cane, while he put his hand on his pulse point. The boy continued to strike his ankles until Thābit told him to stop. Then he called for a cup and brought out from his sleeve a handkerchief containing a drug which he mixed with a little water in the cup. When he opened the butcher's mouth for him to drink the butcher swallowed the drug. At this, shouts and yells to the effect that the physician had revived a dead man rang out in the house and in the street. Thābit barred the doors and made them secure. The butcher opened his eyes and Thābit fed him the convalescent food, sat him up, and sat beside him for a time. When the caliph's men came calling for him he left with them and there was a great tumult with people running to and fro about him until he entered the caliph's palace. When he came before the caliph, the caliph said to him: 'Thābit, what is this Christ-like act of yours I am hearing about?' He said: 'Master, I used to pass by this butcher and noticed him slicing up liver, pouring salt on it, and eating it. At first I was disgusted at this, then I realized that he would suffer a stroke so I began to observe him and compounded a drug for the stroke,* which I kept with me at all times. When I passed by today, heard the shrieks, and asked whether the butcher had died, the people said yes, he died suddenly last night. I knew then that he had suffered the stroke. When I went to him I found he had no pulse so I struck his ankles until the movement of his pulse returned. I made him drink the drug and then he opened his eyes. Afterwards I fed him a light dish. Tonight he will eat a loaf of bread and some francolin,* and tomorrow he will emerge from his house.

When my grandfather, Thābit ibn Qurrah, died—may the mercy of God be upon him—a firm friend of his elegized him in these verses:

Ah, everything but God is mortal. Someone who has gone to foreign parts
 may be expected to return, but he who has died is gone for ever.
I see that those who have gone from us, after having pitched their tents with us,
 are like travellers who lodge on earth, coming in the evening and spending the night.
We announce the death of all the philosophical sciences:
 their light was extinguished when they said, 'Thābit is dead!'
Those who practise them are dismayed because of losing him;
 with him a firm (*thābit*) cornerstone of science has gone.
Whenever they went astray they would be guided on their path by 5
 someone expert in giving a decisive judgement, unearthing the truth.
When death came to him his medical knowledge did not avail,
 nothing that could speak nor anything dumb of what he possessed.
The suddenness of his demise did not let him enjoy wealth;
 Many a livelihood arrives and is lost.
If it were possible for death to be dispelled,
 brave protectors would have defended him against it,
Trusted friends, who loved him sincerely;
 but there is no one who can turn what God decrees.
Abū Hasan,* do not go far!* All of us 10
 are in shock because of your death, crushed by grief.
Can I ever hope that the truth about any uncertainty will be revealed,
 now that your body is buried and your voice silenced?

Your fine exposition would dispel blindness;
when you were speaking every loquacious person was silent.
It was as if you, when asked a question, were scooping from a sea of knowledge,
and, if asked to begin speaking, were hewing from a rock.
Not a single person would seek me in matters of knowledge
who poured out the vessel of knowledge, after your death.
So many loving friends have benefited from you, 15
while people other than you who tried to surpass you would stumble.
I am surprised that on an Earth who made you disappear there would not
forever be established (*yathbuta*) a Thābit like you.
You refined yourself so that nobody could hate you;
when death murdered you there was no one who gloated.
You excelled to the point that there was no one who could deny
your merit except a slandering liar.
Gone is the beacon of science, who was sufficient;
now none remain but erring, blundering people.

Among the sayings of Thābit ibn Qurrah are:

There is nothing more harmful for the older man than to have a skilful cook and a beautiful young servant girl since he will take an excess of food and become ill, and will engage in sexual intercourse to excess and become senile.

The repose of the body is in minimum food, and the repose of the soul is in having few sins, the repose of the heart is in having few concerns, and the repose of the tongue is in keeping speech to a minimum.

Thābit ibn Qurrah wrote 144 books including:

On the causes of the formation of mountains.

Why seawater is salty.

On the pauses in the stasis which occurs between the two opposing movements of the arteries. He composed this book in Syriac. A student of his translated it into Arabic which Thābit revised. After Thābit had composed this book he sent it to Is'hāq ibn Hunayn who approved of it immensely and wrote in his own hand on the last page commending Thābit.

On joint pain and gout.

On using the celestial globe.

On the stones which occur in the kidneys and the bladder.

On the whiteness appearing on the body.

On how the physician should question patients.

A treatise on smallpox and measles.

An answer to a question about the Hippocratics and how many they numbered.

A treatise on the yellowness which occurs in the body, its types, causes, and treatment.

A treatise on choosing the best time for impregnation.

A book demonstrating that weights suspended separately from a single pillar have the same effect as when the weight is combined into a single weight and fixed equally to the entire pillar.

On sundials.

Designs for automata.

A treatise on magic number grids.

A treatise on striking fire by means of two stones.

Among the nine extant books of Thābit ibn Qurrah in Syriac about his religion are:

A treatise on rites, and rules, and laws.

A treatise on shrouding and interring the dead.

A treatise on the doctrines of the Sabians.

On ritual purity and impurity.

A treatise on the reason people speak in riddles.

A treatise on animals fit and unfit for sacrifice.

3. SINĀN SON OF THĀBIT IBN QURRAH
(ninth to tenth century)

Sinān ibn Thābit followed his father in knowledge of the sciences, devotion to them and in mastery of the art of medicine. His aptitude for astronomy was far-reaching. He was in the service of the caliphs al-Muqtadir, al-Qāhir, and al-Rādī as a physician.

Thābit ibn Sinān, his son, says in his chronicle:*

I remember when the vizier, ʿAlī ibn ʿĪsā al-Jarrāh, sent a note to my father, Sinān ibn Thābit. This was in a year of a great many diseases and at that time my father was in charge of the hospitals in Baghdad and elsewhere. In the note the vizier said: 'I have been thinking—may God extend your life—of the state of those who are in the prisons and that, because of their great number and the harshness of their abodes, they must be susceptible to diseases while they are prevented from acting to their own benefit or from seeking advice from physicians about their symptoms. It is necessary then that physicians be assigned to them who will visit them every day and that drugs and sherbets be brought to them. The physicians should go around all the prisons treating the patients there and alleviating their illnesses with the requisite drugs and sherbets. I also suggest that convalescent meals be provided for those who require them.'

This my father did for the rest of his life.

He received another note which said: 'I have been thinking of those people who live in the provinces and that there must be ill people there who are not overseen by any physician since the provinces are empty of physicians. Proceed—may God extend your life—and send physicians and a chest of medicines and sherbets to circulate in the provinces and stay in every locality as long as is required and treat any patients there before moving to the next place.'

My father did this until his colleagues reached Sūrā, the majority of whose population were Jews. He wrote to the vizier informing him that his colleagues in the provinces there were asking permission to remain and treat those people but that he did not know how to answer them since he had no knowledge of the vizier's view on the Jews. My father also informed the vizier that the policy of the hospital was to treat Muslim and non-Muslim alike and he asked him to lay down a rule for him in this matter for him to act upon.

The vizier replied saying: 'I have understood your letter—may God honour you—and there is no disagreement among us that medical treatment of *dhimmī*s* and animals is something correct. However, it should be proposed and acted upon accordingly that humans should be treated before animals and Muslims before *dhimmi*s. And if after treating the Muslims there is anything left over it should be used for the next group. So act—may God honour you—according to this and write to your colleagues to this effect. And advise them to travel in the villages and places in which there are many epidemics and raging diseases, and if they cannot find people to guard them on their journey then they should cease to travel until the way becomes safe, and if they do this then they will not be attacked, God willing.'

In 306/918, my father advised al-Muqtadir to arrange for a hospital in his name and this he did and it was arranged for him at the Damascus Gate. He named it the Muqtadirī hospital and spent on it of his own money 200 dinars every month.

His son also says:

In 319/931, al-Muqtadir became aware that a commoner had died due to an error from a physician. It was ordered that all the practising physicians be prevented from working except for those my father Sinān ibn Thābit had examined and written a note in his own hand stating to what extent he was allowed to practise the art. So they went to my father and he examined them and gave permission to each one of them to practise as was appropriate in the art. In the area of Baghdad they numbered some 860 men not including those who were exempt from examination due to them being known to be advanced in the art or those who were in the service of the ruler.'

Twenty books were written by Sinān ibn Thābit.

4. THĀBIT SON OF SINĀN (*d.* 973–4)

Thābit ibn Sinān was a physician of merit who followed his father in the art of medicine. In the chronicle he wrote in which he mentions the occurrences and events which took place during his time, he says that he and his father were in the service of the caliph, al-Rāḍī. Subsequently he says of himself that he served al-Muttaqī, al-Mustakfī, and al-Mutīʿ. I have found this chronicle in his own hand and he demonstrates his merit therein.

He also says in his chronicle that:

When the former vizier, Ibn Muqlah, was arrested by al-Rāḍī in 324/935, he was beaten with whips and his signature was taken for the amount of 1,000,000 dinars. Then he was handed over to the vizier, al-Khasībī, who gave over questioning of him to Abū Thughrah, and gave responsibility for demanding money to another man at whose hands Ibn Muqlah experienced terrible abominations, hangings, beatings, and rackings.

What I—Thābit ibn Sinān—personally witnessed of this was when al-Khasībī charged me one day with visiting Ibn Muqlah to examine him for a complaint and said that should Ibn Muqlah require bloodletting, it should be done in my presence. So I visited him and I found him stretched out on a threadbare mat with a filthy pillow under his head and naked except for an undergarment. And I saw that his body from his head to the tips of his toes was all the colour of an aubergine and none of it had been spared. And I saw that he was extremely short of breath because of his injured chest. I told al-Khasībī that he was in dire need of bloodletting and he said to me: 'He must be subjected to hard demands so what shall we do with him?' I said: 'I don't know, except that if he is left without bloodletting he will die and if blood is let and he is subject again to abominations then he will also expire.' So al-Khasībī said to Abū Thughrah: 'Go to him and tell him that if he thinks that he will find some respite if his blood is let then he thinks mistakenly. Let his blood but he must remember that the demands will continue.' Then he said to me: 'I would like you to go with him.' I asked to be excused from this but he refused, so I went in with Abū Thughrah and he delivered the message. Ibn Muqlah said: 'If this is the case then I don't wish to undergo bloodletting while I stand before God.' So we returned to al-Khasībī and told him what he had said. He said to me: 'What can you do and what is your opinion?' I said: 'I think that his blood should be let and he should be given respite.' He said: 'Make it so!'

So I returned to Ibn Muqlah and his blood was let in my presence and he was given respite for the day and his situation became more bearable but he expected the abominations to continue the next day and was terrified out of his wits.

Then it so happened that on that day al-Khasībī had cause to go into hiding and Ibn Muqlah remained in respite and no one made any demands of him and he was unexpectedly relieved of his enemy. He returned to his senses and was given a guarantee for what he was owed. Prior to this Ibn Muqlah had handed over to al-Khasībī some 50,000 dinars, and just men had been made to swear that he had sold his estates and those of his sons and all his personal effects to the ruler.

In his book Thābit ibn Sinān also says:

When Ibn Muqlah's hand was cut off, the caliph al-Rāḍī summoned me at the end of the day and commanded me to enter in upon Ibn Muqlah and treat him. So I went to him on the day his hand was cut. I found him to be confined in the cell which was in the arbour court with the door locked. I entered and found him sitting on the base of one of the pillars of the cell and his colour was as grey as the lead he was sitting on. He had become very weak and was in a state of utmost anguish at the throbbing pain of his arm.

I saw in the cell that a dome of canvas had been erected for him with two arches in which there was a prayer mat, and cushions of Tabaristan. About the prayer mat there were many dishes of fine fruits. When he saw me he wept and bemoaned his state and what had befallen him and the shock he was experiencing. I found that his forearm had become severely swollen and that on the cut there was a rough, dark blue cloth bound with a hemp rope.

I addressed him as was necessary and calmed him and untied the rope and removed the cloth under which I found, upon the cut, animal dung. So I ordered that the dung be removed from it. Then, I found that the top of his forearm above the cut had been bound with hemp rope and that it had immersed itself into his arm because of the swelling and his forearm had begun to turn black. So I told him that the rope would be untied and that, in place of the dung, camphor would be put and that his forearm would be anointed with sandalwood and rosewater and camphor. He said: 'Master, do what you think fit.' Then the servant who was with me said: 'I need to ask permission of our master for this.' And he went in to ask permission. When he came out he had with him a large chest filled with camphor and he said: 'Our master has given his permission for you to do as you see fit and he commands that you treat him kindly and provide care for him and remain with him until God grants him well-being.' So I untied the rope and emptied the chest on the place of the cut and anointed his forearm. And so he lived and found relief and the shock subsided.

I asked him whether he had eaten and he said: 'How can I stomach food?' So I arranged for food to be brought and it was brought but he refused to eat, so I was gentle with him and fed him morsels with my own hand and he ate about twenty dirhams of bread and about the same amount in meat of pullets. Then he swore he couldn't swallow another thing. And he drank some cold water and his spirit revived and so I left and the door was locked upon him and he remained alone. In the morning a black servant was permitted to enter and serve him and was confined with him. I went back and forth to see him for many days and he became afflicted with gout in his left leg so I let his blood. He used to be in pain from his right hand which was cut off and from his left leg and would not sleep at night because of the severity of the pain.

Then he regained his health and whenever I visited him he would begin by asking about his son and when I told him he was well he became most calm. Then he would mourn for his own self and weep for the loss of his hand and would say: 'With my hand I served the caliphate three times for three caliphs, and with it I wrote out the Qur'ān twice,* and it is cut off like the hand of a thief! Do you remember when you said to me: "You are experiencing the last of the calamity and relief is near?"' I said: 'Indeed.' He said: 'You see what has befallen me!' I said: 'Nothing remains after this. Now you should expect relief, for you have been treated as none of your peers have been treated. This is the end of the adversity and after the end there is nothing but relief of the burdens.' He said, 'Don't, for this trial has attached itself to me such as to take me from one situation to the next until it leads me to perdition just as the hectic fever attaches itself to the members of the body and does not leave the sufferer until it leads him to his death.' Then he quoted this verse:

> Whenever part of you has died, lament another part!
> For one part of a thing is closely related to another.

And it was as he said. And when Bajkam* closed in on Baghdad, Ibn Muqlah was moved from that place to somewhere even more concealed and no one had any news of him and I was prevented from seeing him. Then his tongue was cut out and he remained in confinement for a long time. Subsequently he was overcome by uncontrollable diarrhoea, and he had no one to treat him or serve him until I heard that he was drawing up water for himself with his left hand and pulling the well-rope with his left hand and holding it in his mouth and his state was utterly wretched until such a time as he died.

5. SĀʿID IBN BISHR IBN ʿABDŪS
(early eleventh century)

Ibn ʿAbdūs was at first a bloodletter in the hospital at Baghdad, after which he occupied himself with the practice of medicine and so distinguished himself that he became one of its great proponents and prominent personalities.

I quote from the physician Ibn Butlān on the reason the most skilled physicians changed the regimen for diseases such as partial paralysis, facial paralysis, lassitude, and the like, that had been treated with hot medicines, to a cold regimen, and in doing so went against the prescriptions of the ancients:

The first person in Baghdad to become aware of this method, to alert others to it and treat patients accordingly was Ibn ʿAbdūs the physician, may God show him mercy. He would treat patients by letting their blood, and giving them cooling and moisture-inducing remedies and would not allow patients to take food. This regimen was very successful. He became the chief physician upon whom kings depended for their regimens. In the hospital, the regimen of the patients was changed from the hot salves and pungent medicaments to barley water and the juice of seeds, and this method worked wonders.*

Ibn ʿAbdūs composed for one of his contemporaries a work on hypochondria and its treatment.

6. IBN BUTLĀN (eleventh century)

Ibn Butlān, a Christian of Baghdad, was proficient in many books of philosophy and the like. He was a contemporary of the Egyptian physician Ibn Riḍwān, and the two of them exchanged extraordinary correspondence and shocking and astonishing writings. Neither of them would compose a book nor form any opinion without the other responding to it and exposing the folly of his opinion. I—Ibn Abī Usaybiʿah—have seen examples of their correspondence and their criticisms of one another. Many incidents took place between Ibn Butlān and Ibn Riḍwān during that time, including some entertaining stories that are not without useful lessons.

Ibn Ridwān was swarthy and was not handsome. He composed a treatise on this subject in reply to those who had faulted him for his ugliness. He claims to demonstrate in this treatise, that the physician of merit does not need to have a beautiful face. Most of Ibn Butlān's criticisms of Ibn Ridwān were of this type. Hence, he says of Ibn Ridwān:

> As soon as the midwives saw his face
> they turned, regretful, from the room,
> And whispered (to avoid disgrace),
> 'We should have left him in the womb.'

He also used to nickname Ibn Ridwān the Crocodile of the Jinn.

Ibn Butlān journeyed from Egypt to Constantinople where he remained for a year. During this time many pestilential diseases occurred.

I quote the following from what he wrote in his own hand:

One of the famous calamities of our time was that which occurred when the comet* rose in Gemini in 446/1054. By the autumn of this year fourteen thousand souls were buried in the Church of St Luke after all the other burial grounds in Constantinople had been filled. By midsummer of 447/1055 the Nile had not risen and most of the inhabitants of Fustat (Old Cairo) and Damascus died along with all the foreigners except those whom God spared. The devastation then went on to Iraq and destroyed most of its inhabitants and it suffered ruin from the blows of aggressor armies. This continued until 454/1062. In most of the lands people suffered from melancholic ulcers and swellings of the spleen, and there was a change in the pattern of paroxysms during fevers and the normal system of crises was disturbed. Consequently the ability to predict was affected.

After this, Ibn Butlān continues:

Because this comet rose in the sign of Gemini, which is the ascendant of Egypt, the pestilence occurred in Fustat, with the Nile failing to rise when the comet appeared in 445/1053,* and Ptolemy's warning of woe to the people of Egypt came true. When Saturn descended into the sign of Cancer the ruin of Iraq, Mosul, and al-Jazīrah was complete, and the lands of Fars, Kirmān, the lands of the West, the Yemen, Fustat, and Syria became desolate. The kings of the earth were in disarray, and war, inflation, and pestilence proliferated. Ptolemy's words that when there is a conjunction of Saturn and Mercury in Cancer the world will be in upheaval proved true.

Ibn Butlān mentioned that the great pestilences affected learning because of the loss of the scholars during his time. 'The lanterns of learning were snuffed out and, after their passing, the minds of men remained in darkness.'

Ibn Butlān wrote thirteen treatises including:

A compendium for monasteries and monks.

On the purchase of servants and inspecting slaves and maidservants.

A tabular guide to health. *

The banquet of the physicians, which he composed for an emir. Ibn Butlān writes in his own hand at the end of it, 'I, the compiler, being the physician at the monastery of the munificent king Constantine on the outskirts of Constantinople, completed copying this at the end of September 1365.'

*An essay using logical methods on objections to those who held that the hatchling is hotter than the pullet.** This he composed in Cairo in 441/1049.

7. IBN SHIBL (*d.* 1081–2)

Ibn Shibl was born and raised in Baghdad and was a philosopher and theologian of merit, a talented littérateur, and a fine poet. He died in Baghdad in 474/1081. Among his many poems is a fine poem on philosophy which shows the strength of his knowledge of philosophical disciplines and theological mysteries.

Another poem is:

Meet the guest of worry with fortitude; you will make him depart,
for worries are guests that feed upon souls.
A misfortune will not increase without (subsequently) decreasing
and whenever one is in a tight spot one will be relieved.
So revive your soul with distraction and it will be content with it;
relief may come any moment.

and

Guard your tongue and do not speak openly about three things,
if you can: your secret, your wealth, and your (religious) views,
Since for these three you will be afflicted by another three:

One who calls you an unbeliever, an envious person, and one who calls you a liar.

8. IBN SAFIYYAH (twelfth century)

A certain person of Iraq said that al-Mustanjid was a harsh, vigilant, and murderous caliph. His vizier was Ibn al-Baladī. Now in the realm there were some great emirs the foremost of whom was Qutb al-Dīn Qāymāz* who was Armenian in origin. He had become very powerful and risen in station until he had gained control of the land and assumed the governance of the realm, without competitor. He had married off his daughters to the great emirs of the realm, but between him and the vizier, Ibn al-Baladī, there was a quarrel.

Then it happened that the caliph fell ill and his doctor was Ibn Safiyyah, the Christian. Ibn al-Baladī, the vizier, warned the caliph to fear the rising power of Qutb al-Dīn and the emirs, and when the doctor, who sought advancement with the emir, came to know something of this he related the conversations to him and this continued for some time.

When the caliph fell ill he decided to arrest Qutb al-Dīn and his associates. When the physician Ibn Safiyyah learned of this he went to Qutb al-Dīn, informed him, and said, 'The vizier has done such-and-such, so have him for breakfast before he has you for dinner!'

Meanwhile, the caliph's illness worsened and he was distracted from the plans he had made to arrest the emirs. Qutb al-Dīn came to the decision to kill the caliph and then be left free to do away with the vizier and he decided along with Ibn Safiyyah, the physician, to prescribe hot baths for the caliph.

Ibn Safiyyah visited the caliph and advised hot baths but the caliph, who felt himself to be a little weak, refused. Then Qutb al-Dīn and some of his associates attended the caliph and said, 'Master, the physician has advised hot baths.' The caliph said, 'We see fit to postpone them.' Then they overturned his decision and placed him in the bathhouse which had been fired for three days and nights and barred the door for a time until he died.

Qutb al-Dīn and the emirs exhibited great grief and went to the caliph's son and appointed him as caliph on their own terms and pledged allegiance to him and he was given the name al-Mustadīʾ.

Some time passed and the new caliph continued to feel resentment for what they had done. He had appointed a new vizier, while the physician, Ibn Safiyyah, continued in his service. Then the caliph and his vizier began to act despotically and independently of Qutb al-Dīn Qāymāz and whenever Ibn Safiyyah learned of anything he would relate it to Qutb al-Dīn.

One night the caliph, al-Mustadīʾ, summoned Ibn Safiyyah and said to him, 'Physician, there is someone I detest the look of and I wish to dismiss him but in a graceful, not disgraceful way.' Ibn Safiyyah said to him, 'Then we will prepare for him a potion both powerful and effective which he will drink and you will be rid of him as you wish.' Then he went and compounded a potion as he had described and brought it back at night. The caliph opened it, examined it, and said, 'Physician, swallow this potion so we can test its efficacy.' The physician recoiled from that and said, 'Master, remember God, remember God, with regard to me!' The caliph said, 'Whenever a physician oversteps his bounds and traverses his limits he will fall into the likes of this. There is no escape for you except by the sword.' So the physician swallowed the potion which he had himself compounded and fled from perdition to perdition. Then he left the caliph's palace and wrote a letter to the emir Qutb al-Dīn informing him of the events and saying to him, 'After me it will be you!' Then he died.

As for Qutb al-Dīn, he was determined to bring down al-Mustadīʾ but God, glory be to Him, made his plot backfire. His wealth was plundered and he fled Baghdad for his life and went to Saladin in Syria, but the ruler refused to accept him. He returned by way of the desert to Mosul but fell ill on the way and died after reaching Mosul.

9. IBN AL-TILMĪDH (*d.* 1165)

Ibn al-Tilmīdh was the foremost of his age in the art of medicine and in its practice. This is indicated by what is known of his writings and his glosses on medical books, as well as the great many people I have seen who bear witness to this. He was practitioner in the ʿAdudī hospital in Baghdad until his death.

At the beginning of his career he had travelled to Persia* and been in service there for many years. He was good at writing and would write in a proportioned script of the utmost beauty and correctness.

He was expert in the Syriac and Persian tongues and deeply versed in the Arabic language and wrote delightful and meaningful poetry although most of his extant poems consist of only two or three verses each and I have found only a few full odes of his. He also wrote many good epistles and I have seen a great volume of his, containing letters and correspondence. Most of his family were secretaries, and Ibn al-Tilmīdh's father was a physician of merit and renown.

Ibn al-Tilmīdh and Awhad al-Zamān* were both in the service of the caliph al-Mustadīʾ. Awhad al-Zamān was better than Ibn al-Tilmīdh in the philosophical disciplines. Ibn al-Tilmīdh, however, had more insight into the art of medicine and was renowned more for it. The two hated each other and were enemies, but Ibn al-Tilmīdh was possessed of more intellect and a better nature than Awhad al-Zamān.

An example of this is that Awhad al-Zamān once wrote a note in which he accused Ibn al-Tilmīdh of things someone like him could not possibly have done. Then he bribed a servant to secretly throw the note in the path of the caliph. This alone displays great evil. When the caliph found the note he took it very badly indeed at first and intended to bring down Ibn al-Tilmīdh. Then he came to his senses and was advised to investigate and to get the servants to confess who had accused Ibn al-Tilmīdh. When the caliph discovered that Awhad al-Zamān had written the note, he became greatly angered and put his life, his wealth, and his books into the hands of Ibn al-Tilmīdh. Ibn al-Tilmīdh, however, who had such a noble nature and great charity, did not encroach upon any of this, but Awhad al-Zamān was banished from the caliph and lost his station.

I—Ibn Abī Usaybiʿah—quote from the physician ʿAbd al-Latīf al-Baghdādī who says:*

Ibn al-Tilmīdh was good company, of noble morals, had generosity and *humanitas*, produced works of renown in medicine, and came to correct conclusions. An example of this is when a woman was brought to him on a stretcher and her family didn't know whether she was alive or dead. It was wintertime so he ordered that she be undressed and that cold water be poured over her frequently and continuously. Then he ordered that she be carried to a warm room which had been fumigated with aloeswood* and perfume and she was wrapped in all kinds of furs for a while. Then she sneezed, moved, sat up, and left, walking, with her family towards her house.

An example of Ibn al-Tilmīdh's decent character was that the back of his house adjoined the Niẓāmiyyah law college in Baghdad and whenever a scholar became ill he would take him in and attend him in his illness, and when he recovered, the physician would give him two gold dinars and send him on his way.

Among the interesting stories about Ibn al-Tilmīdh is that one day he was with the caliph al-Mustaḍīʾ. Ibn al-Tilmīdh was at that time very advanced in age and when he rose to get up he leaned on his knees. The caliph said to him, 'You have become old, Ibn al-Tilmīdh!' He said, 'Yes, and my flasks are smashed.' The caliph thought about what Ibn al-Tilmīdh had said and realized he had said it for a reason. So he enquired and was told that the imām had gifted Ibn al-Tilmīdh an estate called Qawārīr ('Flasks') which he had held tenure of for a time. Then, three years previously, the vizier had taken hold of it. The caliph was amazed at Ibn al-Tilmīdh's good manners and the fact that he hadn't brought the matter up before nor set about requesting the estates. The caliph then commanded that the estates be returned to Ibn al-Tilmīdh and that, in future, none of his possessions were to be encroached upon.

The house of Ibn al-Tilmīdh in which he lived in Baghdad was in the Perfume Market whose gate is adjacent to the Willow Tree Gate by the caliph's Great Palace in the street leading to the banks of the Tigris.

Ibn al-Tilmīdh had a son but he did not comprehend the art of medicine and was, in all respects, far removed from the station of Ibn al-Tilmīdh. A man related to me:

Once we were guests of Ibn al-Tilmīdh's son. An attendant brought a costly dining table upon which the son spread an exquisite white cloth, a vessel in which was vinegar, and choice endives. Then he said, 'Eat, in the name of God.' So we ate a small amount since we were not accustomed to such food. Then he stopped eating and said, 'Servant, bring the washbasin.' The servant brought a silver washbasin and a large piece of soap and poured water for him while he washed his hands and lathered the soap. Then he wiped his mouth, face, and beard until his face was covered with soap and he was all white with lather, looking at us. One of our number couldn't restrain himself from laughing. It was too much for him and he rose and left. The son said, 'What is wrong with him?' So we said, 'Master, he is a little light-headed and this is normal for him.' He said, 'If he were to stay with us we would cure him.' Then we bade him farewell and left, begging God to save us from such ignorance as his.

A certain man of Iraq related that Ibn al-Tilmīdh had a friend whose son died. He was a man of etiquette and knowledge but Ibn al-Tilmīdh did not send his condolences. When they met later the man scolded him for not sending his condolences for his son for the sake of their friendship. But Ibn al-Tilmīdh said, 'Do not blame me for this, for by God, I deserve more condolences than you since your son has died yet the likes of my son is still alive.'

Ibn al-Tilmīdh died in Baghdad on the 28 Rabīʿ I 560/12 February 1165 at the age of 94. He died as a Christian and left behind great wealth, much property, and books of peerless quality all of which was inherited by his son who lived on for a time. Then Ibn al-Tilmīdh's son was strangled in the courtyard of his house during the first portion of the night and his property was taken and his books were carried away by twelve porters. Ibn al-Tilmīdh's son had become a Muslim before his death and it is said he was an old shaykh nearing 80.

Among the extensive poetry of Ibn al-Tilmīdh are the following poems:

> With two glasses I spent my lifetime
> and on them I always relied:
> A glass filled with ink
> and a glass filled with wine.
> With one I established my wisdom
> and with the other I removed the worries of my breast.

> One glass quenches the flame of thirst,
> A second helps to digest one's food,
> The third helping of wine is for joy,
> And one's mind is driven out by adding another cup.

> The sword of your eyelids is superior to
> keen swords in their sheaths:*
> The latter can kill but cannot
> restore souls, warding off death,

Whereas your eyes, looking askance, kill me
but revive me with a furtive glance, quietly.

Do not think that the blackness of his mole is a defect
of Nature, or that it was made to appear by mistake:
Rather, the pen, depicting the curve of his brow,
made a full stop on his cheek.*

Passion has pared me down like a penknife, and your avoidance
has made me waste away so that I have become thinner
than yesterday,*
I cannot be seen until I see you: dust particles are visible
only when the sun is at the horizon.

My newly sprouted beard once was her excuse for being with me;
now it is grey and has become her excuse for rejecting me.
How strange: something that one day led to love,
then changes and, the next day, leads to rejection.

He said as a riddle on the shadow:

A thing of bodies yet not itself embodied;
at times it moves, or it is still.
It is completed at the moments it comes into being or perishes,
but when it is alive there is a waning:
When lights are clearly visible, it is visible to the onlooker,
but when they disappear, it is invisible.

You have slurped a lot of eggs so that
your prick might stay erect for a long time.
What cannot stand with the help of your own 'eggs'*
will not stand with the help of someone else's eggs.

Ibn al-Tilmīdh wrote eighteen books including *The medical formulary** in twenty chapters whose renown and circulation amongst the people is greater than his other books.

10. AWHAD AL-ZAMĀN (ABŪ L-BARAKĀT AL-BAGHDĀDĪ) (*d.* after 1164)

Awhad al-Zamān ('the Unique One of his Time')—that is, Abū l-Barakāt al-Baghdādī—was born in Balat then settled in Baghdad. He was a Jew who subsequently became a Muslim and was in the service of the caliph al-Mustanjid. His compositions are of the utmost good quality and he had a far-reaching interest in the sciences and a great talent for them.

When Awhad al-Zamān started to learn the art of medicine Saʿīd ibn Hibat Allāh* was a distinguished teacher in the art of medicine but would not allow a Jew to study with him at all. Awhad al-Zamān would sit in the shaykh's courtyard so that he could hear all that was being read with him and all the discussions which went on. After a year or thereabouts a question arose from the teacher and his students could not find an answer. When Awhad al-Zamān realized this he entered and put himself at the shaykh's service and said, 'Master, if you permit I will speak about this question.' The shaykh said, 'If you have any knowledge about it then do so.' So he answered with some words of Galen and said, 'Master, this was already discussed on such-and-such a day of such-and-such a month at such-and-such a session and it has remained in my mind since that day.' The shaykh was astounded at his intelligence and said, 'It is not lawful to refuse knowledge to a person such as this.' And from that time on the shaykh kept him close and he became one of his greatest students.

A rare anecdote told about Awhad al-Zamān's treatment of patients occurred when a patient in Baghdad was struck with melancholia and believed that he had an amphora on his head which remained there at all times. When the man walked he would avoid places with low ceilings and would walk very carefully and not allow anyone to approach him lest the amphora lean or fall from his head. He remained thus afflicted for some time and experienced hardship because of it. A group of physicians treated him but to no beneficial effect. Then his case came to Awhad al-Zamān who thought that nothing remained to cure him but illusion so he said to the man's family, 'When I am in my house then bring him to me.' Then Awhad al-Zamān ordered that one of his servants, after the patient had entered and upon his signal, should strike above the patient's head with a large club as if he wanted to break the amphora which the patient thought was on his head. He asked a second servant, who had prepared and taken up an amphora to the roof of the house, to quickly throw that amphora down to the ground when he saw the first servant strike above the head of the melancholic. When Awhad al-Zamān was at home and the patient had come he began to converse with him and reproached him for carrying the amphora. Then he signalled his servant without the patient knowing and the servant approached and said, 'By God, I must break that amphora and relieve you of it!' Then he waved his club and struck above the man's head with it. At the same time, the other servant threw down the amphora from the roof, shattering it with a tremendous crashing. When the patient saw what had been done to him and saw the broken amphora he gasped because they had broken it, but did not doubt that it was the one that had been on his head, as he thought, and the illusion had such an effect that he was cured of that malady.

This is a great type of treatment and a number of the ancient physicians such as Galen and others had similar cases of treatment by illusion, and I—Ibn Abī Usaybiʿah—have mentioned a great deal of this in another book.

One of Awhad al-Zamān's students related the following:

One day a man came with whitlow,* except that the abscess was broken and pus was flowing from it. When Awhad al-Zamān saw this he quickly amputated the man's finger at the joint. We said, 'Master, it would have been enough to treat him as others treat such a case and preserve his finger.' We reproached him, but he didn't utter a word. On the next day another man came with exactly the same condition. Awhad al-Zamān motioned to us to treat him saying, 'Do with him as you see fit.' So we treated him with that which whitlow is treated but the area expanded and the nail and the first of the phalanges of the finger was lost. We tried all possible treatments despite which his finger deteriorated quickly and the matter ended up with amputation. And so we understood that above every person of knowledge there is someone with greater knowledge.

It is said that Awhad al-Zamān became a Muslim when one day he attended the caliph. All those present rose except the chief judge who was present but did not think he should rise with the others since Awhad al-Zamān was a *dhimmī*.* Awhad al-Zamān said, 'O Commander of the Faithful, if the chief judge does not agree with the others because he believes I am not of his religion then I will enter Islam in the presence of our Master and will not let him belittle me in this way.' Then he became a Muslim.

A shaykh and acquaintance of mine who was, at the beginning of his life, a Jew who lived in the Jewish quarter of Baghdad near to the house of Awhad al-Zamān and when he was a youngster used to enter his house, told me that Awhad al-Zamān had three daughters and did not leave behind a male child although he lived for nearly eighty years.

A qadi or judge I knew told me that there was enmity between Awhad al-Zamān and Ibn al-Tilmīdh. After Awhad al-Zamān had become a Muslim, he frequently used to renounce the Jews and curse them and insult them. One day, in the chamber of one of the great and good, and with a group of people present including Ibn al-Tilmīdh, the subject of the Jews came up and Awhad al-Zamān said, 'May God curse the Jews.' At this Ibn al-Tilmīdh said, 'Yes, and the sons of the Jews!' Awhad al-Zamān was dumbfounded and knew that Ibn al-Tilmīdh had alluded to him, but he did not speak.

11. HIBAT ALLĀH IBN AL-FADL (*d.* 1163)

Hibat Allāh was a Baghdadi born and raised who used to practise the art of medicine and was counted among those who are known for the art. He also used to be an oculist, but was, above all, a poet. He was a man of many anecdotes, foul-tongued, and authored a collection of poetry.

Hibat Allāh and the emir and poet named Haysa-Baysa, that is 'Pell-Mell', hated and reviled each other. From time to time they would call a truce but then continue as of old. Pell-Mell was so named because, once, the army in Baghdad had been preparing to leave the city to confront the Seljuq sultan. This was in the time of the caliph al-Muqtafī, and the people were full of talk and excess activity because of this and the emir said, 'Why is it I see the people at pell-mell?' and so he was nicknamed with this name, and the one who attached this epithet to him was Hibat Allāh. Pell-Mell always sought exaggerated eloquence and obscure language in his discourses and his epistles.

Al-Dakhwār told me that Pell-Mell, the poet at Baghdad, wrote the following letter to Ibn al-Tilmīdh asking him to send over some eye-salve made with lead:

'I would apprise you, O learned, solicitous, skilful, precious, and experienced medic—may weal abide with you always, and may ruin stumble blindly and never find you—that I am overcome and sense in my oculi a tendling, neither like the slege of a scorpion, nor like the prick of a neeld, nor yet like the slite of a nathdrack, but rather like the scorching of gledes. Hence, I, from vigilae to vespers, do not know the nocturnal from the diurnal, nor can I distinguish between the serenous and the nubilous. Indeed, often I fall, legs akimbo; at times I am angry and foul-tempered, at others I curl up, and sometimes I stretch out, all of this with a great sighing, and rasping of breath, and sorrow. I can neither flee nor hurry, neither sirmounte nor subiugue. So send at once some lead eye-medicine to salve my disease and slake my thirst.'

When Ibn al-Tilmīdh read the letter he rose at once and took a handful of burnt lead eye-medicine and told one of his companions, 'Take it to him quickly so we don't have to bear reading another one of his letters.'

Pell-Mell wrote several notes to the caliph al-Muqtafī when he requested the village Baʿqūbā from him, including:

The first: 'Lo, herewith the heralds of affection, bearing a letter of laudation, trilled by the drover of a hope, in fulfilment of the abode.'

The second: 'I gallop the steeds of praise in the courtyards of glory, as one who gallops a dashing high horse without goad or effort, petitioning the conclusion of the matter, by your honour.'

The fifth: 'The fifth of the servants petitioning for the downpours of generosity, from the greatest sanctuary, dulcet of rhyme, running like a fleeing she-camel in a burning desert, guiding on the journey, making ease of hard ground; the best decision would be to fulfil her hopes!'

Among the poetry of Hibat Allāh is:

A gazelle who is fond
only of minted gold coins:
He is never pleased with my expressions
coined in verse or prose.

He also said:

You who fear being lampooned:
you may feel safe against being touched by it.
You, with that reputation of yours among people,
are like shit: untouchable.

12. IBN TŪMĀ (*d.* 1223)

Ibn Tūmā was a Christian of Baghdad and was a distinguished physician and a great notable.

I was told that Ibn Tūmā was the physician of the official, Najm al-Dawlah al-Sharābī,* and that his station rose until he became his vizier and secretary. Then he entered the service of the caliph, al-Nāsir, and was among the physicians who visited the caliph during his illnesses. Later he attained the caliph's full favour and was given responsibility for a number of areas of service and oversaw several *dīwān*s and secretaries.

I—Ibn Abī Usaybiʿah—say that Ibn al-Qiftī* related the following about the life of Ibn Tūmā saying:

He was a wise man and a physician, good at treating patients, frequently correct in his diagnoses, successful in his treatments in most cases, and had great good fortune in this regard. He was honourable and trustworthy and advanced during the reign of al-Nāsir until he attained to a rank equal to the viziers. The caliph entrusted him with custody of the property of his courtiers and would deposit it with him. Ibn Tūmā was a good arbiter and had beautiful manners and helped many people in need and by his arbitration prevented many evils.

Towards the end of his reign al-Nāsir's sight became weak and he was distracted most of the time because of great sorrows which had overwhelmed his heart. After he had become unable to inspect the petitions he summoned a woman of Baghdad who was known as Lady Nasīm and took her into his inner circle. Her handwriting resembled the caliph's and so she would sit with him and write letters and notes assisted in this by a eunuch named Rashīq. Al-Nāsir's condition worsened and the woman began to write the letters as she thought fit in which, at times, she would be correct but at others she would err, aided all the while by Rashīq.

It so happened that the vizier wrote an enquiry and by return received the woman's answer which was clearly defective. The vizier was taken aback. Then he called for Ibn Tūmā and asked him to explain the situation. Ibn Tūmā told him of the caliph's loss of sight and the distraction and what the woman and the servant were doing with regard to the letters.

At this the vizier ceased work on most of the matters which came to his attention. The servant and the woman, who had opportunistically sought to bring about certain worldly goals for themselves, realized this and guessed that Ibn Tūmā had made the vizier aware of the situation. Rashīq plotted with two soldiers in the caliph's service to assault Ibn Tūmā and kill him.

One night they watched Ibn Tūmā until he came to the house of the vizier then left to return to the caliph's palace. The two men followed him until he reached the gate of the dark Ghallah Lane where they fell upon him with their knives. After falling to the ground because of the pain of the knife wounds, Ibn Tūmā fled with the killers following him. Someone saw them so Ibn Tūmā shouted 'seize them' but they came back and killed him and wounded the lamp bearer who was with him. Ibn Tūmā was carried dead to his house and buried there that very night. Some men were sent to his house to guard the deposits which he had kept for the ladies, servants, and courtiers.

A search was made for the killers and their arrest was ordered. Early next morning they were brought out to the execution place, their bellies were slit open and they were crucified on the gates of the slaughterhouse opposite the Ghallah Gate where Ibn Tūmā had been stabbed. This murder was on the eve of Thursday 28 Jumādā I in 620/29 June 1223.

13. KAMĀL AL-DĪN IBN YŪNUS (*d.* 1242)

Kamāl al-Dīn ibn Yūnus taught at the school in Mosul and taught all the sciences. He composed works of the utmost good quality and continued to live in Mosul until he died—may God show him mercy.

It is said that Ibn Yūnus knew magic. An example of this was related to me by a qadi or judge and student of his:

Once, an emissary came to the governor of Mosul, from the emperor the king of the Franks* [Frederick II]. The emissary had with him questions about astrology and the like and sought the answers from Ibn Yūnus. The governor sent a message to Ibn Yūnus informing him of this and saying that he should wear beautiful clothes and prepare a splendid salon for the emissary, knowing that Ibn Yūnus used to wear rough clothes and had no knowledge of the things of the world.

The qadi continues:

I was with him when the emissary of the Franks arrived. We looked and saw that the room was adorned with the finest Byzantine carpets and with a group of slaves and servants in fine clothes. The emissary entered and Ibn Yūnus greeted him and wrote the answers to all the questions. When the emissary had gone, all that we had seen before vanished. So I said to Ibn Yūnus 'Master, how wonderful were the splendours and the servants we saw a short time ago!' He smiled and said, 'That is science!'

I—Ibn Abī Usaybiʿah—say that when my uncle Rashīd al-Dīn ʿAlī ibn Khalīfah was a young man he intended to travel to Mosul to study with Ibn Yūnus because of what he had heard about his knowledge and eminence. When his mother, my grandmother, learned of this she wept and begged him not to leave her. She was very fond of him and he could not go against her so he cancelled the journey.

CHAPTER 11

PHYSICIANS IN THE LANDS OF THE PERSIANS

In this chapter one finds physicians from 'the lands of the ʿAjam', the Arabic word denoting non-Arabic speakers in general, but often, and here, referring to Persians. Many, probably all, of these were bilingual and they wrote most or all of their works in Arabic. The four entries in this edition include some famous names. Abū Bakr al-Rāzī, known in the West as Rhazes, a physician, alchemist, philosopher, and free-thinking Muslim who rejected revealed religion, was the author of many works including a multi-volume encyclopaedia on medicine entitled al-Hāwī *(*The comprehensive [book]*), which became known in the Latin West as* Continens [liber]*. He should not be confused with Fakhr al-Dīn al-Rāzī (d. 1209), the scholastic theologian and Qur'anic exegete, who also has an entry. Even more well known than these is, of course, Ibn Sīnā (Avicenna) who receives a long entry despite being, as Ibn Abī Usaybiʿah notes, 'too famous to need a mention and his virtues too outstanding to need to be set down in writing'.*

1. ABŪ BAKR AL-RĀZĪ
(Rhazes or Rasis) (*d. c.*925)

Abū Bakr Muhammad ibn Zakariyyā al-Rāzī was born and raised in Rayy but in his 30s had travelled to Baghdad and stayed there for a time. From a young age al-Rāzī was keen on the intellectual disciplines and studied them as well as literature. He was also a poet. He began to study the art of medicine when he was older.

A Persian physician writes in his book on hospitals:

The reason al-Rāzī began to study the medical art was that when he first came to Baghdad, the City of Peace, he visited the ʿAdudī hospital there. There he was fortunate to meet the hospital's pharmacist, a venerable man, whom al-Rāzī questioned about drugs and who had first discovered them. The pharmacist answered that the first drug to be known was *sempervivum** and that this occurred when Philo* the descendant of Asclepius once had an inflamed swelling on his forearm which caused him severe pain. During his treatment of it he felt the need to go out to the riverside and his servants bore him to a riverbank where this plant happened to be. Philo placed some of the plant on his swelling to try to cool it and because of this his pain subsided. He continued to do this for a long time and repeated it the next day and then it was cured. When the people saw how quickly he had been cured and realized that it was due to this drug they named it 'Life of the World' in Arabic.

On the other hand, a certain person has said:

Abū Bakr al-Rāzī was amongst those who took part in the building of the ʿAdudī Hospital and the ruler ʿAdud al-Dawlah* consulted him about the location where the hospital should be built. Al-Rāzī instructed some of his servants to hang pieces of meat in every quarter of Baghdad and noted the one in which the meat had not gone bad or putrefied quickly and advised that the hospital be built there, and that is its location.

I—Ibn Abī Usaybiʿah—say that according to my information, al-Rāzī lived before the time of ʿAdud al-Dawlah, and he frequented the hospital at a time before ʿAdud al-Dawlah came to renovate it. Al-Rāzī also authored a book describing the hospital and containing all the details of the conditions of the patients who were treated there.

Ibn Juljul says:

Abū Bakr al-Rāzī was in charge of the hospital at Rayy for a time before he began to practise in the ʿAdudī hospital.

Ibn al-Nadīm, the Baghdadi bookseller and author, writes in his *Fihrist* that an old man from Rayy who was asked about al-Rāzī said:

He was an old man with a head as large as a casket. He used to sit with his students beside him and their students beside them and even more students beside them. Someone would come and describe his symptoms to the first person he met and if they didn't know the answer he would pass on to another. Al-Rāzī was honourable and esteemed, amenable to the people, and kind to the poor and the sick. He would even give them generous stipends and tend to them himself. He was constantly writing and copying and whenever I visited him he would always be writing either a draft or a fair copy. He suffered from cataracts in his eyes due to the excessive eating of broad beans* and he became blind at the end of his life.

One anecdote about his innovative diagnoses and excellent deductions is related thus:

A certain reliable physician related that a young man of Baghdad came to Rayy spitting up blood, something which had afflicted him during his journey. He summoned al-Rāzī and showed him and described to him his symptoms. Al-Rāzī took his pulse, examined his urine, and enquired about his condition from the start but could see no indication of any wasting disease or ulcer and could not identify the cause. Al-Rāzī asked the man to give him time to consider the matter at which the man panicked and said, 'I am done for since even this very skilled physician does not know the cause,' and his condition worsened. Then al-Rāzī had an idea and returned to the man and asked him about the waters he had drunk on his journey. When the man told him that he had drunk from swamps and pools, it occurred to al-Rāzī's sharp and intelligent mind that a leech had been in the water and had entered his stomach and that the discharge of blood was due to its effect. He said to the man, 'I will come in the morning and treat you and I will not leave until you are cured. However, there is one condition, and that is that you order your servants to obey me in all I command them to do to you.' The man agreed and al-Rāzī left. Then al-Rāzī saw to it that two large tubs of green algae were gathered which he brought with him to the man the next morning, saying, 'You must swallow everything in these two tubs!' The man swallowed a little and then stopped. Al-Rāzī told him to swallow more and when the man said that he couldn't, al-Rāzī told the servants to take him and place him on his back. They did so and opened his mouth and al-Rāzī began to force the algae down the man's throat pressing it hard and demanding that he swallow it. He threatened to beat him, until he had made him swallow, against his will, one of the tubs in its entirety. And all the while the man was crying out for help. Then the man said, 'I am going to vomit,' at which point al-Rāzī stuffed more algae into his throat until the man vomited. When al-Rāzī examined the vomit, he found a leech in it, that, since it was by nature attracted to the algae, had left its place and attached itself to it. When the man vomited, the leech emerged along with the algae. And the man got up, his health restored.

A student of medicine heard al-Rāzī the physician recount the following story, after his return from being with the emir of Khorasan who was treated by him for a severe illness:

On my way to Nishapur I passed by Bistam which is half way between Nishapur and Rayy. The governor of the city received me, took me to his home and provided every service. He asked me to look at a son of his who had dropsy and took me to an apartment which had been set aside for him where I saw the patient but I had no hope of curing him. So I said some diverting words in the presence of the patient but when I was alone with his father he asked me to tell him the truth, and so I told him that his son had no hope of living, saying, 'Allow him to do as he wishes, for he will not live.'

After a year I left Khorasan and passed by Bistam again where the man received me on my return. When I met him I felt extremely ashamed and was sure that his son had died. Since I was the one who had foretold his death, I feared he might find me burdensome. The governor took me to his house but I did not see anything to indicate that his son had died, although I was reluctant to ask him lest I renew his grief. One day he said to me, 'Do you recognize that young man?', pointing to a handsome youth, healthy, strong, and sanguine, who was standing with the servants attending us. When I said that I didn't he said, 'It is my son who you said would die when you went to Khorasan!' I was astonished and said, 'Tell me the cause of his recovery.'

The man said to me, 'When you left me, my son realized that you had told me he would not live and he said to me, "I have no doubt that this man, the greatest physician of his time, has told you I will not live. I only ask that you keep away the servants for they are my peers and if I see them in good health and I know that I am to die it will cause a fever in my heart which will bring my death on even quicker. So let one of my nurses attend to me." I did as he asked, and every day food for the nurse was brought along with what my son requested but not according to any particular diet. After some days, a dish of *madīrah** was brought for the nurse to eat which she left so my son could see it and then went about her chores. She told me that when she returned she saw that my son had eaten most of what was in the bowl with only a little remaining, the colour of which had changed.'

The old woman recounted to him, 'So I said to him, "What is this?" He said, "Don't go near the bowl," and he pulled it towards himself, saying, "I saw a great viper which came out from its place and slid towards the bowl, ate from it then regurgitated, after which the colour of the food became as you see. I thought to myself that I am dying and I do not want to suffer great pain and I will never have a better chance than this. So I crawled to the bowl and ate from it what I could so that I might die quickly and be relieved. When I could eat no more I returned to my place until you came." '

She continued, 'I could see the *madīrah* on his hands and mouth so I shrieked, but he said, "Don't say anything to anyone, but bury the bowl and its contents in case someone eats it and dies or an animal eats it, bites a person, and kills him." So I did as he said.'

The father continued, saying, 'Then the nurse came out to me and when she had told me everything I was shocked and went to see my son but found him asleep. So I said he was not to be woken until we see what happens. He woke at the end of the day having sweated profusely and asked to go to the bathroom. When he was taken there his bowels opened and that night and the next day he passed more than one hundred bowel movements, which only increased our despair. After a few days of this and little food he asked for some chicken which he ate and his strength began to return. Our hopes of his recovery became stronger and we kept him in isolation and his strength increased until he became as you see him now.'

I—al-Rāzī—was amazed at this, but mentioned that the Ancients had said that when a person suffering from dropsy eats the flesh of an old and long-lived snake, hundreds of years old, he will be cured. 'But had I told you that this was his cure you would have thought I was making false promises. And in any case, how can you tell the age of a snake even if we could find one? So I kept silent.'

Another person said:

The vizier was once a guest at al-Rāzī's house where he ate some food which was as delicious as can be. Afterwards the vizier contrived to buy one of the maids who cooked for al-Rāzī thinking that she would cook similar food for him. When she made the food it was not the same and when the vizier asked her about this she said that the food was the same but all the cooking pots at al-Rāzī's house were of gold and silver. The vizier then imagined that this was the reason for the quality of the food and that al-Rāzī had acquired knowledge of alchemy. When the vizier summoned al-Rāzī and asked him to teach him what he knew of alchemy, al-Rāzī could tell him nothing and denied all knowledge of it, so the vizier had him secretly garotted.

Al-Rāzī was a contemporary of Is'hāq ibn Hunayn and others of the time. He became blind towards the end of his life due to cataracts, and when he was advised to have his eyes couched,* he said, 'No, I have seen enough of the world and am weary of it,' and refused.

Aphorisms of al-Rāzī include:

> Certainty in medicine is an unattainable goal, and the treatment of patients according to what is written in books, without the skilful physician using his own judgement, is fraught with danger.
>
> Life is too short to understand the effect of every plant growing on earth, so use the most well known for which there is a consensus and avoid the unusual. Confine yourself to what you have tried and tested.
>
> When Galen and Aristotle are in agreement on a matter then it is correct. When they disagree, it is very difficult for the intellect to grasp the truth of the matter.
>
> The physician, even though he has his doubts, must always make the patient believe that he will recover, for the state of the body is linked to the state of the mind.
>
> Physicians who are liars and fakes, youth who are inexperienced, those who are careless, and those who are debauched—they are lethal.
>
> The student of medicine who studies with many doctors runs the risk of falling into the errors of each one of them.
>
> If the physician is able to treat a patient using foodstuffs rather than medicines, then he has been truly fortunate.

An example of al-Rāzī's poetry:

> Upon my life, I don't know, even now that decay announces
> an imminent departure, where I shall travel,
> Or where the spirit will dwell after its exiting
> from the dissolved frame and the decayed body.

Among the 226 treatises of al-Rāzī are:

*The comprehensive book** which is al-Rāzī's greatest and most magnificent book on the art of medicine. However, al-Rāzī died before he was able to publish the book.

Doubts and contradictions which are in the books of Galen.

On smallpox and measles. In fourteen chapters.

Book for those without a physician. In it he lists illnesses one by one and states that it is possible to cure them using commonly found drugs. It is also known as *The poor man's book of medicine*.

On the foods and fruits which should be taken at the beginning of a meal and those which should be taken at the end.

An apology for those who play chess.

On the wisdom of backgammon.

The Mansūrī book. Al-Rāzī composed this for the emir and governor of Khorasan.* It is concise and brief while comprising points and rarities of both the theory and practice of the art of medicine in ten discourses.

On the fact that no non-intoxicating drink exists which delivers all the good effects of intoxicating drink on the body.

On the fact that, for all the arts, there does not exist a practitioner recognized for his art, and this is particularly so for medicine, and on what causes ignorant physicians, the general public, and women in the cities to be more successful in some of their treatments of diseases than the learned, and on excusing the physician for this.

On what causes Abū Zayd al-Balkhī to have symptoms of a common cold when he smells roses in springtime.*

2. AL-SHAYKH AL-RAʾĪS IBN SĪNĀ
(Avicenna) (*d.* 1037)

Ibn Sīnā, though too famous to need a mention and his virtues too outstanding to need to be set down in writing, he himself has recorded and described his own life and times in such a way that others need not do so. Hence, we confine ourselves here to what he has said of himself, as well as that which Abū ʿUbayd al-Jūzjānī, the shaykh's companion, told of his life.

Ibn Sīnā himself said:

My father was a man from Balkh from whence he moved to Bukhara. My father married my mother and settled, and my mother gave birth to me and then to my brother. I was taken to a teacher of the Qur'an and a teacher of literature, and when I reached the age of 10 I had mastered the Qur'an and a great deal of literature such that I occasioned amazement.

My father was amongst those who accepted the missionary of the Egyptians and was considered one of the Ismaili sect, and so was my brother. Words such as philosophy, geometry, and Indian arithmetic would roll off their tongues. My father sent me to a man who used to sell vegetables and was proficient in Indian arithmetic so that I might learn from him.

I studied law. Then I began to study the *Isagoge*.* I read other books for myself and the commentaries until I was proficient in the science of logic, the book of Euclid, and the *Almagest*. I then sought to learn medicine and began to read the books which have been composed on that subject. Medicine is not a difficult science and consequently I excelled in it in a very short time so that the eminent physicians began to study the science of medicine under me. As well as this I still attended studies in the law and debated about it, and I, at that time, was 16 years of age.

For the next year and a half I dedicated myself to learning and study. During this time I didn't sleep a single night in its entirety, nor did I occupy myself in any other way during the day. I would frequent the Congregational Mosque and pray and supplicate to the Creator of the Universe until that which was incomprehensible became clear, and that which was difficult became simple.

I used to return at night to my house and place the lamp before me and occupy myself with reading and writing. And whenever I was overcome by sleep or felt some weakness I would turn to drink a cup of wine until my strength came back. Then I would return to my reading, and whenever the slightest sleep overtook me I would dream of those problems so that the particulars of many problems became clear to me during sleep. And so it was until I had mastered all sciences and had become conversant with them as far as is humanly possible.

It so happened that the ruler of Bukhara fell ill with a disease which vexed the physicians so they mentioned me to him and asked him to send for me. Thus I gained the distinction of entering into his service. One day I asked his permission to enter the library. Permission was given and I saw and read books which many people have never even heard of and which I had never seen before or since. And when I reached the age of 18, I completed my studies of all those sciences, and although, at that time, I had a greater capacity for memorizing learning, it is now for me more fully matured. Otherwise it is the same learning and nothing since has come to me as new.

Then my father died and my situation became subject to vicissitudes and I accepted some government posts. It became necessary for me to leave Bukhara and move to Gurgānj where I was presented to the emir. At that time I wore the garb of a law student with a hood wrapped under my neck,* and they allotted to me a generous monthly stipend. Then it became necessary for me to move again and again. Finally, in Jurjan, I met Abū ʿUbayd al-Jūzjānī.

Abū ʿUbayd al-Jūzjānī, Ibn Sīnā's companion, continues, saying, 'This, then, is what the shaykh Ibn Sīnā related to me in his own words and from here on I witnessed the events of his life':

He moved to Rayy and then to Qazwīn, and from there to Hamadan. Then Shams al-Dawlah summoned him, as then he was afflicted with colic. The shaykh treated him until God restored him to health. At that court he won many robes of honour and returned home after forty days and nights, having become a confidant of the emir.

The emir went up to Qirmisīn to make war on ʿAnnāz and retreated towards Hamadan, after which the shaykh was asked to assume the position of vizier. However, the troops mutinied against him, fearing for their own positions so they surrounded the shaykh's house and took him to prison and raided his estates and took everything he owned and asked the emir to kill him. The emir protected him but agreed to banish him from the land in order to appease the troops, so the shaykh withdrew for forty days. The emir Shams al-Dawlah, once again struck with the colic, sought the shaykh and was utterly apologetic to him, who occupied himself treating the emir. He remained there in a position of honour and esteem and the viziership was given to him for a second time.

I myself—Abū ʿUbayd al-Jūzjānī—requested that the shaykh explain the books of Aristotle but he said that he had no time for that but if I agreed he would compose a book which would include everything he considered to be correct with regard to these sciences. Then every night students would gather in his house and I would take a turn reading from his book *The cure*, and someone else would take a turn teaching from *The canon*. When we had finished, singers of all stripes would attend us and the paraphernalia for a drinking party was prepared, which we partook of. Teaching usually took place at night since there was no time for it during the day due to the shaykh's service with the emir. This state of affairs went on for some time.

Shams al-Dawlah was struck again with colic which became severe and he died. Ibn Sīnā withdrew and sent letters in secret to ʿAlāʾ al-Dawlah* seeking to enter his service. Meanwhile I asked him to complete his book *The cure*, and he would write fifty folios every day. Then the vizier, Tāj al-Mulk, became suspicious of him, seized him, and took him to a fortress called Fardajān. He remained there for four months before Tāj al-Mulk brought him to Hamadan.

It then occurred to the shaykh Ibn Sīnā to go towards Isfahan so he, his brother, two servants, and myself set out from Hamadan disguised as Sufis* until we reached Tabarān at the gates of Isfahan, having suffered many hardships on the way. We were met by the shaykh's friends and the emir ʿAlāʾ al-Dawlah's courtiers and the shaykh was provided with clothes and fine mounts and was housed in a quarter known as Gūn Gunbad in a house which had everything required in the way of utensils and furnishings. The shaykh attended the court of ʿAlāʾ al-Dawlah where he was treated with the honour and esteem he deserved. ʿAlāʾ al-Dawlah then designated the eve of Fridays for learned debates in his presence with all other categories of learned men. This included the shaykh, who could not be defeated in any branch of learning.

The shaykh gained a great amount of experience in the remedies that he applied. One example of these is when the shaykh himself was struck with a headache one day. He conceived the notion that some disease-matter was trying to descend towards the membrane of his head and that the occurrence of some sort of swelling was unavoidable.

He ordered that a great quantity of ice be brought, crushed, and wrapped in a cloth and that his head be covered with the cloth. This was done so that the area gained strength and was protected from allowing the disease-matter to enter, and so the shaykh recovered.

The shaykh Ibn Sīnā was powerful in all of his faculties but, of his appetitive faculties, the faculty for sexual intercourse was most powerful and predominant. He frequently exercised this so that it affected his constitution. One day he was struck with colic and because he was so intent on being cured, he administered, in one day, eight enemas to himself so that part of his bowel was lacerated and ulcerations appeared on him. He was obliged to travel with ʿAlāʾ al-Dawlah and he showed signs of the fits which can accompany the colic. Still he kept on treating himself and administering enemas because of the ulcers, and to prevent the colic. One day he ordered that two *dāniq*s of celery seed be taken and mixed with the ingredients for the enema seeking to allay the flatus. However, a certain physician who himself used to go and treat the shaykh put in five dirhams of celery seed, I know not whether deliberately or in error for I was not with him, and the shaykh's ulcers were exacerbated due to the sharpness of the seeds. The shaykh also took mithridate* for the fits, but one of his servants put in it a great amount of opium and gave it to him and he consumed it. The reason for this was that they had embezzled a great amount of money from the shaykh's treasury and wished to see him dead to escape the consequences of their actions. The shaykh was taken in this state to Isfahan where he continued to treat himself. He was so weak that he couldn't walk and he continued to treat himself until he was able to walk and attend the court of ʿAlāʾ al-Dawlah. But still he did not restrain himself and indulged in intercourse often. He did not fully recover from his illness but would relapse and then recover every so often.

When ʿAlāʾ al-Dawlah sought to take Hamadan the shaykh went out with him but that same illness returned once again on the way and when they reached Hamadan the shaykh realized that his strength had collapsed and that it would not be sufficient to stave off the disease. He left off treating himself and used to say, 'The person who used to look after my body is incapable of doing so and now no treatment will be of any use.' He remained in this state for some days until he passed over to the precincts of his Lord. He was 53 years of age, his death being in 428/1037. He was born in 375/985.

This is the end of Abū ʿUbayd's account of the life of the shaykh Ibn Sīnā—may God show him mercy. His grave is beneath the city walls on the south side of Hamadan. It is also said that his body was taken to Isfahan and was buried at a place near the Gūn Gunbad gate.

When Ibn Sīnā died one of his contemporaries said of him:

This Avicenna, mankind's foe,
died wretchedly of constipation.
He was not cured by his own *Cure*
nor salvaged by his own *Salvation*.

The following advice is from the sayings of the shaykh Ibn Sīnā:

And one should know that the most virtuous motion is prayer and the most appropriate stasis is fasting and the most beneficial piety is charity and the most pure inner state is forbearance and the most vain endeavour is ostentation.

And as for wine, one should flee from drinking it for pleasure, nay even curatively or medicinally.

One should not neglect the precepts of the religion and one should respect the divine codes and be regular in physical acts of worship.

May one make a covenant with God that one will adopt this course and abide by this way, and God is the Liege of those who have faith, and He is our sufficiency and the best of trustees.

A vizier wrote to the shaykh Ibn Sīnā complaining of pustules on his forehead:

The protégé of our master, the shaykh, and his friend,
the seedling of his benefaction, nay, the product grown
from his favours,
Complains to him—may God give him lasting life!—
about the traces of pustules that appeared on his forehead.
So be so kind as to grant him the eradication of the ailment,
thus earning
the gratitude of the Prophet together with that of his
offspring.

And Ibn Sīnā replied:

May God cure and banish the complaint he has
on his forehead and make him healthy in His mercy!
He should let loose the sucking leeches who will sip
some blood from the back of his neck, exempting
him from cupping.
Meat he should shun, except light meat; nor should he
let his wine come near him.
He should daub his face with rosewater,
mixed with pressed willow, when he sleeps.
He should not tighten the button, choking him,
nor shout loudly in anger.
This is the treatment, and he who acts accordingly will see
good results and it will take care of his ailment.

Ibn Sīnā wrote a great deal of poetry, including:

Here is grey hair! Its onset had to come.
Now cut it, or dye it, or cover it!
Are you upset by the dew when it falls copiously?
Are you distressed by the sea at its shore?
So often the branch of youth that you were, pleased you
when in leaf; but the leaves had to fall.

and of wine:

Divinity descended in its humanity
as the sun descends in the solar houses.
Someone who was fond of it said
something similar to what the Christians say about Christ:
It, the cup, and what is mixed with it [i.e. water]
are united, like Father, Son, and Spirit.

Ibn Sīnā wrote 101 books, according to our findings, including the following:

The cure, in which the author summarized all four sciences. He composed the parts on physics and metaphysics in twenty days at Hamadan.

The canon of medicine. He composed parts of it at Jurjān and Rayy and completed it at Hamadan. He had hoped to compose a commentary and case notes.

The salvation. Composed on the way to Sābūr Khwāst while in the service of ʿAlāʾ al-Dawlah.*

The book of colic. Composed while imprisoned in the fortress at Fardajān. It is not extant in a complete form.

The book of cardiac drugs. Composed at Hamadan and dedicated to the august and noble Sharīf Abū l-Hasan ʿAlī ibn al-Husayn al-Husaynī.

A treatise on the endive.

On regimen and provisions for soldiers, slaves, and garrisons, and on land taxes.

On showing the falsehood of judicial astrology.

On dream interpretation.

3. FAKHR AL-DĪN AL-RĀZĪ (*d.* 1210)

The learned authority Fakhr al-Dīn, also known as the son of the preacher of al-Rayy, was a leading person among the savants of more recent times. His supremacy has become widely known, and his writings and disciples have spread throughout the world. Whenever he rode out, three hundred law students and others would walk in his retinue. The sultan Muhammad [II] Khwārazm Shāh would come to visit him. He had much knowledge of literature and composed poems in Persian and Arabic. His body was plump and of middling stature, and he wore a long beard. His voice was impressive. He delivered sermons in his native town of al-Rayy and in other towns, and the people came to him from all lands and regions irrespective of their scientific demands and the versatility of their occupations, and each of them ultimately found what he desired for.

The following saying by Fakhr al-Dīn was related to me: 'By God, I very much regret that I have to give up studying during meals, for time is precious.'

A contemporary of mine told me:

I was in the town of Herat when the shaykh Fakhr al-Dīn, coming from the town of Bamyan, entered the city with great pomp and a large retinue. On his arrival, the sultan of Herat came to meet him and received him with great honour. The latter would later erect a pulpit with a prayer rug for him in the front part of the dais of the mosque in Herat, so that he could sit in this place during celebrations and the masses could see him and listen to his words. On that particular day I was present with a group of other people. There was a large crowd attending this meeting. Fakhr al-Dīn was in the front part of the dais, flanked on each side by two rows of his Turkish Mamelukes* leaning on their swords. The sultan came to him and greeted him, and the shaykh invited him to sit at his side.

The shaykh discoursed with much grandeur and eloquence on the soul when just at this moment a dove appeared who circled around in the mosque, pursued by a hawk that was on the verge of capturing it. The dove flew from one side to another until it became exhausted, and then approached the dais where the shaykh Fakhr al-Dīn was sitting. It flew in between the two rows of soldiers, towards the shaykh, and was lucky to be rescued. A poet spontaneously wrote a poem about what had happened and then immediately asked Fakhr al-Dīn's permission to present it, to which he agreed. It reads:

> She came to the Solomon* of this age in her distress,
> while death was looming from the wings of a raptor.
> Who informed the grey dove that your place
> is a sanctuary and a refuge for those who fear?

Shaykh Fakhr al-Dīn was delighted with the poem, called the poet over to him and asked him to sit next to him. After the meeting, he sent him a full robe of honour and a large sum of money. He remained his benefactor forever.

Another man told me the following:

Fakhr al-Dīn's father, a learned authority, was a native of al-Rayy. He studied jurisprudence, specializing in the controversies of the different schools of Islamic law and in its theoretical foundations, until he became a distinguished authority, and was almost without rival. He taught in al-Rayy and also delivered sermons which, because of their excellent content and his great eloquence, were attended by a large crowd. He thus became well-known to all the people in the region. He wrote a number of works on the theoretical foundations of Islamic law and on sermons. He left two sons behind; one of them is Fakhr al-Dīn and the other, the elder, is Rukn al-Dīn.

The latter had a smattering of knowledge regarding jurisprudence and the different schools of Islamic law but he was thoughtless and very much unbalanced. He would always follow his brother Fakhr al-Dīn wherever he went. He would defame his brother, and also those who took heed of his books and words. He would say: 'Am I not older and more learned than he is and better acquainted with the polemics of the different law schools and the theoretical foundations of law? Why, then, does the crowd shout "Fakhr al-Dīn, Fakhr al-Dīn," but I never hear them shout "Rukn al-Dīn"'?. He would slander his brother, so that the people became astonished at him and many of them would mock him.

Whenever something like this reached Fakhr al-Dīn, it distressed him greatly. He did not like his brother being in such a situation, with no one listening to what he had to say. He was always kind to him and repeatedly suggested that he perhaps should take up residence in al-Rayy or elsewhere, where he could visit him and spend time with him to the best of his ability. Yet whenever he posed such a question, his brother's behaviour became worse and the situation would remain exactly the same.

This went on the way it did until Fakhr al-Dīn saw the sultan Khwārazm Shāh. He informed him about the situation of his brother and what he had to suffer, and requested from him that his brother be housed in a certain place. He urged upon the sultan that his brother would not be able to leave this place, or be transported from it, and that he would be provided with everything necessary for his livelihood. Thereupon, the sultan placed Rukn al-Dīn in one of his castles and assigned to him a fief yielding 1,000 dinars a year. He remained there until God decided upon his fate.

The learned authority Fakhr al-Dīn was corpulent, of average height, but had a broad chest, a massive head, and a thick beard. When he died at the height of his life, the hair of his beard had turned grey. He often spoke about death and longed for it, begging God to have mercy upon him. He would say: 'I have achieved all that is humanly possible in the sciences, but I have got so far that now I only wish to meet God, exalted be He, and to glance at His noble countenance.'

Fakhr al-Dīn left two sons behind. The elder of the two studied the sciences and acquired a certain insight in it, while the younger one possessed extraordinary natural talents and an extraordinary intellect. Fakhr al-Dīn often praised his intelligence, saying: 'If this son of mine survives, he will become more learned than I.' When the learned authority Fakhr al-Dīn died, his children remained in Herat. The younger son subsequently adopted the name of his father.

The vizier ʿAlāʾ al-Mulk al-ʿAlawī took over the vizierate under the sultan Khwārazm Shāh. ʿAlāʾ al-Mulk was a distinguished person, who was an expert in the sciences and in literature and he composed poetry in Arabic and Persian. He married the daughter of Shaykh Fakhr al-Dīn. When Genghis Khan, the ruler of the Mongols, crushed and vanquished the sultan Khwārazm Shāh, killed the majority of his soldiers, and Khwārazm Shāh himself was missing, the vizier, ʿAlāʾ al-Mulk, went to see Genghis Khan and sought refuge with him. When he arrived there, Genghis Khan received him with honours and made him one of his courtiers.

Then the Mongols occupied Persia, and they destroyed its castles and cities and slaughtered all the citizens in it, sparing no one's life. ʿAlāʾ al-Mulk, when perceiving that a part of the Mongol army was headed for Herat to destroy the city and kill all its inhabitants, approached Genghis Khan and asked him to assure the protection of the children of Shaykh Fakhr al-Dīn and to have them brought honourably into his presence. Genghis Khan granted this request and promised safe conduct for them. When the soldiers were approaching Herat, they announced that the children of Fakhr al-Dīn, having been granted protection, should stay in a secluded place, where they would be safe. Shaykh Fakhr al-Dīn's house in Herat was one of the largest, most beautiful, and most richly decorated houses.

When the children of Fakhr al-Dīn heard of the announcement, they confidently remained in that house, but were joined there by an enormous crowd, which included inhabitants of the town, relatives, state officials, local notables, many lawyers, and others, who thought they would be safe because of their relationship to Fakhr al-Dīn's children and particularly by their presence in that house. The Mongols, after entering the town, killed everyone they encountered and when they ultimately got to the house, called upon Fakhr al-Dīn's children to make themselves known. When they saw the children, they led them aside, and then massacred all the other people in the house.

They brought the children of Shaykh Fakhr al-Dīn from Herat to Samarkand because the ruler of the Mongols, Genghis Khan, was there at the time, and so was ʿAlāʾ al-Mulk. I—Ibn Abī Usaybiʿah—say: 'I do not know what happened to them thereafter.'

I—Ibn Abī Usaybiʿah—add: Shaykh Fakhr al-Dīn resided most of his life in al-Rayy, but he also went to the region of Khwārazm, where he was taken ill. He subsequently died in Herat as the result of the remainder of an illness. He was ill for a long time, but died on the 1st of Shawwāl 606/29 March 1210, and was transferred to the presence of his Lord, May God, exalted be He, have mercy upon him.

Fakhr al-Dīn composed over sixty works including the following books:

The great commentary [on the Qur'an], entitled *Keys to the Unknown*, in twelve volumes, in his minute handwriting, not including the first sura, to which he devoted *A commentary on sura Fātiha*, in one volume.

On geomancy.

Religions and sects.

The book of syrups.

4. AL-SAMAWʾAL (*d. c.*1174)

Al-Samawʾal was an expert in the mathematical sciences and possessed knowledge of the art of medicine. He originated from North Africa, resided in Baghdad for a while and afterwards moved to Persia, where he remained for the rest of his life. His father, too, had a smattering of the philosophical sciences.

I have copied the following from the handwriting of ʿAbd al-Latīf al-Baghdādī:

This young man of Baghdad, al-Samawʾal, was a Jew who converted to Islam. He died, while still at a young age, in Marāghah. He attained a high degree of expertise in computation and surpassed all his contemporaries in it. He had an extremely keen mind and reached a peak of proficiency in the science of algebra. He lived in Diyarbakir and Azerbaijan.

Ibn al-Qiftī said:

This al-Samawʾal, on arriving in the East, set out for Azerbaijan and entered the service of the house of al-Bahlawān* and their emirs. He lived in the city of Marāghah and there begot children, who like him followed the path of medicine. He then moved to Mosul and Diyarbakir, embraced Islam, and became a true Muslim. He wrote a book, in which he demonstrated the shortcomings of the Jews, the untruth of their claims regarding the Pentateuch* and the passages in there that furnish evidence of its abrogation. He did a proper job in assembling this material.

He died in Marāghah around 570/1174.

Among the eight treatises by al-Samawʾal is his *Book of the refutation of the Jews*, as well as an illustrated book on the right-angled triangle, an *Instructive book on medicine*, and a book on sexual potency.

CHAPTER 12

PHYSICIANS OF INDIA

This chapter is very short, comprising only six Indian physicians in the original, some of whose works were translated into Arabic. Three are included here. One of them, Mankah, is reported to have come to Iraq and have treated the caliph Hārūn al-Rashīd; a contemporary of his, Sālih ibn Bahlah, may have been born in Iraq, judging by his Arabic first name.

In terms of international relations, that of the Middle East with the West dominates in Ibn Abī Usaybiʿah's book. However, its relationship with the lands to its east, though only very briefly illuminated here, shows Indian physicians in a favourable and well-respected light.

1. SHĀNĀQ (CĀNAKYA)
(*c.* fourth century BC)

A CELEBRATED Indian physician, he was the possessor of many treatments and much experience in the art of medicine and a master of various sciences and wisdom, as well as having been proficient in astrology. He was also an eloquent speaker, and stood high in the estimation of Indian kings. The following is an example of his eloquence, taken from a work of his entitled *Sifted jewels*:

'O Prince, beware the pitfalls of time, and fear the mastery of days and the anguish of inevitable death. Know that actions earn recompense, and therefore fear the consequences of time and days, for they are uncertain. Beware then of them. Fate is unknowable. Time is inconstant: fear then its change. It may return with adversity: fear then its assault. It is swift to surprise, and there is no security against its reversal. Know that the man who does not treat himself for the sickness of misdeeds committed during the days of his life will be further than ever from recovery in that abode wherein there are no remedies. The man who in time past has subdued and abased his senses for the betterment of his soul has clearly shown his excellence and made plain his nobility of nature. He who does not control his soul, of which there is only one, will not control his senses, of which there are five, and if he does not control his senses, which are few and tractable, it will be difficult for him to control his advisors, who are numerous and formidable, and then the masses of the people in the outlying regions of the country and the distant parts of the kingdom will be very far from any control.'

Shānāq was the author of several books, notably *On poisons* which was translated from Sanskrit into Persian, and subsequently translated into Arabic for the caliph al-Maʾmūn.

2. MANKAH AL-HINDĪ (MĀNIKYA)
(eighth century)

Knowledgeable about the art of medicine, skilled in treating disease, and moderate in his methods, a philosopher of the Indian sciences, Mankah al-Hindī was also conversant with the Sanskrit and Persian languages. Mankah was a contemporary of Hārūn al-Rashīd, and during the latter's caliphate he travelled from India to Iraq, where he met with the caliph and treated him.

One account relates:

Hārūn al-Rashīd once fell gravely ill. He was treated by a number of physicians, but did not recover. One of his courtiers said to him, 'O Commander of the Faithful, there is a physician in India by the name of Mankah; he is a pious man in their religion and one of their philosophers. If you were to send for him, it may be that God would bring about your recovery at his hands.' Al-Rashīd thereupon dispatched a person to bring him to Baghdad, with a gift to persuade him to undertake the journey. Mankah went to Baghdad and attended the caliph, with the result that Hārūn al-Rashīd was restored to health; he rewarded the physician handsomely and granted him a generous pension.

One day while Mankah was walking in the palace grounds, he saw a huckster who had spread out his cloak on the ground and placed upon it a large and varied array of drugs. The man began to describe a medicinal paste that he had concocted. 'This medicine,' he said, 'is good for a constant fever, a quartan ague, a tertian ague, pain in the back, pain in the knees, abnormal phlegm, haemorrhoids, flatulence, pain in the joints, pain in the eyes, pain in the belly, headache, migraine, strangury, paralysis, the palsy . . .'. Not a single bodily ailment did he omit; his medicine, he said, was a sovereign remedy for every one of them. 'What is that fellow saying?' Mankah asked his interpreter, and the interpreter told him. Mankah smiled. 'One way or the other, the king of the Arabs must be an ignorant man,' he said. 'If the matter is as this fellow said, why has he brought me from my country, separated me from my family, and incurred the expense of my keep, when he has this wonder-worker right here, under his very nose? If the matter is not as this fellow says, why does the king not have him put to death? The law permits the execution of this fellow and anyone like him, for if that is done, only one person will have died, and by his death many will have remained alive, whereas if he is allowed to live, which would be foolishness, he will kill a person a day—indeed, he will kill two, or three, or four individuals every day, as like as not, and that would be to corrupt religion and weaken the kingdom.'

3. SĀLIH IBN BAHLAH AL-HINDĪ
(eighth century)

One of the most distinguished of the learned men of India. He was skilled in Indian methods of treatment, and was influential and far-sighted in the advancement of knowledge. Sālih ibn Bahlah al-Hindī was in Iraq in the time of the caliph Hārūn al-Rashīd.

Ibn al-Dāyah relates an account he heard first-hand from a man who was told it by his patron. In the man's words:

One day the tables had been set for Hārūn al-Rashīd's supper, but Jibrīl ibn Bukhtīshūʿ, his personal physician, was not present. My patron told me that the Commander of the Faithful had ordered him to go and find Jibrīl and bid him attend the caliph at the meal, as was the regular practice. My patron had done so, asking for the missing physician at every suite of apartments where Jibrīl was wont to attend members of the family, but had found no trace of him. Returning, he had informed the caliph of this, whereupon Hārūn al-Rashīd had burst into a torrent of curses against Jibrīl. In the midst of this, in walked Jibrīl himself. 'It would be seemlier,' he said, 'for the Commander of the Faithful to refrain from abusing me in this fashion and instead to weep for his cousin, Ibrāhīm.' The caliph asked him what had happened to Ibrāhīm, and Jibrīl informed him that he had left him near death, and that he would have expired by the time of the night-prayer. Hārūn al-Rashīd was greatly affected at this news: he began to weep, and ordered the tables cleared. Such was his grief that all those who were present were moved to pity for him.

Then his vizier, Jaʿfar,* said, 'O Commander of the Faithful, Jibrīl's medicine is Greek medicine. Sālih ibn Bahlah al-Hindī is no less learned in the art of medicine according to the doctrines of the Indians. If the Commander of the Faithful thinks it advisable, he could have the Indian physician brought here and sent to attend Ibrāhīm. We should then hear his opinion in the case, as we have heard that of Jibrīl.' Hārūn al-Rashīd took this suggestion at once, ordering Jaʿfar to go in search of Sālih ibn Bahlah. Jaʿfar obeyed, and the physician went to Ibrāhīm's chamber, where he saw the sick man and felt his pulse. He then returned to Jaʿfar, who asked him what he had determined, but Sālih replied, 'I will tell no one but the Commander of the Faithful himself.' Jaʿfar tried his utmost to make the physician divulge the information, but he steadfastly refused, and finally Jaʿfar went to Hārūn al-Rashīd. 'Show him in,' said Hārūn al-Rashīd.

Sālih entered and said to the caliph, 'O Commander of the Faithful, you are the imam and the master; no ruling of yours may be overturned by any judge. I call upon you to witness, and all present here, that if your cousin, Ibrāhīm, dies from this illness, every slave belonging to Sālih ibn Bahlah shall be free unconditionally, every beast of burden belonging to him shall be dedicated to charitable purposes, all his property shall be distributed to the poor, and all his wives shall be trebly divorced.' Al-Rashīd replied, 'That was a rash promise to make, O Sālih, in a matter that must ever be hidden from mortal man.' 'Not at all, O Commander of the Faithful,' rejoined Sālih, 'for a matter is hidden from mortal man only if no one has knowledge of it and if indications are lacking. I said what I said on the strength of clear knowledge and unmistakable indications.'

Hārūn al-Rashīd was greatly cheered at this. He began to eat and drink. But at the time of the night-prayer, a letter arrived from the postmaster* in Baghdad, with the news that his cousin, Ibrāhīm, was dead. «*Truly, we belong to God, and to Him we shall return*»,* exclaimed Hārūn, and he began to berate Jaʿfar for advising him to call in Sālih ibn Bahlah, and to revile the Indians and all their medical lore. 'Oh, the shame before God,' he cried, 'that as my cousin lay on his death-bed, I was here drinking wine!' and he called for a measure of wine, which he mixed with water and salt. He then began to drink it and quickly vomited up all the food and drink that he had had in his stomach.

The next morning, Hārūn al-Rashīd went to Ibrāhīm's house, where the servants met him and conducted him to a chamber next to some of Ibrāhīm's sitting-rooms. In the chamber, a carpet had been spread out on the left and another on the right, set with chairs, hassocks, and cushions, while between the two carpets the floor was strewn with pillows. Hārūn al-Rashīd stood leaning on his sword and said, 'It is not fitting, in a house of mourning, to sit with the bereaved family on anything more elaborate than mats. Take away these carpets and pillows!' The carpets were removed by the chamberlains, and Hārūn sat on the mats on the floor. This was the regular practice of the Abbasids from that day forward, in contrast to their previous custom.

Sālih ibn Bahlah stood before Hārūn al-Rashīd, and no one spoke a word to him. Finally, when the company could smell the scent of incense from the braziers, Sālih shouted, 'O God! O God! that the Commander of the Faithful should condemn me to divorce my wife, that he should take her from me and marry her to another, when I am her legitimate husband and entitled to her favours, and that she should become the wife of a man for whom she is not lawful! O God! O God! that the Commander of the Faithful should deprive me of my happiness, when no sin attaches to me! O God! O God! that he should bury his cousin alive! for by God, O Commander of the Faithful, he has not died. Only allow me to go in and see him.' Again and again he repeated these frantic words, and finally the caliph granted him permission to go alone into the room where Ibrāhīm lay.

They began to hear a sound as of a body being slapped with an open hand. Then the sound ceased, and they heard a cry of 'God is most great!' Sālih emerged into their midst, repeating 'God is most great!' and then he turned to the caliph and said, 'Come, O Commander of the Faithful, and I will show you something that will astonish you.' Hārūn al-Rashīd, accompanied by my patron, the eunuch Masrūr the Elder* and another, followed him into the room where Ibrāhīm lay. Sālih then took out a needle that he had with him and stuck it into Ibrāhīm's left hand, between the thumbnail and the flesh, whereupon Ibrāhīm withdrew his hand and brought it close to his body. 'O Commander of the Faithful,' said Sālih, 'does a dead body feel pain?' 'No,' replied the caliph. 'If it should be the wish of the Commander of the Faithful that his cousin should speak to him at once,' said Sālih, 'he shall do so.' 'Pray proceed,' said the caliph. 'O Commander of the Faithful,' said Sālih, 'I fear that if he recovered consciousness to find himself wrapped in a burial shroud smelling of camphor and aromatic burial ointments, his heart would fail him and he would truly die; I should have no means of reviving him. But let the Commander of the Faithful order the shroud removed and his cousin taken and washed again until the smell of the aromatic substances is gone, and then dressed in clothes like those he wore when he was in good health, perfumed with his usual scent, and finally carried to a bed such as he was accustomed to sit upon and sleep in, I shall treat him in the presence of the Commander of the Faithful, and he will speak to you then and there.'

Hārūn al-Rashīd ordered my patron to do as Sālih ibn Bahlah had suggested, and he obeyed. The caliph, accompanied by Sālih, the two eunuchs and my patron, then went to the place where Ibrāhīm lay. Sālih called for some sneezewort* and a bellows from the pharmacy, and proceeded to blow some of the sneezewort into his nose. After he had been doing this for the sixth part of an hour, the body stirred, Ibrāhīm sneezed, sat up before Hārūn al-Rashīd and kissed his hand. The caliph asked him what had happened, and he answered that he had had such a sleep as he did not recall ever having had before. It had been a refreshing sleep, he said, only he had dreamed that a dog had come rushing at him, and when he had attempted to fend it off with his hand, it had bitten him on the left thumb. He had then awoken, but could still feel the pain, and he showed al-Rashīd the place where Sālih had pricked him with the needle.

Ibrāhīm lived for a long time after this adventure. He married al-ʿAbbāsah,* the daughter of the caliph al-Mahdī, and became governor of Egypt and Palestine. He died in Egypt, and his grave is located there.

CHAPTER 13

PHYSICIANS WHO WERE PROMINENT IN THE WESTERN LANDS AND SETTLED THERE

The Muslim West (the Maghrib), including al-Andalus or Muslim Spain, is represented here. Among the Andalusians there are several who later acquired both fame and Latinized names in Europe, including the physician and poet Ibn Zuhr (Avenzoar) and the philosopher and great commentator of Aristotle, Ibn Rushd (Averroes), whose renown in the Latin West outgrew his reputation in the Muslim world. The former is one of a dynasty of medical practitioners, the Ibn Zuhr family, five of whose biographies are given, starting with Abū Marwān ibn Zuhr (d. 1077), through his son, grandson (Avenzoar), great- and great-great grandson who died in 1205.

One of Ibn Abī Usaybiʿah's more unusual sources also appears in this chapter. Abū Marwān al-Bājī, a religious scholar from Seville, met Ibn Abī Usaybiʿah in Damascus on his way to Mecca in early summer of 1237. He provided him with unique information about Andalusian physicians, especially Sevillians, whose biographies were not in the sources available to Ibn Abī Usaybiʿah. Consequently, perhaps, this chapter has the greatest number of entries—eighty-eight in the full version.

1. IS'HĀQ IBN ʿIMRĀN (*d.* before 908)

IS'HĀQ IBN ʿIMRĀN was a well-known physician and a renowned scholar. He was known as 'Instant Poison' because, for all his knowledge, he was frequently unsuccessful in his treatment, and those who made use of his services died.

Ibn Juljul, states:

Is'hāq ibn ʿImrān was a Muslim native of Baghdad who went to Ifrīqiyah during the reign of Ziyādat Allāh,* the ruler of Kairouan. It was Ziyādat Allāh who summoned him, but Is'hāq imposed on him three conditions of which only one was left unfulfilled: upon his arrival, the emir was to send a camel to transport him, provide 1,000 dinars for his expenses, and write a safe-conduct in his own hand so that he could return to his homeland whenever he desired.

Medicine in the Maghrib started with Is'hāq ibn ʿImrān. He was a skilful physician, distinguished for his skill in compounding drugs and proficient in the differentiation of diseases. He resembled the ancients in his knowledge and in the excellence of his talent. Is'hāq settled in Kairouan for some time and composed a number of books.

Discord arose between him and Ziyādat Allāh, and they became increasingly distanced from each other, to such an extent that Ziyādat Allāh had the physician crucified. Is'hāq had asked him for permission to travel to Baghdad, but the emir did not grant his request. Is'hāq used to observe Ziyādat Allāh eating, and would tell him: 'Eat this, leave this.' Then a young Jew from al-Andalus came before Ziyādat Allāh. When Ziyādat Allāh asked him to join them, he obeyed at once. One day Is'hāq was observing him while eating and said: 'Leave that, do not eat it,' but the Jew said: 'He is very strict with you.' Ziyādat Allāh had a respiratory disease: he suffered from shortness of breath. The steward had brought him some curdled milk, and he wanted to eat it, but Is'hāq told him not to do so, while the Jew held that there was nothing to be said against it.

Ziyādat Allāh agreed and ate the curdled milk, but during the night he suffered from such difficulty in breathing that it left him at the point of death. He had Is'hāq summoned. When asked whether he had a remedy, the physician replied: 'I told him not to eat the curdled milk, and he did not accept my advice. I do not have any remedy.' The attendants said to Is'hāq: 'Take 500 *mithqāl*s and treat him.' But he refused until the sum had been raised to 1,000 *mithqāl*s. This he accepted and asked the servants to bring ice, which he ordered Ziyādat Allāh to eat until he was full. When the emir vomited it along with all the milk, which had turned into cheese due to the cold of the ice, Is'hāq said: 'O emir, had this milk entered the ducts of your lungs and stuck there, you would have died from constriction of breathing; but I exerted all my ability and extracted it before it reached there.' However, Ziyādat Allāh said: 'Is'hāq has sold my life at public auction. Discontinue his allowance!'

When his allowance was discontinued, Is'hāq went to a spacious place in one of the squares of Kairouan and installed himself there with a chair, an inkwell, and sheets of paper. He would write prescriptions every day, thereby earning several dinars. But someone told Ziyādat Allāh: 'You have made Is'hāq rich.' The emir then ordered Is'hāq arrested and hauled off to prison, but the people followed him there. That night, Ziyādat Allāh had the physician brought to the palace, where Is'hāq told some stories and made some critical remarks that infuriated the emir. He ordered Is'hāq's forearms bled, and the blood flowed out until he died. Then the emir ordered that the corpse should be crucified, and it remained on the cross for such a long time that a bird made its nest inside it.

One of the things Is'hāq said to Ziyādat Allāh that night was: 'For a very long time, I have been giving you medicines that confuse your mind.' Ziyādat Allāh, who was already insane, was seized by melancholy and died.

Is'hāq ibn ʿImrān is the author of thirteen works, including:

On melancholia.

On the causes of colic and its various types, with an explanation of its remedies.

A book in which he collected the sayings of Galen on beverages.

A discourse on the whiteness of purulent matter, the sediments in urine, and the whiteness of semen.

2. IS'HĀQ AL-ISRĀʾĪLĪ (*d. c.*932)

Is'hāq al-Isrāʾīlī was a proficient, eloquent, and wise physician, well-known for his skilfulness and knowledge; he was a good writer, and high-minded person.

He was originally from Egypt, and initially worked as an oculist. Then he moved to Kairouan following Is'hāq ibn ʿImrān, with whom he studied. He worked as physician for the Imam al-Mahdī,* the ruler of Ifrīqiyah. In addition to his expert knowledge in the art of medicine, Is'hāq al-Isrāʾīlī was proficient in logic and also engaged in other disciplines.

Is'hāq lived a long life, over one hundred years, but he did not take any wife, nor did he beget any son. Once he was asked, 'Wouldn't you be happy if you had a child?' And he said: 'If I ever finish the book *On fevers*, then no!'—by which he meant that the perpetuation of his memory was better achieved with a book than with a son. It is reported that he also said: 'I have written four books that will keep my memory alive better than any son.'

It is related in a history of the Fatimid dynasty* that Is'hāq al-Isrāʾīlī said:

When I came from Egypt to meet Ziyādat Allāh, I found him encamped with the army in Laribus.* He had sent me 500 dinars to encourage me on the journey, and when he received notice of my coming, I was admitted to his presence at the very moment of my arrival. I greeted him as befitted his rank and performed the reverences due to rulers. Then I saw that his assembly had little dignity, and that love for pleasantries and for everything that raises laughter was dominant there. One of the men first addressed me by saying:

Would you say that saltiness purges?

Yes, I said.

And would you claim that sweetness purges?

Yes, I said.

Then saltiness is sweetness and sweetness is saltiness.

But sweetness purges with pleasure and convenience, whilst saltiness purges with roughness, I replied.

He went on trying to get the better of me and indulging in sophistry, and when I realized what he was doing I said:

Would you say that you are a living being?

Yes, he replied.

And that the dog is a living being?

Yes, he said.

Then you are the dog and the dog is you.

Then Ziyādat Allāh burst forth with great laughter and I realized that his love for jest was greater than his love for earnestness.

Is'hāq al-Isrāʾīlī himself said:

When the missionary* of the Imam al-Mahdī arrived in Raqqādah, he took me into his service, and once came to my house. He had a stone in his kidneys and I treated him with a remedy that contained burned scorpions. I was sitting that day with a group of Kutāmah Berbers;* they asked me about different kinds of diseases, but every time I answered they were unable to understand my words. I said to them: 'You are but a herd of beasts. None of you are human except in name only.' Word of this reached the missionary, and when I went to see him he said: 'You have addressed our brothers the believers of the Kutāmah in an inappropriate way. By the Noble God, if you did not have an excuse due to your ignorance of their state and of the extent to which the knowledge of the truth and the truthful people has reached them, I would certainly have you beheaded.'

And Is'hāq said, 'And at that point I saw a man who was concerned with the seriousness of what befell him, and who had no place in him for jest.'

Is'hāq is the author of twelve works including *Fevers*. No book on this matter is better than this one. Ibn Ridwān wrote about it: 'I say that this book is useful and the work of a virtuous man; I have applied many of its contents and have found that there is nothing to add to it. In God is the victory and the help.'

3. IBN MULŪKAH (tenth century)

Ibn Mulūkah, a Christian, lived in the days of the emirs ʿAbd Allāh I and ʿAbd al-Rahmān III. He made medicines with his own hands and performed bloodletting himself. At the door of his house there were thirty chairs for seating patients.

4. TUMLŪN (tenth century)

Tumlūn excelled greatly in medicine, surpassing in it all other people of his time, but did not attend at court; when he was summoned to do so, he asked to be excused, appealing to the emir in person, and finally was excused. There was not a single person of note in his time who did not need his services.

Ibn Juljul says:

A man told me: 'I was with the vizier, ʿAbd Allāh ibn Badr, at a time when his son Muhammad was afflicted with ulcerations all over his body. There were many physicians present, including Tumlūn. Each one of them said something about the ulcers, but Tumlūn remained silent. The vizier said to him: 'Don't you have anything to say? I have seen that you remain silent.' He replied: 'I have an ointment that will heal these ulcers within a day.' The vizier was impressed and ordered him to bring the ointment. He brought it and spread it over the ulcers, and they healed that very night. The vizier gave him 50 dinars while the other physicians left empty-handed.

5. YAHYĀ IBN IS'HĀQ (tenth century)

Yahyā ibn Is'hāq was an intelligent and learned physician, clever at treating diseases, and possessed of great practical skill. He lived at the beginning of the reign of ʿAbd al-Rahmān III, who appointed him vizier. He was entrusted with the supervision of some provinces and the tax agencies and, for some time, he was also military commander of the city of Badajoz. Yahyā stood high in the estimation of ʿAbd al-Rahmān III, who granted him a position of trust and allowed him to treat the women of the harem and the household servants. He wrote a book on medicine in five volumes following the method of the Byzantines. Yahyā had embraced Islam, but his father was Christian.

Ibn Juljul says that someone trustworthy told him the following anecdote about Yahyā ibn Is'hāq, related by a servant who said:

My master had sent me with a letter for Yahyā ibn Is'hāq. I was sitting at the door of his house, at the Gate of the Walnut Tree, when a rustic man came riding towards me on a donkey, shouting as he approached, until finally he stopped at the door of the house. 'I beseech you, go to the vizier,' he said, 'and tell him that I am here.' Alerted by the man's cries, Yahyā came out carrying with him the reply to the letter I had brought. He asked the man: 'What has happened to you?' 'O vizier,' the man replied, 'my penis is swollen, I have not been able to sleep for many days, and I feel that I am about to die.' 'Uncover it,' said Yahyā. The man did so, and it was certainly swollen. Then Yahyā said to the sick man's companion: 'Look for a smooth stone.' The man found one and brought it. Then Yahyā said: 'Take it in your hand and place the penis over it.'

And when his penis was firmly placed on the stone, Yahyā closed his fist and hit the penis with such force that the man swooned. Then the pus flowed out. As soon as the pus that had caused the inflammation had been evacuated, the man opened his eyes and then he urinated, for the urine flowed immediately. Then Yahyā said to him: 'You can go, I have cured your disease. You are a wicked man who sodomized his beast. A grain from its fodder that was there obstructed the orifice of your penis and caused its inflammation, but the grain came out with the pus.' And the man replied: 'I must confess to you that I did that.' This is a proof of Yahyā's fine intuition and good and reliable nature.

Ibn Juljul says:

My informant also treasured a curious anecdote about a case of the caliph ʿAbd al-Rahmān III's treatment. He said that ʿAbd al-Rahmān III suffered once from an earache when the vizier, Yahyā ibn Is'hāq, was still military commander of Badajoz. He received treatment, but the pain did not abate, and he sent a messenger to fetch Yahyā. Yahyā asked why the messenger had come, and he replied that the Prince of the Believers was suffering from an earache that had defied the skills of the physicians. On his way to the palace, Yahyā turned aside and stopped by the Christian quarter, where he enquired about a learned man, and was directed to an elderly monk. When Yahyā asked him whether he had had any experience treating earache, the old monk replied: 'Use warm pigeon blood.' When Yahyā met the Prince of the Believers he treated him with warm pigeon blood; and as soon as the blood was poured into his ear, the earache vanished. This was a prodigious accomplishment that demonstrates Yahyā's careful examination and perseverance in acquiring knowledge.

6. IBN JULJUL (tenth century)

Ibn Juljul was an excellent doctor with experience in treatment and great skill in the art of medicine. He lived in the days of Hishām III and served him as physician. He was clever and was interested in the efficacy of simple drugs and explained their names as found in the book of Dioscorides of Anazarbus, making clear their secrets and shedding light on their obscure meanings.

At the beginning of his book he states:

The book of Dioscorides was translated in Baghdad during the Abbasid caliphate in the days of al-Mutawakkil. It was translated from Greek into Arabic, and that first translation was examined by Hunayn ibn Is'hāq, who corrected and certified it. But the Greek names for which an Arabic term was not known were left in Greek.

Ibn Juljul says:

This book of Dioscorides came to al-Andalus in the first translated version. The people profited from all that could be understood from it until the days of ʿAbd al-Rahmān III, who at that time was the ruler of al-Andalus. Romanos II, the Emperor of Constantinople, presented him with splendid gifts, in the year 337/948,* I think, and among those presents there was a copy of the book of Dioscorides, in its original ancient Greek, illuminated with marvellous Byzantine illustrations of plants.

Romanos wrote to ʿAbd al-Rahmān III: 'You will not profit from the book of Dioscorides unless you have someone with knowledge of Greek, who will recognize the characteristics of those drugs. If there is someone able to do this in your land, then, O king, you will enjoy the benefits of the book.'

Ibn Juljul continues:

At that time, no one among the Christians of Cordova was able to read Greek, and the book of Dioscorides was kept in ʿAbd al-Rahmān's library without being translated to Arabic. Thus, the book remained in al-Andalus while the first translation from Baghdad circulated among the people. In his answer to Romanos, ʿAbd al-Rahmān III asked him to send someone able to speak Greek and Latin to teach some slaves to become translators, and the emperor sent a monk who arrived in al-Andalus in 340/951.

At that time, there were a number of physicians in Cordova who were interested in investigating the names of the drugs in the book of Dioscorides for which the Arabic terms were still unknown. The most eager was the Jewish physician, Hasdāy Ibn Shaprūt, who sought to be close to ʿAbd al-Rahmān. The monk won Ibn Shaprūt's favour and preference and explained to him the names of the drugs in Dioscorides' book that had previously been unknown. Thus Ibn Shaprūt was the first physician in Cordova to prepare the Great Theriac* following the precise explanation of the botanical information in the book.

Ibn Juljul is the author of four works including *Classes of physicians and sages*,* composed in the time of al-Muʾayyad.

7. ABŪ L-SALT UMAYYAH (*d.* 1134)

Abū l-Salt Umayyah was a native of Denia, in eastern al-Andalus, and one of the most outstanding scholars in the domain of medicine and in other disciplines. His well-known legacy includes a number of famous books. He excelled in medicine beyond anything attained by any other physician, and acquired a level of culture that few other educated persons have matched. His knowledge of mathematics was unique, and he was versed in both the theory and practice of music, for he was a good lute player himself. As a raconteur, he was witty, eloquent, and profound. Moreover, he composed beautiful poetry.

Abū l-Salt travelled from al-Andalus to Egypt and lived for some time in Cairo before going back to his homeland. While in Alexandria he was imprisoned.

Ibn Raqīqah told me in Cairo, in 632/1234, the story behind Abū l-Salt's imprisonment in Alexandria:

A boat loaded with copper that had been sailing to Alexandria had sunk not far from there, and no way of raising it could be found, owing to the depth of the sea. Abū l-Salt thought and pondered upon the matter until he came up with an idea. He went to al-Afdal ibn Amīr al-Juyūsh* and told him that if the necessary equipment could be procured, he would be able to raise the ship from the bottom of the sea to the surface, despite its weight. Al-Afdal was delighted and asked him to proceed. He provided all the equipment that Abū l-Salt had requested, investing a great sum of money.

When the equipment was ready, Abū l-Salt had it placed in a big boat with the same dimensions as the boat that had sunk. Making her fast with twisted ropes made of raw silk, he had a number of experienced divers swim down and tie the ropes to the submerged boat. He had used geometric shapes to design a device, worked from the boat in which they were standing, that would raise the wreck, and he instructed his crew what to do with it. When they operated it, the silk ropes were drawn towards them little by little and rolled around the wheels that they had in their hands, and the submerged boat appeared before them and rose almost to the surface. But then the ropes broke, and the ship fell and sank back to the bottom of the sea.

Abū l-Salt had acted in good faith when he designed his invention to raise the boat, but fate was not on his side. Al-Afdal became furious with him because of all the money that he had invested in the device, which was now lost, and, although he did not deserve it, the vizier had him arrested. He remained imprisoned for some time, until some notables interceded for him and he was released.*

Abū l-Salt—may God have mercy upon him—died on a Monday at the beginning of Muharram 529/October–November 1134 in the city of Mahdia, and was buried in Monastir.

When Abū l-Salt was about return to al-Andalus, an Egyptian poet wrote verses in Cairo and sent them to him in Mahdia. In them he describes his affection for the physician and the days they had spent together in Alexandria. They begin with the following lines:

Is there no recovery from my illness after separation from you?
It is a poison, but the antidote lies in meeting you.
O sun of excellence that has set in the west, though its light
shines over every country in the east:
May the first spring rain (ʿ*ahd*) water a time (ʿ*ahd*) when I knew you, its memory (ʿ*ahd*)
restored in my heart by a promise (ʿ*ahd*) and a covenant that will not be lost,*
Renewed by a recollection that is sweet, as when
a little turtledove coos, hidden by leaves of the trees.
You have a generous character, 'haute couture',
whereas most other people's characters (*akhlāq*) are shabby (*akhlāq*).
I have been weakened, Abū l-Salt, since your abodes have become
remote from mine, by worries and yearnings.
When it is hard for me to extinguish them with my tears
they occur while they burn between my eyelids.
Clouds, urged onward by a sighing, that is drawn
through my collarbones and chest, by a gasping.

Abū l-Salt wrote a great deal of poetry. He described a grey horse with the following verses:

A grey horse: like a shooting star in the morning it came,
moving in gold-woven horse-cloths.

Someone envying me, having seen it behind me
as a spare mount, on its way to battle, said:
'Who has bridled dawn with the Pleiades
and has saddled the lightning with a crescent moon?'

On fleas, he wrote:

Many a night of never-ending darkness,
Its evening distant from its dawn,
Like the night of a yearning slave of love,
Has lengthened, in its darkness, my sleeplessness.
The creatures that best love to harm other creatures
Think my blood is more delicious than vintage wine.
They gulp it down without ever sobering up,
Not omitting a morning drink because they had an evening drink.
If I were to spend the night above the top of Capella
It would not stop them from visiting me,
Like lovers coming at night to their beloved.
They know more about veins than Hippocrates,
Such as the median arm vein and the basilic vein.
They cut the veins with a thin lancet
Of their snout, sharpened and pointed,
Like a skilled and gentle physician.

He also wrote:

Youthful passion mixed the water of youth with its fire
from the roses of his cheeks and the myrtle of his cheek-down.
He is an idol who contains all the novelties of beauty,
so as to gain possession of my heart, in a chain of captivity.
The full moon is contained by his buttons; a twig is
in his belt, and a curved sand-dune fills his loin-cloth

8. ABŪ MARWĀN IBN ZUHR (*d.* 1077)

Abū Marwān Ibn Zuhr excelled in the art of medicine, was experienced in its practical aspects and became well known for his dexterity. His father, the jurist Muhammad, was one of the jurists and experts in Hadith* in Seville.

The qadi or judge Sāʿid al-Andalusī says:

Abū Marwān Ibn Zuhr travelled east and lived in Kairouan and Egypt, where he practised medicine for a long time. He then returned to al-Andalus and went to the city of Denia. When Abū Marwān presented himself at court, the emir Mujāhid showed him great honour and invited the physician to enter his service, which he did, enjoying the ruler's favour during his reign. Abū Marwān made a name for himself in Denia for his pre-eminence in medicine, and his fame reached all corners of al-Andalus. He had unique ideas in the field of medicine. One of them was his rejection of steam-baths, because he was convinced that they decomposed the body and corrupted the humours.

This is a view that contradicts ancient and modern opinions and was considered a mistake by both the experts and non-experts; what is more, when the bath is administered as it is required and gradually,* it is an excellent exercise and a beneficial practice because it opens the pores, removes impurities and reduces the excess of chyme.

I—Ibn Abī Usaybiʿah—continue: Abū Marwān Ibn Zuhr moved from Denia to Seville and remained there until his death. He left behind impressive wealth in Seville and other places, including both urban properties and rural estates.

9. ABŪ L-ʿALĀʾ IBN ZUHR, SON OF ABŪ MARWĀN (*d.* 1130)

Abū l-ʿAlāʾ ibn Zuhr was well known for his skills and knowledge, and for the precision of his treatments. There are anecdotes about the remedies he administered to the sick, and about his ability to determine their states and the pain they suffered without hearing it from them, but merely by inspecting their urine phials and feeling their pulse.

Abū l-ʿAlāʾ lived in the time of the 'Veiled Men', also known as Almoravids.* He prospered under their rule, acquiring great prestige and enjoying a good reputation, but he had begun to practise as a physician when he was young, in the days of al-Muʿtadid, king of Seville. He also applied himself to the study of literature and was an excellent writer. It was in his time that Ibn Sīnā's *Qānūn* (Avicenna's *Canon*) arrived in the West.

The Egyptian, Ibn Jumayʿ, wrote:

A certain merchant brought a copy of Ibn Sīnā's book from Iraq to al-Andalus. It had been executed extremely beautifully, and he presented it as a gift to Abū l-ʿAlāʾ ibn Zuhr as a way of ingratiating himself with him, because he had never seen the book before. But when Ibn Zuhr examined the *Qānūn* he criticized and rejected it. He did not deposit it in his library, but decided instead to cut it into strips and to use them to write prescriptions for his patients.

Another historian wrote:

Despite his young age, the name of Abū l-ʿAlāʾ ibn Zuhr was a byword for excellence, and he was praised as a master of knowledge. Good fortune paved his path towards success, and Fate was not content with bestowing on him merely a lowly standing. He surpassed all virtuous men in knowledge and purity of descent, and he was also the most magnanimous and liberal of persons. His only fault was being impulsive and prone to use obscene language, but is there anyone with all the qualities of the perfect man in complete harmony?

Abū l-ʿAlāʾ ibn Zuhr was buried in Seville outside the Gate of Victory. He was the author of the following poems.

On love he wrote:

You who shoot at me with arrows that have no aim
but my heart, and have no substitute for it;
Who makes me ill with eyelids that are filled with sickness,*
yet are healthy (it is their nature to nurse and be sick):
Grant me if only with an apparition from you that will visit me at night;
for sometimes an accident may fill the place of a substance.

After hearing that Ibn Manzūr, the chief qadi of Seville, said, mockingly, 'Is Ibn Zuhr ill?' he wrote:

They said that Ibn Manzūr expressed, tirelessly, his surprise
at my being ill. I said, someone who walks may stumble.
Galen was ill all the time;
'eating' bribes is something the esteemed jurist does.

Abū l-ʿAlāʾ ibn Zuhr is the author of nine works including *Medical experiences*, which, after his death, [the Almoravid sultan] ʿAlī ibn Yūsuf ibn Tāshfīn compiled in Marrakesh and other places in Morocco and al-Andalus and copied in Jumādā II 526/April–May 1132.*

10. ʿABD AL-MALIK IBN ZUHR (AVENZOAR), SON OF ABŪ L-ʿALĀʾ (*d.* 1162)

ʿAbd al-Malik ibn Zuhr followed in the footsteps of his father, Abū l-ʿAlāʾ, in medicine. He acquired an extensive knowledge of simple and compound drugs, and was skilled in treatment. His fame spread throughout al-Andalus and other lands, and their physicians studied his works. None in his time matched his skills in the varied tasks of the art of medicine.

ʿAbd al-Malik ibn Zuhr served the Almoravids, and he enjoyed his lot of luxury and wealth from his attendance on them, until the advent of the Mahdī, Ibn Tūmart, and the ascension to power of ʿAbd al-Muʾmin, who was recognized as caliph. He engaged ʿAbd al-Malik ibn Zuhr as his personal physician, and gratified him with luxuries and stipends that surpassed all his desires.

A contemporary of mine told me that the caliph ʿAbd al-Muʾmin needed to take purgatives, and that he hated to drink medicines of that kind. But ʿAbd al-Malik ibn Zuhr made it bearable for him: he went to his garden and watered a vine with water into which he had poured the medicine. The water had thus acquired the strength of the purgative with which it had been infused or boiled. As a result, the vine absorbed the strength of the needed purgative and the grapes that grew from it contained the power of the medicine. ʿAbd al-Malik then helped the caliph by giving him a bunch of those grapes and telling him to eat them. The caliph, who trusted his physician, ate them. ʿAbd al-Malik observed the caliph as he ate, and then said: 'O Commander of the Faithful, that is enough; you have eaten ten grapes and they will help you to sit on the toilet ten times.' ʿAbd al-Malik then explained to him the reason for his words, and the caliph went to the toilet the said number of times, finding relief and getting better. In that way ʿAbd al-Malik ibn Zuhr enhanced his status at court.

Another man, originally from Murcia, told me that when he was on his way to the palace of the caliph in Seville, he met ʿAbd al-Malik ibn Zuhr in the bathhouse. There was a man there who was sick with an intestinal disease, his abdomen had swollen and his skin had turned yellow. He was in great pain, and asked for someone who could examine him. Some days later, when ʿAbd al-Malik ibn Zuhr was there, the man asked again, and he examined him. ʿAbd al-Malik observed that the man had an old jug by his head from which he used to drink water. 'Break this jug,' the physician said, 'it is the cause of your disease.' 'No sir, by God,' the man replied, 'for I have no other.' But ʿAbd al-Malik ordered one of his servants to break it, and out of the broken jug came a frog that had grown to a great size inside it. 'That is the end of your disease,' said Ibn Zuhr. 'Look: that is what you were drinking!' And thereafter the man recovered.

The qadi Abū Marwān al-Bājī,* originally from Seville, said to me—Ibn Abī Usaybiʿah:

A trustworthy person told me that there was a wise man in Seville known as the Mouse who was versed in medicine and the author of a valuable book on simple drugs in two volumes. ʿAbd al-Malik ibn Zuhr, who used to eat green figs very often, once offered some of them to that physician called the Mouse, but he would not eat any, or only one fig every year. He told ʿAbd al-Malik that he feared he would develop a superating ulcer—they call it *naghlah** in their dialect—if he ate figs so frequently. To which ʿAbd al-Malik replied: 'There is no reason for such apprehension. And if you do not eat figs, you might suffer from convulsions.'

In the event, the Mouse did die from convulsions; and ʿAbd al-Malik Ibn Zuhr also developed an ulcer in his side and died because of that. This shows how accurate their predictions were.

11. AL-HAFĪD, 'THE GRANDSON' ABŪ BAKR IBN ZUHR, SON OF ʿABD AL-MALIK (*d.* 1199)

Abū Bakr ibn Zuhr (al-Hafīd, 'the grandson' of Abū l-ʿAlāʾ ibn Zuhr), the vizier, was a respectable, wise, cultured, and noble man born and raised in Seville. He distinguished himself in the study of various disciplines and learned medicine with his father, ʿAbd al-Malik, acquiring direct knowledge of its practical aspects. He was of middle height, with a healthy complexion and strong limbs, and even in old age he retained an excellent colour and could move with the same vitality, only he became hard of hearing at the end of his life. He memorized the Qur'an, and studied Hadith and also Arabic literature and language. No one in his time was better versed in the Arabic language, and it is said that he mastered both medicine and literature. He also excelled at composing poetry, and wrote a number of famous *muwashshahahs** that are still sung and are among the best poems in that style. With his solid religious convictions, his determination, and his love for the good, Abū Bakr ibn Zuhr was respected, and he used to speak boldly.

The qadi Abū Marwān al-Bājī told me:

Someone trustworthy told me that a man from the al-Yanāqī family* was a friend of Abū Bakr ibn Zuhr and that they used to meet often to play chess together. One day while they were playing chess, Abū Bakr felt that his friend was not as cheerful as usual and said to him: 'What is on your mind? It seems that you are distracted with something; tell me about it.' 'Yes,' replied al-Yanāqī, 'I have a daughter, and I am marrying her to a man who has asked for her hand, but I need 300 dinars for the dowry.' 'Play, and forget about that,' said Abū Bakr, 'for I have 300 dinars less 5 dinars that you may have.' They played for a while, and then al-Yanāqī took his leave, and Abū Bakr ibn Zuhr gave him the money. But he soon returned and put the 300 dinars less 5 dinars in Abū Bakr's hands. 'What is this?' he exclaimed. 'I have sold some olive trees,' replied the man, 'for 700 dinars, and I am giving you 300 less 5 dinars in return for that which you kindly gave me: you receive this from me, and I still keep 400 dinars.' 'Keep it all for yourself and put it to good use,' said Abū Bakr, 'I did not give you that money so that you would return it to me.' But the man refused. 'By God,' he said, 'at this moment I have no need to take anything from anyone, either as a gift or as a loan.' 'But are you my friend or my enemy?' asked Abū Bakr. 'I am your friend,' the man answered, 'and the one who loves you the most.' To which Abū Bakr replied: 'True friends do not have anything that they would not share if one of them is in need.' But the man would not accept, until Abū Bakr ibn Zuhr exclaimed: 'By God, if you do not accept it we will become enemies, and I will not speak to you ever again.' And the man finally took the money and thanked his friend.

He also said:

The Almohad* caliph, Abū Yūsuf Yaʿqūb al-Mansūr, decided that no book of logic and philosophy should remain in his lands, and many were burnt. He was adamant that no one should engage in those disciplines any more, and that if anyone were to be found studying those arts or in possession of any book dealing with them, he would be severely punished.

A botanist from Seville said:

Abū Bakr ibn Zuhr had taken two pupils to practise medicine with him; they worked together and were affiliated with him for some time. One day when they met Abū Bakr ibn Zuhr, one of them happened to have a small book of logic with him. When he saw it, Abū Bakr exclaimed: 'What is this?' Then he took the book, inspected it, and when he realized that it was a book on logic, threw it away. Even though he was barefoot, he stormed towards his pupils, intending to beat them. They fled, but Abū Bakr rushed after them despite his state, not sparing any insult, while they ran off before him—and they certainly ran a long distance.

Afterwards, the pupils avoided Abū Bakr for some time, not daring to approach him at all. Subsequently, however, they went to visit him, and on that occasion they excused themselves, claiming that the book had not been theirs and that they had seen a young man in the street with the book in question. Intending to subject him to scorn and ridicule, they had taken the book from him, and, forgetting that they had it with them, had gone on to meet their master.

Abū Bakr ibn Zuhr pretended to be deceived by all this, but before forgiving them, he ordered them to memorize the Qur'an, to study Qur'anic commentary, Hadith, and Sharia; and to commit themselves to abide by the requirements of the Sharia without exception. The pupils obeyed his orders. One day when they were with Abū Bakr, he presented to them that book on logic they had brought, and said: 'Now you are free to read this book and others of that kind with me, and to study them.' They were greatly surprised, but this incident demonstrates his great intelligence and virtue.

The qadi Abū Marwān al-Bājī also told me:

Abū Zayd, the vizier of the caliph, al-Mansūr, was an enemy of Abū Bakr ibn Zuhr and envied him. When he saw the prestige and excellence of the physician's position and the high regard that his work commanded, he conspired against him, intending to poison him with the help of one of his associates, putting the poison into some eggs. Abū Bakr was then working with his sister and his niece, both of whom possessed a sound practical knowledge of medicine and knew how to administer remedies. They were especially experienced in the treatment of women and used to take care of the women of al-Mansūr; in fact, they had delivered all the sons of al-Mansūr and his family. When Abū Bakr's sister died, he did not eat any of the poisoned eggs, but his niece passed away with her mother, since no treatment was effective.

The qadi adds: The vizier Abū Zayd was subsequently murdered by members of his own family.

One of the many poems of Abū Bakr ibn Zuhr was recited to me by a man who heard it from Abū Bakr himself about his love for his son:

I have someone like a sandgrouse chick,
    a young one, with whom my heart has been left behind.
My house is far from him; how lonely I am
    without that dear little person and that dear face!
He yearns for me and I yearn for him;
    he weeps for me and I weep for him.
Yearning has become tired between us two,
    from him to me and from me to him!

The qadi Abū Marwān al-Bājī told me that a man who lived in Seville, heard these verses from Abū Bakr himself, towards the end of his life:

I looked into the mirror when it had been polished
    and my eyes could not believe all they saw.
I saw a little old man whom I did not know,
    while I used to know, before that, a young man.
'Where is he,' I asked, 'who once dwelled here?
    When did he depart from this place, when?'
The mirror, thinking me stupid, replied to me, without speaking:
    'That one was here once; this one came afterwards.
Take it easy! This one will not last forever.
    Don't you see how grass withers after it has grown?'
Pretty women used to say, 'Dear brother!' but now
    pretty women say, 'Dear father!'

Abū Bakr ibn Zuhr also recited the poem:*

O cupbearer, the complaint is addressed to you!
We called upon you but you did not listen.

There was a drinking companion I fell in love with;
I drank wine (*rāh*) from his hand (*rāhah*).
Whenever he woke up from his inebriation
He drew the wineskin towards him, leaned back,
And poured me four in four,*
A willow branch, inclining from where he had been straight:
He who loves him spends the night, from excess of passion,
With pounding heart and weakened strength.
Whenever he thinks of separation he weeps;
But why should he weep for something that has not happened?
I have no patience nor fortitude;
O people! They have reproached me, doing their utmost;
They rejected my complaint about what I suffer.
One is entitled to complain about a state such as I am in:
The distress of despair and the humiliation of desire.
Why have my eyes become dim-sighted by a glance?
After seeing you they dislike the light of the moon.
If you wish, listen to my story.
My eyes have become wretched from long weeping;
One part of me wept with me for another part of me.
A hot heart* and tears that flow—
He knows his sin yet does not acknowledge it.
O you who turns away from what I describe:
The love in me for you has grown and thrived.
Let the beloved not think I am pretending!

And:

The speech of a reproacher
Is something that will pass with the winds.
My proper conduct has been brushed aside, I am deprived of my right behaviour
By an open mouth that turned eyes away from chamomile flowers*
That are watered by a mixture of musk and wine
Like bubbles floating
On the surface of pure water.
Who can help me when he, like a full moon, reveals himself in the dark?
I have fallen in love, because of his cheeks, with the moon when it is full,
And I have fallen in love, because of his body, with lissom figures.
Like a tender bough
His is unable to carry a sash.

12. ABŪ MUHAMMAD IBN ZUHR, SON OF ABŪ BAKR (*d.* 1205–6)

Abū Muhammad ibn Zuhr was a handsome and clever man with sound judgement, and extraordinary intelligence. His way of life was virtuous, although he liked luxurious clothes. He took a keen interest in medicine, having been dedicated to the study of all its aspects. Abū Muhammad worked with his father, who helped him to discover the theoretical and practical secrets of that art. The caliph Muhammad al-Nāsir held Abū Muhammad in high regard and bestowed great honours upon him, aware as he was of the physician's vast knowledge and the high rank of his family.

This is one of the most amazing stories that the qadi, Abū Marwān al-Bājī, told me about Abū Muhammad:

One day I was with him and he said to me: I dreamt of my sister last night—she had already died. It was as if I asked her: 'Sister, by God, would you tell me how long will I live?' She replied: 'Two *tapias* and a half'—the *tapia* is a wooden frame used in construction and is known in the Maghrib by that name. Its length is ten spans. And I said to her: 'I asked you a serious question, and you have answered with a joke.' But she replied: 'No, by God, I spoke in earnest; but you did not understand. Is not a *tapia* ten spans, and two *tapias* and a half twenty-five? Then you will live for twenty-five years.'

The qadi continued:

When he told me about this dream, I said: 'Do not interpret anything from that, because it is probably a confused dream.' But he died before the end of that year, being, as he had been told, 25 years old, no more or less. He left two sons, both with noble personal virtues and good family background; one was named Abū Marwān ʿAbd al-Malik and the other Abū l-ʿAlāʾ Muhammad. This latter, who was the younger, took an interest in the art of medicine and became an expert in the books of Galen. They both lived in Seville.

13. IBN RUSHD (Averroes) (*d.* 1198)

The qadi Ibn Rushd was born and raised in Cordova. His virtue is well known, as is his engagement with various disciplines. He excelled in law and in the analysis of discordant Hadiths. Ibn Rushd also devoted himself to medicine. He was an excellent writer with regard to both form and content: his medical *Book of generalities** is the best of his works. Ibn Rushd and ʿAbd al-Malik ibn Zuhr (Avenzoar) were close friends, and when Ibn Rushd wrote this book on medical generalities ʿAbd al-Malik Ibn Zuhr asked him to write a book on particularities as well, so that the two treatises would complement each other and form a perfect compendium on medicine.

The qadi Abū Marwān al-Bājī told me:

Ibn Rushd had a sound intellect, wore shabby clothes, and possessed a strong spirit. He devoted himself to the study of various disciplines, including medicine and philosophy.

Ibn Rushd held the position of qadi in Seville for a time before moving to Cordova to serve the caliph Abū Yūsuf Yaʿqūb al-Mansūr at the beginning of his reign; later, al-Mansūr's son Muhammad al-Nāsir also held him in high regard. Al-Mansūr sent for Ibn Rushd when he left Cordova and covered him with honours and placed him by his side.

Later, however, al-Mansūr punished Ibn Rushd by having him exiled to Lucena—a city not far from Seville once populated by Jews. The caliph also acted against other people of rank and virtue. It seems that the reason behind this was their alleged engagement with philosophy and the disciplines of the ancients.

They lived in exile for some time, until a group of notables from Seville testified in favour of Ibn Rushd, swearing that he had been innocent of his alleged offence, and the caliph al-Mansūr had mercy on him and on the others. This took place in 595/1198.

The qadi Abū Marwān also said:

One of the things that al-Mansūr resented about Ibn Rushd is that, while sitting in one of his assemblies, Ibn Rushd, on his own initiative or perhaps talking to him or discussing some scientific issue, addressed al-Mansūr saying, 'Listen, my friend.' Also, Ibn Rushd once described a book on animals, and when he came to the giraffe, he said: 'I have seen a giraffe in the palace of the king of the Berbers'—i.e. al-Mansūr. These words reached al-Mansūr, who found them offensive, and that was one of the reasons why he sent Ibn Rushd into exile. It is also said that Ibn Rushd apologized, claiming that he had meant to say 'the king of the two continents', but had misspelled it, and the reader had said 'king of the Berbers'.*

Ibn Rushd, may God have mercy upon him, died in Marrakesh at the beginning of 595/early November 1198. Ibn Rushd lived a long life and left a son who was also a physician.

Ibn Rushd is the author of over forty-five works on philosophy and medicine.

14. ABŪ L-HAJJĀJ IBN MŪRĀTĪR
(*d.* between 1213 and 1224)

Abū l-Hajjāj was from Mūrātīr in eastern al-Andalus, a town close to Valencia. He excelled in the art of medicine and acquired extensive experience in that domain. Abū l-Hajjāj led a virtuous life and possessed good judgement. He knew about legal disciplines, studying Hadith and the *Mudawwanah*.* In addition, he was a man of letters and a poet, and he loved licentious literature. Many stories circulated about him.

The qadi Abū Marwān al-Bājī told me:

We were once in Tunis with the ruler, Muhammad al-Nāsir. and the army was suffering as a result of high prices and a shortage of barley. Abū l-Hajjāj composed a *muwashshahah* on al-Nāsir in which he altered a verse that Abū Bakr ibn Zuhr usually included in some of his *muwashshahah*s.

Abū Bakr's verse ran:

> A feast does not consist of wearing a fine robe and a suit
> Or smelling perfume:
> Rather, a feast consists of a meeting
> With one's beloved.

While Abū l-Hajjāj's version was as follows:

> A feast does not consist of wearing a fine robe and a suit
> Made of silk:
> Rather, a feast consists of a meeting
> With barley.

Then al-Nāsir gave him ten dry measures of barley, the price of which at that time was 50 dinars.

Abū l-Hajjāj lived a long life and died of gout in Marrakesh during the reign of the caliph, Yūsuf II.

15. ABŪ BAKR IBN ABŪ L-HASAN AL-ZUHRĪ
(*d.* between 1213 and 1224)

Abū Bakr was the son of a jurist and qadi and was born and raised in Seville. He was generous and gentle, with a good character and a noble soul. He served the governor of Seville, as his personal physician, and used to attend sick people and write prescriptions for them without asking anything in return.

At the beginning of his career Abū Bakr was passionate about chess, and was such an excellent player that he came to be known as 'The Chess-player'. The qadi Abū Marwān al-Bājī told me:

I once asked Abū Bakr about the reasons that had led him to study medicine, and he replied: 'I used to play chess constantly. Few others in Seville played as well as I, to such an extent that I was called Abū Bakr the Chess-player. I became furious when I heard that, and I told myself that I had to turn away from chess and cultivate some other discipline, so that it, instead of chess, would be associated with my name. But I knew that, even if I were to devote my entire life to study law or other humanistic disciplines, I would not reach the required degree of excellence to be nicknamed after it. So I decided to approach ʿAbd al-Malik ibn Zuhr (Avenzoar) and study medicine with him. I listened to his lectures and wrote prescriptions for the sick people who came to consult him, and after that I became famous for my medical skills, and the former nickname that I had hated was forgotten.'

Abū Bakr lived eighty-five years. He died during the caliphate of Yūsuf II, and was buried in Seville.

16. IBN AL-ASAMM (thirteenth century)

Ibn al-Asamm was one of the most renowned physicians of Seville. He was greatly experienced in the art of medicine and had excellent observation skills. There are well-known stories and numerous anecdotes about his knowledge and his ability to know the state of patients, from what kind of afflictions they were suffering, and what they had eaten simply by examining their phials of urine.

A contemporary of mine told me—Ibn Abī Usaybiʿah:

I was with Ibn al-Asamm one day when we saw a group of people who were calling him. Among them there was a man on a beast, and when we approached, we saw that he had a snake in his mouth: its head had gone down his throat, and the part of it that was outside had been knotted with a hemp string to the man's arm. 'What has happened here?' asked Ibn al-Asamm, and they replied: 'He always sleeps with his mouth open. Last night he had eaten cheese before going to bed and this snake came along, licked his mouth and entered it while he was asleep. When the snake felt that someone was coming it panicked and part of it went down his throat, but we grabbed the snake and tied it with this string to prevent it from descending any further. Then we brought him to you.'

When Ibn al-Asamm looked at the man he found him about to die of fear, and said to him: 'Don't worry!'; and, addressing the others: 'You almost killed him.' He then cut the string, and the snake descended into the man's stomach. At this point, Ibn al-Asamm said to the man: 'Now you will heal.' Ordering him not to move, he took some drugs and infused them in boiling water. Then he made the man drink it. Although it was very hot, he drank it. Ibn al-Asamm examined his stomach and exclaimed: 'The snake is dead.' Then he made the man drink from another jar of water in which he had boiled some stuff and explained: 'This will tear the snake to bits with the movements of the stomach.' After two hours he made him drink some water in which he had boiled emetic drugs: the man's stomach heaved and he almost choked on his vomit, but Ibn al-Asamm covered his eyes and he kept on vomiting into a basin until we saw the snake, which had been torn to bits. Ibn al-Asamm ordered him to keep on vomiting until he had expelled all the remains of the snake and his stomach was empty. Then he said: 'Cheer up, for you have been cured,' and the man went away healthy and content after having been at death's door.

CHAPTER 14
FAMOUS PHYSICIANS IN EGYPT

In this chapter, one finds again a few physicians who acquired renown in the West, such as the great Jewish philosopher and religious scholar Maimonides (Mūsā ibn Maymūn in Arabic, Moshe ben Maimon in Hebrew). Born in Cordova he moved to Egypt, where he died in 1204, having been one of Saladin's physicians and one of the teachers of Ibn Abī Usaybiʿah's uncle, Rashīd al-Dīn. The Basra-born polymath, Ibn al-Haytham (Alhazen), famous especially as a mathematician and physicist, and sometimes called 'the father of modern optics', is also present. His entry and that of Ibn Riḍwān comprise two more of the autobiographies related by Ibn Abī Usaybiʿah, the former slightly self-righteous and the latter including a rare natal horoscope as well as advice on living sensibly and prudently.

Encountered for the first time are people whom Ibn Abī Usaybiʿah had met personally, including Maimonides' son, Ibrāhīm, with whom Ibn Abī Usaybiʿah worked at the Nāsiri hospital in Cairo. We also find the first mention of Ibn Abī Usaybiʿah's book in the entry to Abū Saʿīd Muhammad. He writes a letter to Ibn Abī Usaybiʿah in 1269 saying he has come across an (earlier) copy of the book in Cairo, thus providing a terminus post quem *for what is probably the last revision Ibn Abī Usaybiʿah made of his book.*

1. POLITIANUS (*d. c.*802)

POLITIANUS was a famous physician in Egypt and a Christian scholar of the Melkite sect.

Eutychius, Patriarch of Alexandria, says in his book:

Politianus was made Patriarch of Alexandria in the fourth year of the reign of the Abbasid caliph al-Mansūr, remaining in that office for forty-six years until his death,* and was also a physician.

During the days of Hārūn al-Rashīd, the governor of Egypt sent a very beautiful concubine from the Copts* of Lower Egypt to the caliph, who grew to love her very much. One day, she became very ill. The physicians treated her, but were unable to cure her. They said to Hārūn al-Rashīd, 'Have your governor in Egypt, send you one of the physicians of Egypt since they know more about how to treat an Egyptian slave-girl than the physicians of Iraq.' So the caliph ordered the governor to choose one of the most skilled physicians of Egypt. The governor called upon Politianus, told him of the caliph's love for the girl and her illness, and conducted him to Hārūn al-Rashīd. Politianus carried with him some of the coarse cake and small salted fish of Egypt, and when he arrived in Baghdad and attended to the slave-girl, he gave her the cake and small salted fish to eat, whereupon she was cured of her illness.

From that time on, the coarse cake and small salted fish were imported from Egypt to the imperial storehouses in Baghdad. Hārūn al-Rashīd rewarded Politianus with ample wealth and issued a decree ordering that every church that had been taken from the Melkites by the Jacobites should revert back into the possession of the Melkites. Politianus returned to Egypt and reclaimed many churches from the Jacobites.*

2. AL-HASAN IBN ZĪRAK (*d. c.*882)

Al-Hasan ibn Zīrak was a physician in Egypt during the governorship of Ibn Tūlūn and attended him while he resided in Egypt. If Ibn Tūlūn travelled, however, the physician Saʿīd ibn Tawfīl accompanied him. In 269/882 Ibn Tūlūn went to Damascus and from there to the fortifications along the frontier to restore order.* On his way back he passed through Antioch, where he consumed a lot of water buffalo milk and became afflicted with vomiting and diarrhoea. The efforts of Saʿīd ibn Tawfīl to cure him were not successful, and Ibn Tūlūn returned to Egypt, ill and discontented with that physician.

Upon entering Fustat (Old Cairo), Ibn Tūlūn summoned al-Hasan ibn Zīrak and complained to him about Saʿīd ibn Tawfīl's treatment. Ibn Zīrak put his mind at rest about the illness and wished him a speedy recovery. Ibn Tūlūn's ailment eased with rest, peace and quiet, being reunited with his family, and the attentions of al-Hasan ibn Zīrak.

But Ibn Tūlūn maintained a discreet silence about his sexual intercourse with women, as a result of which his condition worsened. At this point he summoned the physicians and threatened them, while concealing from them his improper regimen, his sexual intercourse, and his craving for marinated fish, which one of his concubines secretly brought to him, but no sooner did he eat it than he developed severe diarrhoea.

Sending for al-Hasan ibn Zīrak, he said to him: 'I believe what you prescribed for me today was not correct.' To which al-Hasan ibn Zīrak replied: 'Let the emir, may God help him, summon all the physicians of Fustat to his residence in the early morning each day, so that they can reach a consensus as to what the emir should take that morning. I have administered nothing to you except those things whose composition merits your confidence, and all of them stimulate the retentive faculty in your stomach and liver.' 'By God,' answered Ibn Tūlūn, 'if you do not succeed in the treatment of my illness, I will cut off your head. You are experimenting on a sick person and nothing is to be gained by doing that.'

So, al-Hasan ibn Zīrak departed, trembling. He was an old man and his liver became inflamed through fear and anxiety, which prevented him from eating and sleeping, and he soon developed severe diarrhoea. He was overcome with worry and became delirious, talking irrationally about the illness of Ibn Tūlūn, until he died the following morning.

3. SAʿĪD IBN TAWFĪL (*d.* 882)

Saʿīd ibn Tawfīl was a Christian physician, distinguished in the art of medicine. He was in the service of Ibn Tūlūn as one of his special physicians, attending him both while travelling and while in his capital. Before his death, however, Ibn Tūlūn turned against him. The reason for this was as follows:

As mentioned above, Ibn Tūlūn had gone to Syria and had proceeded to the fortifications along the frontiers to restore order. Upon his return to Antioch, he developed vomiting and diarrhoea from the water buffalo milk, of which he had hastily drunk too much. He summoned his physician, Saʿīd ibn Tawfīl, only to learn that he had gone to a church in Antioch. Ibn Tūlūn became angry, and when Saʿīd appeared, the emir upbraided him for his lateness, but, in disdain, refrained from complaining about what ailed him. Then the following night, the emir's illness worsened and he sent for Saʿīd, who arrived slightly inebriated. Ibn Ṭūlūn said to him, 'I have been ill for these past two days, and you are drinking wine!' Saʿīd replied: 'My lord, you called for me yesterday while I was in church, as is my usual custom, and when I appeared, you told me nothing!' 'Would it not have been appropriate to ask about my condition?' said Ibn Ṭūlūn. 'Being your physician is one thing, my lord,' replied Saʿīd ibn Tawfīl, 'but asking your household about your private affairs is another.' 'What should I do now?' Ibn Ṭūlūn then enquired. 'Tonight and tomorrow do not touch any food,' the physician replied, 'even though you crave it.' 'But I am hungry, by God!' Ibn Ṭūlūn protested, 'I will not be able to last that long!' 'This is a false hunger,' Saʿīd said, 'caused by a coldness of the stomach.'

When the middle of the night came, however, the emir called for something to eat. He was brought pullets, hot roasted meat, and bread stuffed with fowl and cold young goat's meat.* Once he had eaten them, his diarrhoea ceased. So his servant went out and said to Saʿīd: 'The emir ate lamb and roasted meat and his condition has been alleviated.' 'God is now the one to call on for help,' Saʿīd replied, 'for the expulsive force causing the bowel movements had been weakened by his abstention from food, but now the bowel movements will become dreadful.' And by God, just before daybreak the emir had more than ten evacuations. Ibn Ṭūlūn left Antioch with his condition steadily worsening, although his strength was such that he managed to bear it. As he headed for Egypt riding became intolerable, so a cart pulled by men was made for him and on it he was more comfortable. But before arriving at al-Faramā, Ibn Ṭūlūn complained of its discomfort and therefore continued to Fustat by boat. On the deck of the boat a domed tent was pitched for him to lie in.

When Ibn Ṭūlūn alighted in Egypt, he complained about Saʿīd ibn Tawfīl to his secretary who then spoke to Saʿīd reproachfully, 'You are skilful in your profession, and you have no vice except that you are prideful, rather than humble, in serving the emir. Even though the emir speaks the language eloquently, he is foreign in disposition and relies on your guidance. Your approach has alienated him from you. You should be kind to him, be of use to him, be devoted to him, and pay attention to his condition.'

To this Saʿīd replied, 'By God, my service to him would be like that of a mouse to a cat, or a lamb to a wolf. Indeed, I'd rather be killed than attend on him.'

Ibn Ṭūlūn eventually died of this very illness.

Ibn al-Qiftī writes:

At the beginning of Saʿīd ibn Tawfīl's association with Ibn Ṭūlūn, the physician employed a youth of ugly appearance. His name was Hāshim and he tended to Saʿīd's mule and would look after it when Saʿīd was in attendance on Ibn Ṭūlūn. Occasionally when they returned, Saʿīd employed Hāshim in pulverizing drugs at his house, and Hāshim would stoke the fires under the medicinal decoctions. Saʿīd ibn Tawfīl had a very handsome and intelligent son who was well-versed in medicine.

One day, early in their association, Ibn Ṭūlūn instructed Saʿīd to find a physician for the women's quarters, one who would be present during Saʿīd's absence. Saʿīd told him: 'I have a son whom I have taught and trained.' Ibn Ṭūlūn said, 'Let me see him.' Saʿīd brought him to the ruler, and Ibn Ṭūlūn saw a handsome youth possessed of all admirable qualities. 'But he is not suitable for service in the women's quarters,' Ibn Ṭūlūn said, 'for someone of sound knowledge but ugly appearance is required for them.' Saʿīd was afraid that Ibn Ṭūlūn would bring in a stranger who would contradict and oppose him, so Saʿīd took Hāshim and dressed him in a *durrāʿah** and a pair of leather boots, and Ibn Ṭūlūn appointed him to the women's quarters.

A certain physician related the following:

I met Saʿīd ibn Tawfīl who was accompanied by a man who said to him, 'What position did you assign Hāshim?' Saʿīd replied, 'In the service of the women's quarters, because the emir requested an ugly person.' And the man said to him, 'Among the sons of physicians there would have been one who is ugly, but whose education was good and whose lineage was sound and suitable for the position. If he becomes established, he may revert to his base habits and low-class origins.' At these words, Saʿīd laughed heedlessly. In the event, Hāshim established himself securely in the women's quarters by preparing beneficial potions for fatness, pregnancy, maintaining good complexions, and growing luxuriant hair—so much so that the women preferred Hāshim to Saʿīd.

When the physicians assembled before Ibn Ṭūlūn the next morning, as they had done every day since his illness had worsened, Miʾat Alf, the mother of Abū l-ʿAshāʾir* (Ibn Ṭūlūn's grandson), said: 'The physicians have assembled, but Hāshim is not present, and, by God, my lord, none of them can compare to him.' So she brought Hāshim out and when he came before the emir, Hāshim looked directly at him and said: 'The emir has been neglected, resulting in his present condition. May God not reward the one who is responsible for this matter.'

Ibn Ṭūlūn said to him, 'What is the right course of action, O blessed one?' 'You should take a small dose, in which there is such and such,' Hāshim replied, and he listed nearly one hundred items that have a binding effect at the time they are taken, but are harmful later because they deplete the body's strength. So Ibn Ṭūlūn took the medicine and abandoned what Saʿīd and the other physicians had made for him.

When it caused constipation, Ibn Ṭūlūn was pleased and thought his recovery was complete. Then he said to Hāshim, 'Saʿīd ibn Tawfīl had prohibited me this past month from having even a morsel of porridge though I craved it.' Hāshim said, 'My lord, Saʿīd was mistaken. Porridge is nourishing and has a good effect on you.' So it was brought in a large bowl, most of which he ate. Then he lay down to sleep. The porridge stuck fast stopping the diarrhoea, so he imagined that his condition had improved.

All of this was concealed from Saʿīd ibn Tawfīl. When he appeared, Ibn Ṭūlūn questioned him: 'What have you to say about porridge?' Saʿīd replied: 'It weighs heavily on the organs and the emir's organs need something that will lighten them.' Ibn Ṭūlūn said to him, 'Spare me this foolishness! I have already eaten it and it has proved beneficial to me, God be praised.'

Fruit having arrived from Syria, Ibn Ṭūlūn asked Saʿīd ibn Tawfīl's view on quinces. Saʿīd answered, 'Suck on them on an empty stomach and empty bowels, for then they will be of benefit.' When Saʿīd had left him, Ibn Ṭūlūn ate some quinces, but the quinces encountered the porridge and pushed it and so caused another bout of diarrhoea. Ibn Ṭūlūn summoned Saʿīd and said, 'You son of a whore! You said that quinces would be beneficial to me but the loose bowel movements have returned.' Saʿīd got up to examine the stools and returned to him, saying: 'This porridge, which you said I was wrong to prohibit, remained in your intestines, which, because of their weakness, were unable to digest it until the quince pushed it through. I did not prescribe that you eat the quinces, rather I advised sucking on them.' Then he asked how many quinces he had eaten, to which Ibn Ṭūlūn replied 'two.' Saʿīd said: 'You ate the quinces to satisfy yourself and not as a course of treatment.'

'You son of a whore!' Ibn Tūlūn retorted, 'you sit there making sport of me while you are perfectly healthy and I am seriously ill.' Then he called for whips and gave Saʿīd ibn Tawfīl two hundred lashes and had him led around on a camel, with a crier proclaiming, 'This is the reward of one who was trusted but was disloyal.'

The emir's associates plundered Saʿīd's house, and he died two days later. That was in Egypt in the year in which Ibn Tūlūn died in the month of Dhū l-Qaʿdah.

4. AL-TAMĪMĪ (*d.* after 980)

Al-Tamīmī spent his early years in Jerusalem and its vicinity. He had an excellent knowledge of plants, including their forms and what has been said about them. Al-Tamīmī had extensive experience in the preparation of salves and simple drugs. In addition, he investigated the Great Theriac* and its composition, and he compounded many versions of it with a very sure hand. Later in life he moved to Egypt and resided there until his death, may God have mercy upon him.

He was a private physician to Ibn Tughj,* the ruler of the city of al-Ramlah, and Ibn Tughj was fond of him. The physician had made a number of salves, medicinal perfumes, and fragrant fumigatories for Ibn Tughj to protect him from pestilential diseases, and he recorded the recipes for them in his writings.

Ibn al-Qifṭī relates that, later, al-Tamīmī entered the service of the vizier, Ibn Killis, for whom he composed a large work in a number of volumes, entitled *Material of survival by cleansing corrupted air and precautions against the harm caused by pestilence*. This was during the reign of al-Muʿizz in Cairo.

It is said that al-Tamīmī related the following story concerning his father:

My father, may God be pleased with him, told me that once he got excessively drunk and that this impaired his reasoning abilities to the point that he fell from a considerable height to the floor of an inn where he was staying. He was not conscious, and the innkeeper attended to him and carried him to his own living quarters. When my father awoke and rose, he felt pain and weakness in several parts of his body. He did not know what might have caused the pain, so he rode off, attending to matters until noontime. Then he returned to the inn and said to the innkeeper, 'I have intense pain and weakness in my body and I do not know the reason for it.' 'You should thank God you are still alive,' said the innkeeper. 'Why?' my father asked. 'Don't you know what happened to you yesterday?' said the innkeeper. 'You fell from the highest place in the inn onto the floor while you were drunk.' 'From where?' my father asked, and the innkeeper showed him the place. When he saw it, pain and throbbing began immediately, so much so that he could not endure it. He began to shout and groan until a physician was sent for, who opened a vein and bled him. The physician also bound up his bruised joints with bandages. It was many days before my father recovered and the pain had ceased.

I—Ibn Abī Usaybiʿah—say that there is a similar story. A certain merchant who was on one of his trips in the desert with his fellow-travellers went to sleep at a site which was along the road while his companions were sitting around him. A snake emerged from somewhere and, happening upon the man's foot unexpectedly, bit him and then slithered away. The merchant woke up in terror from the pain, clutched his foot and groaned. One of the others said to him, 'Nothing has happened except that you extended your foot quickly and brushed against a thorn,' and he pretended to remove the thorn, saying, 'No injury remains.' The pain subsided after that, and they all continued on their way. After a while, they returned to that same place and stopped to rest there. His companion said to him 'Do you know what actually caused the pain which befell you in this place? A snake bit your foot and we saw it but did not tell you.' Immediately, an intense pain began in the man's foot and penetrated into his body until it approached his heart and he lost consciousness. Then his condition steadily worsened until he died. The reason for that was that fears and psychological events can have a strong effect on the body. When the man realized that his injury had been a snake bite, he was affected by that idea, and the remaining poison in his body at that spot spread through his body. When it reached his heart, it killed him.

Ibn al-Qifṭī continues:

When al-Tamīmī was in his native city of Jerusalem he undertook the study of medicine and the principles of compounding medicines.

He prepared a theriac which he called the Saviour of Souls. Concerning this, al-Tamīmī says, 'This is a theriac that I prepared in Jerusalem from a limited number of ingredients. It is very effective antidote to the harm of poisons, whether swallowed or inserted into the body through the poisonous bite of vipers and serpents, various kinds of venomous snakes, yellow scorpions and such like, as well as centipedes. It is also a proved antidote to the sting of tarantulas and lizards and has no equal.' He gives its ingredients and the manner of preparing it in his book, *Material of survival*.

Among al-Tamīmī's writings are three works on theriac, including an epistle to his son on the preparation of the Great Theriac, with a warning against drugs used for it erroneously, a description of the correct plants, the times for collecting them, their method of kneading them, and an account of the antidote's usefulness and its proven application.

5. IBN AL-HAYTHAM (Alhazen) (*d. c.*1039–41)

Ibn al-Haytham was originally from Basra, but moved to Egypt, where he remained until the end of his life. He was of an excellent character, highly intelligent, and expert in various branches of learning. None of his contemporaries was his equal in the mathematical sciences. He was constantly absorbed in study, prolific as an author, very ascetic, and dedicated to doing good. He summarized and commented upon many of the works of Aristotle. Similarly, he condensed many of Galen's books on medicine and became an expert in the fundamentals of the art of medicine, as well as its rules and general principles. Even though he did not practise medicine and did not have training in medical treatment, his many compositions were very useful. He also had good penmanship and a sound knowledge of the Arabic language.

Ibn al-Qiftī had the following to say about Ibn al-Haytham:

News of Ibn al-Haytham and his mastery of knowledge reached al-Hākim, the ruler of Egypt, who was himself inclined towards great learning. For that reason, al-Hākim desired to meet him personally. Subsequently, al-Hākim was informed that Ibn al-Haytham had said, 'If I were in Egypt, I would conduct work on the Nile that would result in making it useful at all times, during both its rise and fall, for I have been informed that the Nile flows down from an elevated location that is at the border of Egypt.' This made al-Hākim more eager than ever to meet Ibn al-Haytham, and he secretly sent him a sum of money and urged him to come to Egypt. Accordingly, Ibn al-Haytham travelled to Egypt, and when he arrived there, al-Hākim came out to receive him. They met in a village near Cairo known as al-Khandaq just outside the gates of al-Muʿizz's city* (Cairo), and al-Hākim had lodgings prepared for him and ordered that he should be shown hospitality and respect. Ibn al-Haytham stayed there until he had rested, at which point al-Hākim asked him to proceed with his promised works on the Nile.

Consequently Ibn al-Haytham set out, accompanied by a group of skilled workers specializing in construction who would help him with the engineering works he had in mind. When, however, he had travelled the length of the country and seen the remains of monuments from antiquity from when it had been inhabited by a past civilization*—and these were of the utmost perfection of construction and engineering and included celestial designs, geometric patterns, and marvellous paintings—he realized that what he had intended to do was impossible. For those who had preceded him in ancient times had not lacked the knowledge that he himself possessed, and if the project had been feasible, they would have carried it out. Ibn al-Haytham's scheme was in ruins and his enthusiasm failed. He went to the place known as 'the cataracts', south of the city of Aswan. It is an elevated place from which the waters of the Nile flow down, and here he observed the Nile, studied it, and scouted both banks. As a result, he learned first-hand that his plan could not be executed in a suitable way as he had intended, and he realized the error of what he had promised. Ashamed and prevented from doing what he had intended, Ibn al-Haytham returned and apologized to al-Hākim, who appeared to accept this.

Subsequently, al-Hākim appointed him a government official. Ibn al-Haytham, however, held his post only out of fear and not out of desire, for he realized he had made an error in accepting office since al-Hākim was extremely capricious, spilling blood without cause and for the slightest reason due to some imagined pretext. Ibn al-Haytham pondered the matter, but could not think of a way to get out of his predicament except by displaying madness and disturbance of mind. So he undertook to do just that, and word of it spread abroad. As a result, al-Hākim and his deputies confiscated his property. Al-Hākim assigned a guardian to attend to him and look after his affairs, and Ibn al-Haytham was bound and put under house arrest. This arrangement continued until Ibn al-Haytham learned of the caliph's death, after which he quickly demonstrated his sanity and returned to his former condition. He moved from his home and took up residence in a pavilion near the entrance of al-Azhar Mosque, one of the mosques of Cairo. There he stayed, living the life of an ascetic, content and satisfied. The property that had been confiscated was returned to him, and he occupied himself with composing, copying, and teaching. Ibn al-Haytham had a particularly precise style of handwriting that he turned to account by transcribing a large number of treatises on mathematics.

Ibn al-Qiftī also says that a Jewish scholar told him in Aleppo:

I have heard that Ibn al-Haytham would copy three books a year in his particular field of interest—Euclid, the 'Intermediate books', and the *Almagest*. It would take him one year to copy them. When he had completed the work of transcribing them, someone would come to him and pay him 150 Egyptian dinars for them. This became a set price over which he would not bargain or negotiate, and it was his means of subsistence for the year. He continued to do this until he died in Cairo at the end of 430/September 1039* or shortly thereafter.

I—Ibn Abī Usaybiʿah—have transcribed the following from a treatise in Ibn al-Haytham's own handwriting on the knowledge of the ancients, one that he was composing and working on at the end of 417/January 1027, when he was 63 lunar years of age:

Since my childhood, I have continually reflected on the varying beliefs of people, each group of people holding fast to what they believed. But I was doubtful about all of it and was certain that there was only one truth and that the differences are in the approaches towards the truth. When I had completed my acquisition of intellectual matters, I devoted myself to seeking the origin of truth, and I directed my conjectures and my concentration towards the attainment of that by which the falsities of dubious opinions can be disclosed and the errors of complicated, deluded ideas dispelled. I directed my resolve towards the attainment of the belief nearest to God, His praise be extolled, leading to God's approval and obedience to Him and piety. I was in the same position as Galen who in the seventh book of his treatise *On therapeutic method*, addresses his pupil: 'I don't know how it came about—whether I may say it was by wondrous coincidence, or by inspiration from God, or by sheer madness, or through some other means—that I since my youth have despised ordinary people and had little regard for them and ignored them, preferring truth and the acquisition of knowledge. I am certain in my own mind that people cannot acquire anything better in this world, nothing that would bring one closer to God, than these two matters.'

I thoroughly investigated a variety of opinions and beliefs, and the several religious sciences, but I found no profit in that, nor did I perceive in it a clear path to the truth nor a sure road to indisputable belief. Rather, I came to realize that I would reach the truth only through doctrines whose constituent elements were perceptible and whose forms were rational. These I found only in Aristotle's exposition of logic, physics, and metaphysics, which constitute the essence and nature of philosophy, where Aristotle begins by drawing general, particular, universal, and intrinsic conclusions.

When I—Ibn al-Haytham—realized that, I devoted all my efforts to studying the philosophical disciplines, which comprise three branches of learning: mathematics, physics, and metaphysics. I therefore concentrated on the fundamentals and principles which govern these three fields and their consequences, and I arrived at a good understanding of them in all their depths and heights. Then I considered the nature of mankind—its association with decay and its susceptibility to destruction and depletion, and how in the vigour of youth a person is able to master the required concepts underlying these basic principles more readily, but when that person reaches old age and the time of senility, his constitution becomes inadequate, his strength weakens, his reasoning power fails, and there is deterioration in his ability to do what he had routinely done before.

From these three fundamental subjects (mathematics, physics, metaphysics) I explained in detail, summarized, and condensed in an orderly way what I was able to understand and discern. I have drawn upon their assorted contents to compose works that clarify and reveal the obscurities of these three fundamental domains right up to the present time, which is Dhū l-Hijjah 417/January–February 1027.

As long as I live, I will devote all my energy and all my strength to such endeavours with three aims in mind: first, to benefit the person seeking truth and influence him during my lifetime and after my death; second, as an exercise for myself in these matters to confirm what my reflection on these disciplines has formulated and organized; and, third, to create for myself a treasure-house and provision for the time of old age and period of senility.

In doing this I have followed what Galen says in the seventh book of his treatise *On therapeutic method*: 'In all my writings, it has been and remains my intention to do one of two things: either to benefit someone through something useful and profitable, or to benefit myself through mental exercise, by which I enjoy myself at the time of my writing it and at the same time make a store-house for the time of old age.'

I will explain what I have written concerning these three fundamental subjects so that people will realize my position with regard to the pursuit of truth and in order to disassociate myself from the likes of the foolish, common people, elevating myself to resemble or emulate the pious and the best of mankind.

I—Ibn al-Haytham—have composed twenty-five treatises in the domain of the mathematical sciences, while in the natural sciences and metaphysics, I have composed forty-four books.

Then I—Ibn al-Haytham—attached a list of everything I had composed regarding the knowledge of the ancients to an essay in which I made clear that all worldly as well as religious matters are consequences of the philosophical sciences. This essay was the final one in my numbered discourses on these disciplines, being the seventieth. That total excludes a number of epistles and compositions that have passed from me into the hands of certain people in Basra and al-Ahwāz, for which the originals have been lost, and which I have been prevented from copying by preoccupation with worldly affairs and the disruption of travel, as frequently happens to scholars. A similar misfortune befell Galen, and he even mentions this in one of his books, where he says: 'I had composed many books whose originals I had handed over to a group of my colleagues but work and travel prevented me from copying them until they had become dispersed.'*

If God prolongs my life and gives me time, I will compose, comment upon, and summarize from these disciplines many more ideas that repeatedly occur to me and prompt me to disclose them. But God will do what He wills, and He will decide what He wants, and in His hands is the ultimate power over all things, for He is the initiator and the restorer.

What it is essential to know is that what I composed and abstracted from the learning of the ancients was done with the intention of addressing the most learned wise men and the exemplary intellectuals. It is like the one who says:

Many a dead one has become alive by dint of his knowledge
and many a survivor has already died by dint of his ignorance and error.
Therefore acquire knowledge so that you may live forever
and count survival in ignorance as nothing.

These are two verses by Abū l-Qāsim, son of the famous vizier ʿAlī ibn ʿĪsā al-Jarrāh, may God be pleased with them both, which are inscribed on his tomb.

I—Ibn Abī Usaybiʿah—say that the date of Ibn al-Haytham's writing of this epistle was Dhū l-Hijjah 417/February 1027, and this is the end of his autograph, may God have mercy upon him.

There is a catalogue (*fihrist*), which I found, of the books written by Ibn al-Haytham up to the end of 429/October 1038. This includes ninety-two works.

6. AL-MUBASHSHIR IBN FĀTIK (active 1048)

Al-Mubashshir ibn Fātik was a notable emir of Egypt who was accounted one of its most eminent scholars. He was constantly engrossed in study and was devoted to the pursuit of virtue, to meeting with the people of Egypt, debating with them, and putting to use what he learned from them. Ibn al-Haytham was one of those with whom he associated and from whom he learned a great deal about mathematical astronomy and the mathematical sciences. He was a prolific writer and I—Ibn Abī Usaybiʿah—have found many books in his handwriting concerned with the writings of the ancients. He acquired a very large number of books, most of which still exist, although their pages have become discoloured due to immersion in water.

The shaykh Ibn Raqīqah, related the following account to me, Ibn Abī Usaybiʿah, while in Egypt:

The emir, al-Mubashshir Ibn Fātik, desired to attain knowledge and had a large library. He was constantly occupied only with study and writing, and most of the time, even when engaged in travel, he could not bear to forsake his books, for he believed that they were more important than anything else that he possessed. He had a wife of high status who was also from the ruling class. When he died, may God have mercy upon him, his wife, accompanied by her slave-girls, betook herself to his library. She was resentful of the books because of the time he had spent with them, all the while neglecting her. While lamenting him, she and the servant girls threw the books into a big pool of water in the courtyard. The books were subsequently retrieved from the water, but by then most of them had become waterlogged. And so that is the reason why most of the books of al-Mubashshir ibn Fātik are in their present condition.

Al-Mubashshir ibn Fātik is the author of four works including *Choicest maxims and best sayings*.*

7. IBN RIDWĀN (*d.* 1061)

Ibn Ridwān was born and raised in Egypt where he studied medicine.

In his autobiography, he gives a detailed account of his medical studies and the circumstances in which he came to undertake them. He says in that connection:

It is appropriate for every person to take up the profession most suitable for him. The art of medicine is closest to philosophy with regard to obedience to God, the Mighty and Glorious. The astrological signs at my birth indicated that medicine should be my profession. Moreover, a life of merit is more pleasing to me than any other. I undertook the study of medicine when I was a boy of 15 years, but it will be best to relate to you the entire story:

I was born in Fustat at a locality situated at 30° latitude and 55° longitude.* The ascendant was, according to the table of the astronomer, Yahyā ibn Abī Mansūr, Aries at 5°36', and the mid-point of the tenth house was Capricorn at 5°28'. The positions of the planets were the following: the Sun was in the sign of Aquarius at 5°32', the Moon in Scorpio at 8°15', with a latitude of south 8°17'. Saturn was in Sagittarius at 29°, Jupiter in Capricorn at 5°28', Mars in Aquarius at 21°48', Venus in Sagittarius at 24°20', and Mercury in Aquarius at

19°. The Lot of Fortune* was in Capricorn at 4°5' and its opposite in Cancer at 22°10'. The Dragon's Head* was in Sagittarius at 17°11', the Dragon's Tail in Gemini at 17°41'. Vega* was in Capricorn at 1°22' and Sirius in Cancer at 5°12'.* [See Appendix 5, Figure 6 for a modern diagram illustrating the birth horoscope of Ibn Riḍwān.]

At the age of 6, I began to devote myself to study, and when I was 10, I moved to the capital, Cairo, and concentrated even more on learning. When I reached the age of 14, I undertook the study of medicine and philosophy but since I had no money to support myself, studying was difficult and troublesome for me. Sometimes I made a living by practising the art of astrology, at other times that of medicine, and at yet others by teaching. I continued in that way, exerting all my efforts in learning, until I was 32, at which time I was well-known as a physician. What I then earned from practising medicine was not only sufficient but left me a surplus that has lasted even until the present day, when I am at the end of my fifty-ninth year. With the revenues left after my expenses, I have acquired properties in this city of Cairo which—if God decrees that they remain secure and allows me to attain old age—will yield enough to enable me to subsist.

Since the age of 32 until the present day, I have been keeping a memorandum-book for myself, amending it every year up to this present account as I approach my sixtieth year. My daily exertion in my profession is sufficient exercise for me to maintain the health of my body. After that exertion, I rest, and then eat food selected for the preservation of my health. In my conduct, I do my best to be modest, considerate to others, helpful to the dejected, alert to the anxieties of the unfortunate, and of help to the poor. I make it my purpose in all this to enjoy the satisfaction that comes from good deeds and thoughtful sentiments.

But it is essential as well that this brings monetary profit from which I can earn a living. I spend money on the health of my body and the maintenance of my household, neither squandering money nor being stingy. Rather, I keep to the path of moderation, as is prudent at any time. I inspect the furnishings of my household and whatever is in need of repair, I repair, and whatever is in need of replacement, I replace. In addition, I am responsible for what is required in terms of food, honey, olive oil, and firewood, as well as clothing. Whatever remains after my expenses, I spend on various good purposes such as donations to family, associates, neighbours, and maintenance of the household. Accumulated revenue from my properties I set aside for their maintenance and further investment in them, and for times of need. When contemplating a new enterprise in commerce, building, or something else, I assign the matter great importance and, if the project seems likely to be successful, I promptly allocate the appropriate amount, but if it seems unlikely to be successful, I reject it. I inform myself as much as I can regarding worthy enterprises and make the necessary arrangements.

I make certain my clothes are decorated with marks of distinction and are clean. I also use a pleasant perfume. I am quiet, and hold my tongue regarding people's failings. I try not to speak except when appropriate and take care to avoid swearing and criticizing others' opinions. I avoid pridefulness and love of superiority. I reject worries of greed and dejection. If adversity befalls me, I rely on God, exalted be He, and face it reasonably, without cowardice or rashness. When I transact business with anyone, I settle the account without giving or raising credit unless compelled to do so. If someone requests a loan from me, I give it to him, but I do not refuse it if he repays it. My leisure time after having finished my work, I spend in the worship of God, may He be glorified, by focusing upon the contemplation of «*the government of the Heavens and the Earth*»* and praising Him who wisely rules. I have studied Aristotle's treatise *On management** and I aspire to adhere to its prescriptions from morning to night.

During times of solitude I review my actions and sentiments during the day, and I am pleased with what was good or proper or beneficial but distressed by anything that was bad, shameful, or harmful, and I promise myself not to repeat it.

As for my personal amusements, I make my main recreation reflecting upon God, mighty and glorious is He, and praising Him by contemplating «*the government of the Heavens and the Earth*». Men of learning and the ancients have written a great number of works concerning those things. From among them, I prefer to concentrate on five books of *belles-lettres*, ten books on Sharia, the books of Hippocrates and Galen on the art of medicine, four books on agriculture and pharmacology, some technical books, and some books by savants. Other books I either sell at any price I can get or I store in cases, but selling them is better than storing them.

I—Ibn Abī Usaybiʿah—say that the above is the whole of what he relates in his autobiography.

His birthplace was Giza in Egypt, but he grew up in Cairo, where his father was a baker. Ibn Ridwān's house was in the Qasr al-Shamʿ quarter of Fustat, and the house is known to this day* by his name, even though the house itself is nearly gone, with only a small remnant of it left. During Ibn Ridwān's lifetime Egypt was stricken with a great shortage of commodities, most of the inhabitants perished, and many of the survivors fled the country.*

I—Ibn Abī Usaybiʿah—have transcribed from a manuscript in the writing of Ibn Butlān* that this shortage of supplies in Egypt occurred in 445/1053. He says: 'In the following year, the level of the Nile fell and the shortage of food increased, followed by a great pestilence that reached its peak by 447/1055. It was reported that the ruler* supplied 80,000 shrouds for the dead from his own purse and that he lost 800 military commanders. The ruler, however, acquired considerable revenue from the estates of those who died without heirs.'

A copyist from Malaga has related to me—Ibn Abī Usaybiʿah—that the mind of Ibn Ridwān became deranged towards the end of his life. The reason for this change occurred during the period of the famine when he adopted an orphan-girl whom he raised in his house. One day he left her by herself in the house where he had accumulated valuables and gold worth about 20,000 dinars. She took all of it and fled. She was never heard from again, and Ibn Ridwān was not successful in finding out where she had gone. From then on, his mental faculties deteriorated.

I—Ibn Abī Usaybiʿah—say that Ibn Ridwān was often inclined to polemics against his contemporaries, both physicians and others, and many of his predecessors as well. Ibn Ridwān did not have a teacher in the medical arts whom he followed.* On this issue he composed a book in which he argues that learning medicine from books is more satisfactory than learning it from teachers. Ibn Butlān refuted this and other opinions of Ibn Ridwān's in a monograph.

Among the sayings of Ibn Ridwān that I—Ibn Abī Usaybiʿah—have transcribed from a manuscript in his own handwriting is the following:

The physician, according to the opinion of Hippocrates, should be one who encompasses seven qualities:*

1. He should have an excellent moral character, a healthy body, a keen intelligence, and a pleasing appearance, while also being prudent, possessing an excellent memory and having a good disposition.
2. He should be well-dressed, pleasant smelling, and clean in his body and attire.
3. He should respect the confidences of patients and not divulge anything about their illnesses.
4. His desire to cure those who are ill should be greater than his desire for any payment he might request. His desire to treat the poor should be greater than his desire to treat the rich.
5. He should be eager to teach and do his utmost to benefit the people.
6. He should be sound of heart,* modest in appearance and truthful in speech, while paying no attention to anything regarding the women or the wealth that he sees in the houses of the upper classes, let alone meddling with any of them.
7. He should be trustworthy with regard to people's lives and property, neither prescribing any lethal medicine nor giving instructions regarding one, nor any drug causing abortion, and he should provide treatment for his enemy in no less correct a manner as he would treat his friend.

Ibn Ridwān composed over one hundred books including the following:

The useful book on the method of medical learning, * in three parts.

On the fact that Galen did not make a mistake in his statements regarding milk, as some people thought.

Letter providing answers to questions concerning tumours that were put to the author by a shaykh.

Letter concerning the treatment of a boy who had the disease known as elephantiasis and leprosy.

On the refutation of the treatise by Ibn Butlān on the hen and the pullet.

On the mouse.

On the confusions put forward by Ibn Butlān.

On the fact that what Ibn Butlān is ignorant of is certainty and wisdom and that which he is informed about is error and sophistry.

On the fact that Ibn Butlān does not understand what he himself says, much less what others say.

On drawing attention to the senseless jabber in the statements of Ibn Butlān.

On the prevention of harm from sweets to feverish persons.

8. SALĀMAH IBN RAHMŪN (active *c.*1117)

Salāmah ibn Rahmūn was a Jewish physician who was one of the distinguished citizens of Egypt. He was known not only for his excellent achievements in the art of medicine, but also for his knowledge of the writings of Galen and his investigation of their obscure passages. Ibn Rahmūn was also an outstanding scholar in the domains of logic and the philosophical disciplines, and composed a number of works in those fields.

Abū l-Salt Umayyah mentions Salāmah ibn Rahmūn in his 'Egyptian epistle'. Speaking of the physicians whom he met in Egypt [in 1117], he says:

The most characteristic of those whom I saw and would count amongst the physicians was a Jew by the name of Salāmah ibn Rahmūn. He was acquainted with al-Mubashshir ibn Fātik and learned from him something of the art of logic, in which he became particularly distinguished. Subsequently, he endeavoured to teach all the works on logic, natural philosophy, and metaphysics, but he explained them as he saw fit, interpreting and summarizing them with no evidence of having grasped the finer points of knowledge. Rather, he talked excessively and so committed errors; his answers were hurried and frequently erroneous.

Upon my first encounter with him, I asked him about issues that had arisen in the course of his discussion—things that could be understood even by one who did not have any great breadth of knowledge in the field. He answered in a way that exposed his shortcomings, articulated his incompetence, and made clear his lack of imagination and understanding. In his great pretensions and his incompetence at even the easiest of his undertakings, he can be compared to the man described by the poet:*

> He tucks up his robe to wade into deep waters
> but the waves engulf him on the shore.

Abū l-Salt continues:

There was in Cairo a physician from Antioch called Jirjis, who was nicknamed 'The Philosopher' in the way you might call a raven 'Whitey'. Jirjis went out of his way to make fun of Salāmah Ibn Rahmūn and mock him. He used to compose fake medical and philosophical essays, written in the style of language of the common people, which were meaningless and of no use whatsoever. Then Jirjis would see to it that they were placed in the hands of someone who would ask Salāmah about their meaning, whereupon Salāmah would discourse at length about them and explain them anyhow, without caution or care, but rather with haste and indifference to content. Some amusing compositions by him are extant.

The following lines by this Jirjis have been recited to me—Abū l-Salt. This is one of the best lampoons on an ill-starred physician that I have ever heard, and I imagine the lines are indeed by him:

> In Abū l-Khayr's* scales, with his ignorance,
> a virtuous man weighs but lightly.
> His poor patient, through his evil omen,
> is in a sea of perdition without shore.
> Three things enter at the same time:
> his face, a bier, and the man who washes the corpse.

Salāmah ibn Rahmūn is the author of four works including: *On the reason why women in Egypt grow fat when past their youth*.

9. IBN JUMAYʿ (*d. c.*1198)

Ibn Jumayʿ was not only a famous physician and well-known scholar, but also a personage of distinction. Ibn Jumayʿ worked hard at the art of medicine, becoming skilled at treatment and composing works on the subject. He was born and raised in Fustat. He served Saladin, enjoying the ruler's favour and holding an influential position of the highest level during his reign. It was for Saladin, who consulted him in medical matters, that Ibn Jumayʿ compounded the Great Theriac.

I—Ibn Abī Usaybiʿah—say that Ibn Jumayʿ paid particular attention to Arabic language usage and accuracy of expression. He would never give a lecture without having al-Jawharī's Arabic dictionary, *al-Sihāh*, near to hand, and whenever he came across a word which he was not sure he understood, he would look it up, relying on what al-Jawharī said concerning it.

A man from Egypt related the following account to me:

One day, while Ibn Jumayʿ was sitting in his shop* near the candle market in Fustat, a funeral procession passed by him. When he saw the procession, he called out to the mourners, telling them that their beloved was not dead and that if they interred him, they would be burying him alive. They stood looking at him in surprise at his words, unable to believe what he had said. Then one of them said, 'What harm will it do us to test what he says. If it is true, it is what we want. If it is not true, it makes no difference.' They summoned him over and said, 'Prove what you just said to us.' Ibn Jumayʿ instructed them to return home and remove the shroud from the deceased. There he said, 'Carry him to the bathhouse.' He then poured hot water on the body to warm it up, bathed it with warm compresses, and immersed it in water until a small amount of sensation was perceptible and the man moved slightly. 'Rejoice at his return to life!' cried Ibn Jumayʿ. He then continued treating the person until he regained consciousness and felt well. This was the beginning of Ibn Jumayʿ's fame for excellence and knowledge in the medical art, for it seemed that he had performed a miracle. Afterwards, he was asked how he had known that the body being carried covered in shrouds was still alive. 'I looked at his feet,' he replied, 'and saw that they were upright. The feet of those who have died are splayed out. So I surmised that he was alive, and my guess was correct.'

I—Ibn Abī Usaybiʿah—say that there was in Egypt, a famous poet who had a malicious tongue. He composed many satirical poems about Ibn Jumayʿ. The following is one of those that have been recited to me:

> You lied, misspelling, when you claimed
> your father was Jumayʿ the Jew.
> Jumayʿ the Jew (*Jumayʿ al-Yahūdī*) is not your father,
> but your father is 'all the Jews'* (*jamīʿ al-yahūdī*).

I have copied a poem from a manuscript with a eulogy for Ibn Jumayʿ and this is part of it:

He rests between the stones of the earth, and through him
the congregation has become redolent with sweet fragrance.
He had an open face, pure and joyful, smiling,
not gruff of character or like a sullen person.
I used to eulogize him, honouring him,
and now I elegize him, as much as a deprived one can.

Ibn Jumayʿ's eight writings include treatises on the lemon, its syrups and beneficial uses, on rhubarb and its medicinal uses, and on the curvature of the spine.

10. IBN SHŪʿAH (*d.* 1183)

Ibn Shūʿah numbered amongst the notable scholars and very eminent physicians. He was an Israelite, famous for his mastery of the art and knowledge in medicine, eye medicaments, and wound healing. He was good-natured, light-hearted, and given to joking, and he used to compose poetry and play the musical stringed instrument called a *qīthārah*.* Ibn Shūʿah served Saladin in medicine when the latter was in Egypt, and the ruler held him in high esteem.

There was a Sufi* jurist in Damascus, with the nickname al-Najm (the Star). He was sombre in nature, abstemious in his mode of life, and rigid with regard to religion, following the letter of the law. He was acquainted with the father and brother of Saladin.

When Saladin's uncle, Asʿad al-Dīn [Shīrkūh], travelled to Egypt, al-Najm followed him and took up quarters in a mosque. He was sharp-tongued and would slander the inhabitants of the palace, for his way of glorifying God was to insult them. Whenever he saw a non-Muslim riding a mount, he sought to kill him, and for that reason they used to avoid him. One day al-Najm saw Ibn Shūʿah riding and threw a stone at him, which hit his eye and knocked it out.

Among the poetry of Ibn Shūʿah are the following lines composed after al-Najm had knocked out his eye:

Be not amazed that the eyes are dimmed by
the sun's rays: that is a well-known thing.
Rather, be amazed that my eye was blinded because I looked
at al-Najm (the Star), whereas he is a slight, obscure person.

11. MAIMONIDES (AL-RAʾĪS MŪSĀ) (d. 1204)

Maimonides, or al-Raʾīs Mūsā ('The Master' Mūsā or Moses) was a Jew, learned in the customs of the Jewish people and numbered amongst their religious authorities and most distinguished scholars. In Egypt, he was the head of their community, for he was unique in his time in the art of medicine and its practice, versatile in many disciplines and well-versed in philosophy. Sultan Saladin became aware of him and sought his medical advice, as did Saladin's son, al-Malik al-Afdal.

It has been said that Maimonides had converted to Islam while in the Maghrib, memorized the Qur'an, and studied Islamic jurisprudence. Then, when he went to Egypt, he took up residence in Fustat where he reverted back to his former faith.*

The following verses were written in praise of Maimonides:

I see that Galen's medicine is for the body only,
but Maimonides' medicine is for mind and body.
If he were to treat Time with his medical knowledge
he would cure it of ignorance with knowledge.
And if the full moon were to seek his medical advice
the fullness it claims would be fulfilled,
And he would treat, on the day of its fullness, its spots
and cure it, on the day of its invisibility, of its sickness.

12. IBRĀHĪM, THE SON OF MAIMONIDES (*d.* between 1234 and 1241)

Ibrāhīm, the son of Maimonides, was born and raised in Fuṣṭat. He was a well-known physician in the service of al-Malik al-Kāmil, but also from the palace made frequent visits to the hospital in Cairo, where he treated the ill.

I—Ibn Abī Usaybiʿah—met him in Cairo in 631 or 632/1233 or 1234 while I was practising in the hospital* there at that time. I found him to be a tall elderly man, slim of build, charming in company, witty in conversation, and distinguished in medicine. He died in Egypt in the 630s/1232–41.

13. IBN ABĪ L-BAYĀN (d. 1236)

Ibn Abī l-Bayān was a Karaite* Israelite born in Cairo in 556/1160. He was a recognized master in the medical art. Whenever we treated the ill at the Nāsirī hospital in Cairo, I—Ibn Abī Usaybiʿah—witnessed the excellence of his achievements with regard to the knowledge and identification of diseases, the recollection of drug therapies, and his acquaintance with what Galen said concerning them—all of which defies description.

Among the physicians of his time, Ibn Abī l-Bayān was the most capable in compounding drugs and knowing the quantities and weights that are appropriate. So much so that when patients with rare illnesses came to consult him, he would dictate on the spot, in accordance with just what that patient required, a prescription for compound drugs such as pastilles, medicinal powders, syrups, and so forth.

I found that someone said about him:

If an illness is complicated, internally,
Ibn al-Bayān will come with a clear exposition (*bayān*) of it.
So if you are desirous of good health,
then take from him immunity against your disease.

Ibn Abī l-Bayān lived over eighty years, but towards the end of his life his eyesight became weak.

He is the author of two works including a medical formulary of the compound drugs generally used in the hospitals and pharmacies of Egypt, Syria, and Iraq, in twelve chapters. I—Ibn Abī Usaybiʿah—read it under his guidance and corrected my copy with his help.

14. ABŪ SULAYMĀN DĀWŪD (active *c.* 1184)

Abū Sulaymān Dāwūd was a Christian physician in Egypt in the days of the caliphs, who held him in high regard. He excelled in the medical art, being experienced in both its theory and practice. Abū Sulaymān Dāwūd was a native of Jerusalem, subsequently moving to Egypt where he gained extensive knowledge of astrology.

It so happened that Abū Sulaymān Dāwūd learned through astrological readings that Saladin would conquer Jerusalem on a certain day of a certain month in a certain year and that he would enter the city through the Rahmah Gate. Therefore Abū Sulaymān spoke with one of his five sons*—that is, the knight, Abū l-Khayr—for this child had been brought up alongside the king of Jerusalem's leprous son, Prince Baldwin [IV of Jerusalem], and had taught him horsemanship, so that when Baldwin was crowned king, he made Abū l-Khayr a knight.

Unlike his four brothers, who were physicians, Abū l-Khayr became a soldier. Thus, Abū Sulaymān Dāwūd told this son to go as his messenger to Saladin and to report the good news to him that he would be king of Jerusalem at the designated time. Abū l-Khayr obeyed his father's order and set out on his way to Saladin. He reached him on the first day of 580/14 April 1184, while the people were busy celebrating. So Abū l-Khayr went directly to the jurist, ʿĪsā,* who was very pleased to see him, and then went with him to Saladin and delivered the message from his father. The sultan was very pleased indeed and bestowed upon him a splendid reward and gave him a yellow banner and arrow of that colour, saying to him, 'When God enables me to do what you have said, put this yellow banner and arrow above your house and the quarter in which you live will be completely safe because of the protection given your house.'

In the fullness of time, all of what Abū Sulaymān Dāwūd had predicted came true, for the jurist ʿĪsā entered the house in which Abū l-Khayr was living in order to protect it. No one in Jerusalem was spared from imprisonment, killing, or levying of taxes, except the house of Abū Sulaymān Dāwūd. Saladin doubled the amount that Abū Sulaymān's sons had been receiving from the Franks.* Then he wrote a decree valid throughout the entire realm, both land and sea, exempting them from all the regulations applied to Christians, and they have been free from them to this day.

I—Ibn Abī Usaybiʿah—say that Saladin conquered Jerusalem on 27 Rajab 583/2 October 1187.

15. ABŪ HULAYQAH (*d.* after 1250)

Son of Abū l-Khayr and grandson of Abū Sulaymān Dāwūd, illustrious and learned, the physician known as Abū Hulayqah ('the one with the ring'), is peerless in his time in the medical art and the philosophical disciplines, versatile in various fields, excellent at methods of treatment, and careful in administering drug remedies. He is also compassionate with the sick, desirous of doing good deeds, and diligent in matters of the faith to which he adheres.*

I—Ibn Abī Usaybiʿah—have met him many times and observed his excellence in methods of treatment, the pleasant nature of his company and the perfection of his character, which are beyond description.

Abū Hulayqah was born in the fortress of Jaʿbar in 591/1195 and once, it happened that when al-Malik al-Kāmil entered the bath, his father, Abū l-Khayr, gave his son some fruit and rosewater and ordered him to take them to the sultan. He took them, and when al-Malik al-Kāmil came out from the bath, Abū Hulayqah presented the gifts to him. Al-Malik al-Kāmil received them and went into the treasure house, where he emptied the plates of fruit, filled them with bolts of brilliant fabric, and sent the tray to Abū Hulayqah's father. Al-Malik al-Kāmil then took the boy, who was at that time about 8 years old,* by the hand and escorted him to his father, al-Malik al-ʿĀdil I. When al-Malik al-ʿĀdil saw the boy, whom he had never seen before, he said to his son, al-Malik al-Kāmil, noticing the resemblance, 'Is this the son of Abū l-Khayr? Bring him to me!' Abū Hulayqah was placed in front of al-ʿĀdil, who clasped him by his hand and spoke with him a long time. Then al-Malik al-ʿĀdil turned to the boy's father, the knight Abū l-Khayr, who was standing in service along with the attendants, and said to him, 'This son of yours is an intelligent boy. Do not teach him the military craft for I have many soldiers. Yours is a family blessed by God, and I have benefited from your medicine. Send him to his uncle in Damascus to study medicine.'

Abū Hulayqah was not his name until after Sultan al-Malik al-Kāmil made a remark one day, when the future Abū Hulayqah was sitting with physicians at the gate and the sultan ordered his servant to look for him. The servant asked him, 'O Sir, which physician do you mean?' and the sultan replied, 'The one with the (ear)ring,' after which Abū Hulayqah became known by this name from that day on.

One of the accomplishments of Abū Hulayqah was that he thoroughly understood al-Malik al-Kāmil's pulse. So much so that one day the sultan hid himself behind the curtain along with the women patients of his household. Abū Hulayqah took the pulse of each person and gave prescriptions for them, but when he reached the pulse of the sultan, he knew it was him, and so he said: 'This is the pulse of our master, the sultan, and it is healthy, God be praised.' Al-Malik al-Kāmil was greatly amazed and his esteem for Abū Hulayqah rose even further.

Another story about Abū Hulayqah:

Al-Malik al-Kāmil had a muezzin, Jaʿfar, who had a stone that blocked his urethra from which he suffered intensely to the point that he was on the verge of death. He wrote to al-Malik al-Kāmil and informed him of his condition and requested an authorization to go home to be treated. When he arrived at his house there were present all the great physicians of the day. Each of them prescribed something for him, but nothing was of any benefit. So, al-Malik al-Kāmil summoned Abū Hulayqah, who gave Jaʿfar a dose of his theriac. In the amount of time it took for it to reach his stomach, its potency penetrated to the place of the stone and broke it up so that it came out when he urinated, stained from the colouring of the medication. The patient recovered immediately and returned to his service of the sultan and called for the noon prayer. At this time, the sultan was encamped at Giza, Cairo, and when he heard the muezzin's voice he ordered him brought to him. When Jaʿfar arrived, al-Malik al-Kāmil said to him, 'What about the letter that you sent to me yesterday, in which you said you were near death? Tell me what is wrong with you.' The muezzin said, 'O Master, it would have been so had it not been for my lord's physician, Abū Hulayqah, who gave me a theriac that cured me immediately.'

Another story about Abū Hulayqah:

One day, the sultan requested that he prepare for him a sauce to be eaten with a meat stew while on his journeys. The sultan suggested to him that it should strengthen his stomach, stimulate his appetite, and in addition act as a laxative. So Abū Hulayqah prepared a sauce according to the following recipe:

Take one part parsley and half a part each of lemon balm and the pulp of fresh citron, both of which have been soaked in water and salt for several days and then soaked in fresh water. In a brewer's mortar, pound each of them separately until it becomes like an ointment. Then mix all of them in the mortar and squeeze the juice of a ripe green lemon over it and sprinkle a good amount of Andarānī salt* on it for seasoning. Then, put the mixture into small jars, each one holding the amount of a serving, filling the container full because if there is any space left it will spoil. Seal those containers with good oil and store them away.

When the sultan partook of the sauce, he got the desired results, and he praised it greatly. One day, the sultan was travelling to Asia Minor, and he said to Abū Hulayqah, 'Will this sauce keep for a long time? Does it last even a month?' Abū Hulayqah answered, 'Yes, if it is prepared in the way I indicated.' The sultan said, 'Every month make me some in a quantity sufficient to last the month and send it to me at the beginning of every (lunar) month.' Abū Hulayqah continued producing the sauce each month and sending it to him along the difficult mountain passes of Asia Minor. The sultan never failed to use it while travelling and praised it profusely.

I—Ibn Abī Usaybiʿah—say that all of the relatives of Abū Hulayqah are known in Egypt and Syria as the Banū Shākir, due to the fame of his uncle, Abū Shākir,* and his good reputation. All those related to him became known as Banū Shākir, even if they were not his sons. When I—Ibn Abī Usaybiʿah—met the scholar Abū Hulayqah, he had heard that I mentioned the well-known physicians of his family and that I had described their learning and achievements. He was kind and thanked me and I recited to him, extemporizing:

> And why should I not thank those whose excellence
> has travelled east and west?
> In the heaven of lofty deeds there shine from them
> auspicious stars that have never set:
> People whose status among mankind
> ranks as high as a planet.
> They have written so many books on medicine,
> containing every original, amazing idea.
> My gratefulness (*shukrī*) to the Banū Shākir
> will not cease to be with those far and near;
> I have immortalized a lasting glory among them,
> by fine description and friendly praise.

As for the reason of the ring (*hulayqah*) that was attached to the ear of Abū Hulayqah and from which he got his nickname, it is as follows: His father had no male children who lived, except for Abū Hulayqah. When his mother was pregnant with him, his father was advised to have made a silver ring and to give the value of its silver as alms. Abū l-Khayr had arranged to have a jeweller present at the hour in which his child was born, in order to pierce the child's ear and put the silver ring in it. He did so, and God gave the child life. His mother made a pact with her son that he was not to remove the ring, and so it remained. Then Abū Hulayqah married and had a number of male children who died, as had happened in his own family. He thought of the ring and he had one made for his eldest son, Abū Saʿīd Muhammad.

The poetry of Abū Hulayqah includes this poem that he recited to me—Ibn Abī Usaybiʿah—himself, and in the presence of Saladin's brother:

The beloved granted a meeting on a night
when the chaperone was heedless and slept, unaware,
In a meadow that, but for its transience, would resemble
the gardens of Eden in all its attributes.
The birds were singing amid the branches with their voices
and the wine was unveiled in the cups of those who poured it,
While my companion was the bright Moon, in which
the senses were revelling, in all their senses.

When Abū Hulayqah was in Damietta, his father became ill in Cairo. When a letter arrived telling him his father had recovered, he composed the following:

The clouds of bliss rained upon me
since the misery you complained of ceased.
Since I saw your handwriting I have been clothed in happiness;
so what can I do to be duly grateful?

16. ABŪ SAʿĪD MUHAMMAD, SON OF ABŪ HULAYQAH (active in 1269)

Abū Saʿīd Muhammad is unique amongst scholars and the most perfect physician. Born in Cairo in 620/1223, he was given the name Muhammad when he became a Muslim during the reign of Baybars I. God bestowed upon him the most perfect intellect, the best manners, abundant intelligence, and vast knowledge. Abū Saʿīd Muhammad has mastered the medical art so thoroughly and has come to know the philosophical disciplines to such an extent that no one comes close to him in regard to anything he undertakes. Nor do they possess the fine characteristics that are combined in him, including courteous speech and the generous giving of alms to friends and relations, near and far.

I—Ibn Abī Usaybiʿah—received a letter from him while he was in Baybars' military encampment in Shawwāl 667/June 1269.* The letter displayed brilliant refinement, abundant knowledge, Asmaʿian insight,* an Akhzamite nature,* great affection, and immense charity. In the letter, he states that he found in Cairo a copy of this book that I have written on the classes of physicians and that he had acquired it for his library. He exaggerated in the characterization of the book, which shows the generosity of his nature and the nobleness of his origins. At the beginning of his letter there was the following line:

> And I am a man who loves you because of good qualities
> I heard: the ear, like the eye, can fall in love.*

I replied to him in writing with a poem in the same metre and rhyme, the first lines of which are:

> A letter reached me, pleasing with its inscription,
> containing thoughts that shine like the sun,
> The letter of a noble, generous man who is to be lauded,
> with a bright countenance, its light sparkling.

Abū Saʿīd Muhammad has two brothers. One is prominent in the art of ophthalmology and, before reaching the age of 20, had composed a book on eye medicaments for the sultan, al-Malik al-Sālih Najm al-Dīn Ayyūb. The other brother, who is the younger, is accounted a scholar distinguished in the art of medicine.

17. ASʿAD AL-DĪN (*d.* 1237)

Asʿad al-Dīn was one of the most eminent scholars, prominent amongst the erudite, with a keen intellect and great dedication to learning.

I—Ibn Abī Usaybiʿah—first met him in Damascus at the beginning of Rajab 630/April 1233 and found him to be an elderly man of handsome appearance, with attractive grey hair, a perfect build, and a light-brown complexion, charming in conversation and well endowed with honourable virtues. Subsequently I met him again in Egypt where he received me kindly and warmly. He had been a friend of my father's for many years. The death of Asʿad al-Dīn was in Cairo in 635/1237.

CHAPTER 15

FAMOUS SYRIAN PHYSICIANS

This final, longest chapter lists the physicians of al-Shām, the old Arabic name for Syria, loosely corresponding to Greater Syria or what used to be called the Levant. Thirteen of these physicians were known personally to Ibn Abī Usaybiʿah. including his paternal uncle, Rashīd al-Dīn ʿAlī ibn Khalīfah and one of his teachers, known as al-Dakhwār. Their entries are replete with first-hand oral accounts and this is the chapter most revealing of Ibn Abī Usaybiʿah's own life.

Though always flattering and complimentary in his initial introduction, even of people he later criticizes, Ibn Abī Usaybiʿah seems particularly effusive in his use of superlatives for people alive at the time of writing, such as ʿIzz al-Dīn ibn al-Suwaydī—who presumably might read what had been written.

The fourth autobiography included by Ibn Abī Usaybiʿah is here, that of ʿAbd al-Laṭīf al-Baghdādī, a particularly interesting figure, with an impressive talent for memorization, even allowing for possible exaggeration. He was a polymath with an encyclopaedic oeuvre and an independent mind, not afraid to attack colleagues or great authorities and was also a teacher of Ibn Abī Usaybiʿah's father and uncle.

Two further mentions of Ibn Abī Usaybiʿah's book are made, most memorably with the judge, Rafīʿ al-Dīn al-Jīlī, who looks through a copy and sees he is not included, though this is obviously rectified in a later revision. He is also the 'courier' who takes the commissioned, and second-mentioned, copy to Amīn al-Dawlah in 1245.

Strangely absent from this chapter is an entry on Ibn Abī Usaybiʿah's contemporary, the Damascene physician Ibn al-Nafīs (d. 1288), who worked mostly in Cairo and is famous for having posited by logical deduction, rather than by anatomical observation, the movement of blood in the lungs. Whether some offence, rivalry, or simple oversight of* *a younger colleague in a different city was at the bottom of this absence, one can only speculate.*

Though poetry is found in profusion throughout the later chapters, two particularly long and amusing poems by the same author are found in this chapter. The first one describes the various costs of inviting over drinking companions, surprisingly unchanged with time, while the second is a jesting rant on the annoyance of having shoes made that do not fit. Also in this chapter, as well as the previous, are examples of the author's own poetic skills, while the final poem is, fittingly, a poem of praise for Ibn Abī Usaybiʿah's book.

1. AL-YABRŪDĪ (*d. c.*1058)

Al-Yabrūdī was a Jacobite Christian.* He excelled in the art of medicine, being thoroughly acquainted both with its theoretical basis and with its practical application, and was considered one of the most respected and outstanding representatives of that art. He was always busy working, was very fond of studying, and held virtue in high esteem.

Another physician, a Christian from Damascus, told me that al-Yabrūdī was born and spent the first years of his life in Yabrūd, a large village where many Christians live. In that village, al-Yabrūdī, like the other Christian inhabitants, engaged in agricultural work. He also collected wormwood* in an outlying district of Damascus near to his home, loaded it on a pack animal, and brought it to the city, where he sold it to be used as fuel for heating ovens and other such purposes. One day, as he was coming in with a load of wormwood, he saw a person whose nose was bleeding profusely being bled by an elderly doctor on the other side of his body, the side opposite the place from which the blood was escaping. He stopped and watched the doctor and then asked him: 'Why are you bleeding this person, when the quantity of blood escaping from his nose is more than sufficient?' The doctor replied that he was doing so in order to staunch the flow of blood from the nose by drawing the blood to the side of body opposite the place from which the blood was escaping. 'Ah?' said al-Yabrūdī: 'Where I come from, when we wish to divert a stream, it is our practice to dig an outlet in a new direction, but one that is not directly opposite to that of the old bed. The water then ceases to flow in the old bed and passes into the new one. Why not adopt a similar procedure and bleed from the other side?' The elderly doctor did so, and the man's nosebleed stopped. Seeing from al-Yabrūdī's question that he was keen of understanding, the doctor said, 'if you devote yourself to the art of medicine, you will become a good physician.'

A story similar to the preceding one, although not quite the same, is attributed to my wise teacher, al-Dakhwār:

I have heard that there lived in Damascus a bloodletter who was not accounted one of the most skilful practitioners of his trade. It once happened that when bleeding a young man, he cut the artery. He became confused and panicky; he attempted to staunch the blood, but was unable to do so. As a crowd gathered, a young boy appeared at his side and said: 'Uncle, bleed him at the other arm.' Grateful for any advice, the operator bled his patient's other arm. The boy then said, 'Bind up the first incision,' and he did so. When he tightened the bandage, the flow of blood stopped. He then closed the other incision, whereupon the flow of blood was checked and finally ceased altogether.

Some time later, the bloodletter saw the same youth driving a pack animal with a load of wormwood. The bloodletter stopped him and said, 'How did you know what to do?' 'I have sometimes seen my father irrigating his vineyard,' said the youth, 'when a breach opens in an irrigation channel and the water goes gushing out. My father is able to stop it only if he makes another opening that will reduce the volume of water pouring out through the breach. Only then can he close the breach.' At this, the surgeon told him to give up selling wormwood, took him under his wing and taught him the art of medicine. Thanks to this incident, al-Yabrūdī became one of the most celebrated and erudite physicians of his time.

I—Ibn Abī Usaybiʿah—say: al-Yabrūdī corresponded regularly with Ibn Ridwān of Cairo and other Egyptian physicians, asking them various questions on medical matters and engaging them in discussions. He copied a very large number of medical books personally, including in particular the books of Galen, commentaries and compilations of them. Moreover, I have heard that one day al-Yabrūdī was crossing a market in Damascus when he saw a person undertake to eat several *ratl*s of boiled horsemeat, of the quality that is sold in the markets, for a bet. As al-Yabrūdī watched, this person ate far too much, overloading his stomach, and then drank a lot of beer and ice-water, causing his condition to become severely perturbed. Al-Yabrūdī then realized that the man would soon lose consciousness, and if left, he would be in danger of death. He therefore followed the man to his house to see how his condition would develop. A very short time later, his family began to weep and wail, for they thought that he had died. Al-Yabrūdī went to them and said, 'I shall cure him. There is nothing wrong with him.' Then he brought him to a nearby bathhouse, gently prised his jaws open, and poured some boiled water containing a mild emetic down his throat. This brought on moderate vomiting. Al-Yabrūdī then gave him supportive treatment, until he regained consciousness and was restored to health. The family were astonished at what al-Yabrūdī had done and praised the wonderful way in which he came to the man's rescue. This affair became well known and did much to establish his fame. I—Ibn Abī Usaybiʿah—say: this story indicates that al-Yabrūdī, by studying the man's condition and observing what happened to him, had read his symptoms accurately and realized that he could save him if he could treat him in time.

A similar story is related by a Persian physician:

One day I saw a man making a bet with someone that he could eat a certain quantity of carrots. I stayed and watched to see what would happen to him, not because it was my wish to have social intercourse with people of that kind, God no! But I wanted to see what would happen if a lot of food was forced into his stomach. He ate his carrots while sitting on a wall so that he could see everyone standing around him and was able to jest with them.

When he had eaten the greater part of them, I observed that the masticated carrots were coming back into his throat in the form of a stringy, pulpy mass impregnated with saliva. His eyeballs protruded, his breathing stopped, his face turned red, his jugular veins and the veins of his head became engorged with blood, and then his face darkened and turned ashen. He retched more than he vomited, but finally threw up much of what he had eaten.

I understood from this that his breathing had stopped because the stomach was pressing the diaphragm towards the mouth* and preventing it from returning to its state of expansion for respiration. As to the fact that his colour reddened and his jugular and other veins became engorged with blood, I presumed that this was caused by the natural flow of blood towards the head, as happens to someone whose arm is bandaged for bleeding. In the latter case, the natural flow goes in the direction in which it is stimulated to go. As to the fact that his face subsequently darkened and turned ashen, I must presume that the cause of it was the poor temperament of his heart. If he had not vomited as much as he did, if the stomach had continued to press on the diaphragm so that he was prevented from breathing altogether, he would have died of asphyxia, as we have seen in many who have died as a result of vomiting. As to the fact that he retched more than he vomited, I understood that the retching was caused by the severity of the disturbance of the stomach.

It thus appears that when food enters the stomach in large quantities, it causes the stomach to stretch and all its folds to expand, as I once saw when dissecting a beast of prey* live in the presence of the emir. One of those who were present pronounced the animal's stomach to be small. But then I began to pour water into its mouth. We kept on pouring one jugful after another down its throat, until we had poured in some forty *ratl*. Upon examination, I observed that the inner layer of the stomach had stretched until its surface had become as smooth as the surface of the outer layer. I then perforated the stomach, and once the water had come out, the stomach contracted and the folds of the interior returned to their original state, as did the pylorus. As God is my witness, after all this, the animal was still alive.

One author relates an account told to him by a Syrian about a baker, who was making bread in his oven in the city of Damascus, when a man came by, selling apricots. The baker bought some and began to eat them with hot bread. No sooner had he finished than he fell unconscious and appeared to be dead. People flocked around, brought in physicians and searched for signs of life, but finding none, concluded that he must be dead. He was washed and wrapped in a shroud, prayers were recited, and then the man was carried to the cemetery. As the procession was passing the city gate, it met a physician, a man by the name of al-Yabrūdī who was a skilful, intelligent, and wise physician. When he had heard what had happened, he said, 'Put him down so that I can take a look.' The physician turned the man over, looking for signs of life. He then opened the man's mouth and made him swallow something (or, according to another account, administered him an enema), whereupon the food he had eaten was immediately expelled. The man opened his eyes and spoke, and then returned to his shop.

Al-Yabrūdī died in Damascus and was buried in the Jacobite church there.

Al-Yabrūdī is the author of two works, including:

On the fact that the hen is cooler by nature than that of the hatchling or pullet.

2. ABŪ L-HAKAM (*d.* 1155)

The wise and cultured shaykh Abū l-Hakam, a native of Murcia in Andalusia, was a distinguished scholar in the philosophical sciences, and well versed in the art of medicine, besides being noted for his literary erudition and renowned for his poetry. He was good at telling funny stories, made jokes, enjoyed entertainment, and loved to be amused. Many of his poems are dirges for people who were still alive in his time, but his intention was merely jest and buffoonery. He was excessively fond of drinking. He loved play-acting, and when excited, would mime and sing to accompany his performance.

Abū l-Hakam travelled to Baghdad and Basra and then returned to Damascus, where he lived until his death. He died—may God have mercy upon him—in Damascus, when the last two hours of the night of Wednesday 6 Dhū l-Qaʿdah 549/12 January 1155 had elapsed.

Abū l-Hakam would compose defamatory poems against a group of contemporary poets, who had ridiculed him. One of those poets lampooned him in the following satirical verses:

> We have a doctor, a poet, with an inverted eyelid,
> May God relieve us of him!
> Whenever he visits a patient in the morning
> he composes an elegy for him the same day.

Abū l-Hakam was described in this satirical poem as having inverted eyelids for the following reason: one night, Abū l-Hakam left a man's house in a state of intoxication, with the result that he fell down and cut his face. Next morning, visitors kept asking him how he had happened to fall. He thereupon dashed off some verses, placed them near his head, and whenever someone asked him about his condition, he gave them to that person to read. He then took a mirror and looked at the gash under his eyelid left by the fall and recited the following lines:

Wine has left on my cheek
  a wound like a ewe's cunt.
I fell flat on my face,
  my turban flying off,
And have remained disgraced. But for
  the night my privates would have shown.
I know all this
  came from perfect pleasure.
Who can give me another like it,
  even for the price of my beard being shaven off?

Among Abū l-Hakam's poems, is one in an unaffected style that he entitled *The Domestic Scandal*. He describes the damage and costs that may befall someone when he invites over his drinking companions:

Any domestic scandal tends
To happen through one's own best friends.
Now listen to a well-tried man:
He'll tell you how it all began:
All that may come from invitations
And all their diverse tribulations.
Provide the food, provide the fun;
Then suffer all the damage done.

Disliked by all, the Awful Bore 5
Comes first. Then: spongers at the door!
Whatever food may be provided,
The host will be severely chided.
Creep up his mother's **** he may,
From censure he can't hide away.
'Not enough spices!' says one guest,
'It's rather burnt!' declare the rest.
Another says, 'Too little salt!
—I'm merely helpful, finding fault.'
He grabs the food from far and near, 10
Then drinks some water, fresh and clear,

Since 'wholesome water has no peer.'
The next thing he demands is beer,
With ice in summer. When it's cold:
'A fire, if I may be so bold!'
Who needs a tooth-pick? Take a straw:
The mats lie ready on the floor.

And after this there comes the wine,
Delicious, choice; it tastes divine.
One person says, 'It's vinegár!' 15
Another says, 'Defective jar!'
And someone else is now complaining:
He wants a filter, for the straining.
Some large carafes are brought in there,
In which the wine is mixed with care.
Someone cries out, 'But that's still pure!'
And pours more water, to be sure.
'He's got an ulcer,' mocks another,
'O, don't add water! Please, don't bother!'
Fruits, nuts, with any fragrant smell, 20
Go down, it seems, extremely well.
Some fussy person's fancy's tickled
Only by basil and things pickled,
While yet another man supposes
Wine goes with apples and with roses.

The singers' fee may cause some tension,
Their agent may cause apprehension;
A fix you should be quick to handle:
Spread round your cash, for fear of scandal.
Sometimes they get into a swoon: 25
Fear not! They'll have their breakfast soon.
If you invite them in December,
Make sure of stove and burning ember!
From it there flies up many a spark
That on your carpet leaves its mark:
Your once-new carpet now is peppered
With dots like any spotted leopard.
And don't forget the meat: kebab
Or sliced, for ev'ryone to grab.
And when the cold is over, pep 30
Them up with fans and cool julep.

Your drinking-friends come in all sorts:
The wine reveals their favourite sports.
There's one whose forte and whose strength
Is telling stories at great length,
While he is busy masticating.
—Nobody heeds what he's relating.
Forgets himself, speaks out of turn:
They slink away in unconcern.
Another weighs his words with care 35
And gives himself a haughty air.
Another acts the fool. He's after
A cheap but all-embracing laughter.
Someone becomes morose when stewed;
Instead of leaving he gets rude.
Someone as sober as a judge
Arrives, and bears all drunks a grudge.
There's one light-fingered Jim-'ll-fix-it,
Sees something rather nice: he nicks it.
A knife, a flask, a handkerchief, 40
A dicing-bowl fit for a thief.
Now someone pulls (abracadabra!)
A chain right off the candelabra,
'Extinguishing' (he says) 'a wick.'
It is, of course, a little trick.

Don't mind their winks whenever any
Should leave their place 'to spend a penny':
It's slaves and slave-girls they will seek,
To pinch a tit or bite a cheek.
One's hospitality's abused 45
Yet worse: one's wife is being seduced,
One's sister, daughter, or one's son
(Especially a pretty one).
In this one ought to be forgiving,
For, after all, your friends are living;
A man is flesh and blood and bone;
He is no statue or a stone.

And if among them is a glutton
Your banquet isn't worth a button.
Eating is all that he is doing; 50
Heedless of all, he's good at chewing.
Drinking with friends he must decline:

He says he doesn't care for wine.
He buggers sleeping drunks at night,
Consumes their sweets in broad daylight.
Your friends will start an ugly brawl,
But you will suffer, that is all.
They break the cups and bottles each
And ev'ry vessel within reach.
The row spreads to your neighbours, too, 55
Who falsely will belabour you;
Straight to the bailiff they appeal:
Surely, complete is your ordeal.
Thus may a man gain loss of face;
And if the party did take place
On Friday's eve, there's worse disgrace.
If in the fighting blood is shed,
The host may just as well be dead.
If someone tumbles and gets killed,
One merely pays some light wergild;
For drinking in an upstairs room 60
Brings people closer to their Doom.

Think of the harm that's coming from it!
The mats are soiled with bits of vomit.
And then one seeks something to eat,
One's drinking-bout not yet complete.
When you wake up—you've hardly slept—
And now the floor has to be swept,
You will be henpecked by your wife,
In bed and up, always at strife,
Who, when the sun's up, will remind you 65
Of last night's trials, now behind you.
—That is, if they have gone. If not
(They stayed, toped on, slept on the spot),
Then all your hope is now forlorn,
When the sun rises on the morn.

Offer your friends your choicest wine,
And cakes, and heads of sheep and kine;
Pawn chairs and stools, pull out all stops,
Pledge them at the off-licence shops.
But if some guest misses one sandal, 70
You'll be involved in one more scandal;
So tell your boy to guard them well,

Lest your kind comrades give you hell.
Don't mind your losses in this fix.
Provide your lamps with num'rous wicks.
Someone at last wants to strike camp:
He leaves and robs you of your lamp;
With in his hand a full wineskin
To please his friends and next-of-kin.
If oil runs out, give it no thought: 75
Amidst this ruin it is naught.
All costs must by the Host be paid
When in the Balance he is weighed.
Latter-Day Prophets who go dry
Deserve a good punch in the eye.
The debts he owns—a pretty sum—
Prove him to be a stupid bum.

He would be spared all this forever
If he were wise, astute, and clever.
A scandal, quite without a match: 80
He whom it strikes, strikes a bad patch!
At other people's places drinking
Is better, in my way of thinking.
Well, then. Repentance of one's vices
Is always best when there's a crisis.

On the theme of keeping one's secret he said:

I shall shun Laylā, though my love of her is in my heart,
for fear I might provoke a chaperone or grudging enemy.
I shall hide a secret we had between her and me;
for if I said I fucked her I would have disclosed it.

3. SUKKARAH AL-HALABĪ
(twelfth century)

Sukkarah was a Jewish shaykh from the city of Aleppo, a small man, but a skilled medical practitioner, with a long experience in the care of patients.

I have heard the following account:

When the sultan Nūr al-Dīn ibn Zangī was in Aleppo, he kept with him in the citadel a concubine of whom he was particularly fond. This girl once became seriously ill. The sultan had to go to Damascus, but his heart remained with her, and he sent constantly to enquire about her. Her illness persisted, although she received medical treatment from some of the most distinguished physicians. Finally, however, the wise doctor Sukkarah was brought to her. He found that she had very little appetite, had a humoral imbalance, and was not able to move and rise from her bed. He returned to her several times with the other physicians and eventually asked permission to visit her alone.

Once closeted with his patient, Sukkarah said to her, 'My lady, I want you to answer whatever questions I pose to you, and to tell me everything without hiding any facts from me.' She promised to do so, and in return, at his request, assured him that no harm would befall him. Then he asked: 'Where are you from?' 'I am from the Alan,'* she replied. 'The Alan people in their homeland are Christians,' said the physician. 'Tell me, what did you usually eat at home?' 'Cow's meat,' she replied. 'And what kind of wine did you drink there, my lady?' 'Such and such,' she replied. 'Rejoice at your good health,' said the physician, and took his leave.

Sukkarah went back to his house, bought a calf, slaughtered it, and cooked a part of it. He returned to the citadel, carrying with him a bowl containing a piece of boiled meat marinated in milk and garlic and topped with a thin slice of bread. This he placed in front of the girl, saying, 'Eat.' She leaned over to it and began to dip the bread in the milk and garlic and ate until she was full. The physician then took a small vessel from his sleeve and said, 'My lady, this is a drink that will do you good. Take it!' No sooner had she drunk it than she became sleepy. He covered her with a mantle made of squirrel fur, which caused her to perspire heavily, but when she awoke the next morning, she felt well. Sukkarah gave her the same food and drink on the following two days, until her health was fully restored.

The girl treated the physician most generously, giving him a tray filled with pieces of jewellery. He thanked her, but added, 'Nevertheless I would like you to write a letter for me to the sultan, informing him about the nature of your illness and explaining that you were cured by me.' She promised to do so, and wrote to the sultan, telling him that she had been on the verge of death and had been treated by so-and-so, who was the only person who proved able to cure her, while all the other physicians had been baffled by her illness, and asking the sultan to reward him.

When the sultan read the letter, he sent for Sukkarah, showed him great honour and said: 'Ask whatever you want, and I shall give it to you.' 'My Lord,' replied the physician, 'grant me ten *faddān*s of land, five in one village and five in another.' 'It is yours,' replied the sultan, 'and we shall draw up a contract of purchase and sale, so that it will remain yours in perpetuity.' The sultan wrote out the contract then and bestowed a robe of honour upon him.

Sukkarah went back to Aleppo, where he accumulated great wealth. He and his children after him lived there comfortably all their lives.

4. IBN AL-SALĀH (*d. c.*1145)

The learned Ibn al-Salāh was outstanding in the philosophical sciences, being thoroughly acquainted with their finer points and obscure aspects. He was eloquent, expressing himself clearly and writing in an elegant style. He was also a distinguished physician.

Ibn al-Salāh was Persian by origin, being originally from Hamadan, but he had settled in Baghdad. From there, he was invited to the court of Timurtāsh* who showed him great honour. After having enjoyed the ruler's friendship for a time, he travelled to Damascus where he remained until his death. He died—may God have mercy upon him—in Damascus on a Sunday night in 540/1145 or a little later, and was interred in the cemetery of the Sufis* near the river Bānyās on the outskirts of Damascus.

I have copied the following account:

The learned shaykh Ibn al-Salāh travelled from Baghdad to Damascus. Ibn al-Salāh wanted to have a pair of shoes made, of a Baghdadi type called *shamshak*.* When he asked for a good shoemaker who could produce them, he was directed to a man by the name of Saʿdān the Shoemaker, with whom he placed an order for a pair of *shamshak* shoes. When, after some time, the shoes were finished, Ibn al-Salāh discovered that the toes were too narrow and that the shoes were too long and shoddily made. As a result, he complained incessantly.

When Abū l-Hakam, the physician (mentioned above), heard about this, he composed the following poem in jest, putting it into the mouth of the philosopher. In it he mentions numerous technical terms from the fields of logic, natural sciences, and geometry.

My mind cannot describe my tribulation,
And strange will be my case's explanation.
    We shall divulge to you here, since you ask us,
    The pain and shame we suffered in Damascus.
Though as a rule I'm well-informed, I came
Quite ignorant about it, to my shame.
    My feet wore *shamshak* shoes, in Baghdad made,
    But now, the work of cruel Time, decayed.
'I'll find that Fate sends me another pair along,' 5
I thought. But O how wrong I was, how wrong!
    I met a scoundrel who, alas, lived there.
    My God! The misery I had to bear!
'Dear Saʿd,' I said. 'Please help me, be so good!
And you will earn a scholar's gratitude.
    Select a piece of leather, overall
    Well-tanned with vinegar and eke with gall!'
'I'm at your service,' he replied, 'for it's
A duty for each person in his wits!'
    And 20 dirhams on the nail I paid; 10
    Delivery, for two months, was delayed!
I said, when God decreed the work was done,
'Saʿd may have finished what was once begun.'
    He showed me *shamshak* shoes with toes too tight,
    And heels that would kill heel and foot all right!
And backs almost as bad, to soles attached
In utter worthlessness not badly matched.
    A figure hard to solve for cracks, if ever,
    Defeating people powerful and clever.
The heel is downward to the nadir bent, 15
The front makes to the zenith its ascent.
    And their proportions all unsound to boot;
    Their shoddiness had spread to branch and root.
The two shoe-sides do not run parallel:
The one goes up, the other down (to hell!).
    There's so much to find fault: loose stitches, holes,
    And there are awful cuts in strings and soles.
The joining had been necessary, stringent;
But that the shoes came unjoined was contingent.
    A compound syllogism in confusion: 20
    Without a categorical conclusion.
I can't protect my feet with their transversal:
Shoe-shapes like these should not exist, God curse all!

Vague is their genus, Isagogically;
One can't define their species logically.
Their shapes corrupted since origination:
Who, for their badness, will give consolation?
They had, we hoped, potentiality,
But did not come as actuality.
Not 'not quite perfect': they are below par 25
In general and in particular.
Let's not affirm what's truly negative,
Or make bad propositions positive.
Their categories look all wrong to me
Their substance, quality, and quantity.
All propositions wrong, it's evident;
All syllogisms are deficient.
All proofs are wrong, though types may vary:
The gen'ral, positive, and necessary.
The instep, as a dial's gnomon made, 30
Would soon be seen to swerve from sun to shade.
They flip-flop on my feet. It is still dry;
How will I walk in mud, if I'm to try?
I'm baffled by them and at my wits' end,
Saʿdān left me no sense at all, my friend!
In all, it's clear to me: the brain is cracking,
The mind confused, all reason sorely lacking.
How ruinous, a family, as one sees,
So worthy of disgrace and miseries!
If Euclid were alive and saw this figure, 35
He could not solve the thing, for all his rigour.
I swear an oath by God, Giver of breath,
And by the prophets Hūd,* Dhū l-Kifl, Seth,
And by the Suras found in the Qur'an,
Yāsīn, Tāhā, Maryam, The Ants, Luqmān:*
If I won't slip on a declivity
The man shall get away from me scot-free;
And I shall not lampoon the town in verse,
Nor blame a shoemaker, in jest or worse.
It was a friend from Hell that God did send; 40
May He not bless me with another 'friend'!
By his delay my heart was grieved not half!
As much as Moses suffered with that calf.*
Some people pestered Aristotle, when
He did not want to jest with them again.

Hippocrates saw lots of misery,
But never saw a man as sick as me.
    If Galen's foot was bitten by such shoes,
    With balms and salves he'd heal both wound and bruise.
And Ibn Lūqa* rather walked barefoot 45
And to reproaches he gave not a hoot.
    And Abū Nasr al-Fārābī,* too,
    Walked barefoot rather than on such a shoe.
Thus leading scholars never ceased to suffer
Ills undeserved from any stupid duffer.
    And now I am resolved to leave this city
    For home. I'm sorry; it's a pity.
If I were in Baghdad (I wish I'd stayed)
Kind people quickly would come to my aid.
    I would have helpful friends, keen students who 50
    Write down what I dictate that's wise and true.
Could I but fly to it! Can anyone
Help me with this? Alas, it can't be done.
    In Syria I suffered woes uncounted;
    I wish that I had never here dismounted.
Here in Damascus where I live I find
I live with people that are not my kind.
    I swear: neither the rain-stars,* pouring out
    their rain on earth after a lasting drought,
Nor al-Khansāʾ, lamenting her dead sibling,* 55
While tears along her cheeks were dribbling,
    Shed more than I shed tears, when I
    Saw on my feet such footwear all awry!
I'm sick and at the fag end of my tether:
I wish I had no feet left altogether.
    I've not exhausted all deficiencies;
    Please tell me how to shun their evil, please!
The instep was so tight I fear it may
Make my whole body shrink and waste away.
    Those *shamshak* shoes! When I beheld their ill 60
    Design, I knew for certain they would kill,
And make me ill. I'm under no illusion:
No herbs can save me from it, or infusion.
    My death will be announced here; they'll recite:
    'Our plight, here on this earth, is like your plight!'*
Don't be amazed. I suffered hardships sore,
No one has ever borne such misery before.*

5. SHIHĀB AL-DĪN AL-SUHRAWARDĪ (*d.* 1191)

The distinguished and learned authority, Shihāb al-Dīn al-Suhrawardī,* was unparalleled in his mastery of the sciences, with his extensive knowledge of the philosophical arts and grasp of the principles of astronomy. He was extremely intelligent and quick-witted, and possessed an excellent way of expressing himself: he would get the better of any opponent, regardless of the subject under discussion. His knowledge, however, was greater than his common sense.

When al-Suhrawardī arrived in Aleppo he entered into debate with the local experts of jurisprudence. None of them was able to stand against him, and consequently they loathed him bitterly. The sultan al-Malik al-Zāhir summoned him, along with a number of distinguished scholars, jurists, and theologians. They held a long debate, in which al-Suhrawardī displayed effortless superiority over the others, and dazzled everyone with his great knowledge. He made a good impression on al-Malik al-Zāhir, thereby acquiring rank and prestige.

However, the favour shown by al-Malik al-Zāhir merely inflamed the hatred of his rivals, who prepared attested statements alleging that he was an infidel and sent them to Saladin in Damascus. 'If this man stays,' they wrote, 'he will corrupt the faith of your son, al-Malik al-Zāhir; if he is set free and sent away, he will corrupt any region of the country in which he settles.'

Upon receiving this letter, Saladin had his scribe draft a letter concerning the matter and send it to his son, in Aleppo. In the letter, Saladin wrote that al-Suhrawardī surely had to be killed. When al-Suhrawardī heard of this, he realized that he must die and that there was no way for him to escape his fate. Accordingly, he chose to be left in an isolated place and be denied food and drink until he should meet God, exalted be He. This happened at the end of 586/1191 in the citadel of Aleppo, when he was about 36 years old.

It is said that al-Suhrawardī knew much about the art of natural magic and that many people witnessed him performing marvels of this specific kind. A physician told me about this and asserted that he met him once in person outside the Gate of Deliverance or *Bāb al-Faraj*. They were walking in the direction of a great open field together with a group of students and others, and were engaged in conversation about that art, its marvels, and the shaykh's knowledge of it.

Al-Suhrawardī listened as he walked along and then exclaimed: 'How beautiful Damascus is and how beautiful this place!'* We looked and saw in the direction of the east lofty whitewashed palaces built closely together, constructed and ornamented in a most beautiful manner. The enclosure contained large windows, in which were the most beautiful women imaginable. Singing voices and musical instruments were heard. There were intertwining trees and broad rivers were flowing. We had not known this place before and were greatly astonished at it. The crowd was delighted at the view, but also perplexed by what they saw.

The physician continued:

We continued to see it for an hour, but then it vanished and we again viewed what we had long been accustomed to see there. But when I first gazed at this wondrous manifestation it felt as if I had quietly dozed off without anyone noticing. My perception did not seem to be in touch with reality.

A Persian legal scholar told me the following story:

We had left Damascus and were in al-Qābūn, in the company of the shaykh Shihāb al-Dīn al-Suhrawardī, when we encountered a flock of sheep with a Turkmen shepherd. We said to the shaykh: 'O master, we would like to eat one of those sheep.' 'I have 10 dirhams,' replied al-Suhrawardī, 'Take them and buy yourself a sheep.' So we bargained about the price. Then the shaykh said to us: 'Take a sheep and go! I shall stay here and seek to come to terms with him.'

We moved on, while the shaykh stayed, talking with the man and trying to negotiate. After we had walked a little further, he left the shepherd and followed us. The Turkmen came after him shouting, but he paid no attention and did not speak to him. When the Turkmen caught up with him, he grabbed the shaykh's left arm in a fury and cried: 'Where are you going and why did you walk away from me?' Suddenly the shaykh's arm came off his shoulder. The Turkmen found himself holding it in his hand, with blood pouring out from it. The Turkmen turned pale, stood in bewilderment, and threw away the arm, filled with fear. The shaykh turned, picked up the arm with his right hand and rejoined us. The Turkmen kept on looking around at us until he was out of sight. When the shaykh reached us, we saw nothing but a handkerchief in his right hand.*

When al-Suhrawardī was killed, he recited a poem, from which the following lines are taken:

> Say to companions who thought they saw me dead
> and wept for me out of grief when they saw me:
> Do not think that I am dead;
> that dead one, by God, is not I.
> I am a bird and that is my cage:
> I flew from it and it was left vacant, as a security.
> And today I converse with a Host
> and I see God with my own eyes, in bliss.

6. RAFĪʿ AL-DĪN AL-JĪLĪ (*d.* 1244)

The venerable qadi or judge and learned authority Rafīʿ al-Dīn al-Jīlī was one of the most outstanding scholars in the philosophical sciences, dogmatic theology, legal theory, and methodology, the natural sciences and medicine. Having settled in Damascus, he taught as an expert of religious law at the al-ʿAdhrāwiyyah law college. I—Ibn Abī Usaybiʿah—studied philosophy with him for a time. He was eloquent, highly intelligent, and read and studied constantly.

As time went on, however, many people complained about him and, to make a long story short, in the end he was arrested and put to death—may God have mercy upon him. Following an argument between Rafīʿ al-Dīn and the vizier Amīn al-Dawlah, Rafīʿ al-Dīn was placed under guard and brought, under an escort of the vizier's men, to a place near Baalbek, where there was an immense, reputedly bottomless pit, known as the Cave of Afqah. These men were ordered to tie his hands and then push him into the pit. One of the men who was among those present on that occasion told me that when Rafīʿ al-Dīn was pushed into this pit, he was crushed by the fall, but that his clothing appeared to have caught on the side of the cave near the bottom. 'We stayed there for approximately three days,' he told me, 'listening to his moaning and groaning. After some time, it became weaker and weaker and then it stopped, so that we were sure that he was dead. Then we went away.'

I—Ibn Abī Usaybiʿah—say: It is curious to note that the qadi Rafīʿ al-Dīn went over a copy of this book (my history of physicians) in my presence, in which I had not included him.* He looked through it, but stopped when he had finished the account of Shihāb al-Dīn al-Suhrawardī. He was much impressed by it and spoke: 'You have included him, but you have omitted others who were greater than he,' referring to himself. Then he added, 'Al-Suhrawardī's situation was most unfortunate indeed, but at least he died in the end. And God mighty and glorious decreed that Rafīʿ al-Dīn should be put to death like him. Praise the Lord, who determines the fate of His creatures according to His will.' Rafīʿ al-Dīn died in Dhū l-Hijjah 641/May 1244.

When Rafīʿ al-Dīn was appointed as chief qadi in Damascus, I composed a poem to congratulate him, from which the following lines come:

Lasting glory and good fortune and high standing
for all time, and elevation and brilliance,
Through the lasting life of our master Rafīʿ al-Dīn,
man of all-encompassing generosity and of benevolence!
Chief Judge, most exalted master, through whose lofty qualities
scholarship and scholars rise high,
Unique in noble traits, though all of mankind
share some of them.
If any man of eloquent speech wished
to count his noble traits, the eloquent would fall short.
How many enemies attest to his excellence
—and excellence is not (normally) attested by enemies!
He has composed works that clearly express
everything that the ancients did garble.
Through him Jīl has things to boast of among countries;
likewise this generation (*jīl*) is raised through him.
...
Be well, live long, in a lasting life of ease,
as long as a dove sings in its grove!

7. SHAMS AL-DĪN AL-KHUSRAWSHĀHĪ
(*d.* 1254)

This honourable and learned scholar was a native of Khusrawshāh, a small village very near Tabriz. He was a leading scholar, an outstanding philosopher, a model to mankind, and an honour to Islam. Tireless in the pursuit of learning, and a man of great merit and virtue, he was one of the most brilliant disciples of the shaykh Fakhr al-Dīn, the son of the preacher of al-Rayy.

When Shams al-Dīn arrived in Damascus, and I met him, I found him to be an elderly gentleman with pleasant manners and an attractive way of speaking. He was intelligent and very learned. One day when I was at his home a Persian jurist brought him a book written in a very tiny handwriting, one-eighth the size of Baghdadi script,* and in a rather irregular format. After looking at it and thoroughly examining it, he kissed it and laid it down forthwith. Upon my asking him the reason for this, he said, 'This is the handwriting of our master, the imam Fakhr al-Dīn, the preacher—may God have mercy upon him.' I felt great esteem for him because of the respect that he had shown towards his master.

8. SAYF AL-DĪN AL-ĀMIDĪ (*d.* 1233)

The esteemed leading authority and learned scholar Sayf al-Dīn al-Āmidī was one of the most distinguished, erudite, and intelligent men of his time, supreme in his knowledge of the philosophical sciences, the several schools of theology and the principles of medicine. He was a spirited person and was impressive in appearance. He was also an eloquent speaker and an excellent writer as well.

I—Ibn Abī Usaybiʿah—used to meet with him and study under him. This was because of the firm friendship between him and my father. The first time I met him, I had come to his house with my father. He lived in a paved courtyard near the al-ʿĀdiliyyah law college in Damascus. After we had greeted him, he observed the formalities by welcoming us with amiable words. Then we sat down. He looked at us and spoke these exact words: 'I have never seen a father and a son resemble each other more than you do.'

Sayf al-Dīn remained in Damascus until his death—may God have mercy upon him—on 4 Safar 631/9 November 1233. He is the author of nineteen works.

9. IBN AL-MUTRĀN (*d.* 1191)

Ibn al-Mutrān was a leading philosopher and a most erudite scholar. He was Damascus born and bred and his father was also a prominent physician. He had a sharp intellect, spoke eloquently, and studied constantly.

Ibn al-Mutrān was a handsome man, who was particularly fond of luxurious, costly clothes. He served as a physician to Saladin who showed him great favour, so that he enjoyed high status and great prestige. The ruler made him his chamberlain and appointed him in charge of the household, a post for which he paid him extremely well. Saladin—may God have mercy upon him—was a noble and excellent man, who was very generous to those who served him and to everyone who asked him for assistance, so much so that when he passed away, his treasury was found to be empty.* He had complete confidence in Ibn al-Mutrān, and showered the physician with favours and gifts, providing him with opulent means. Ibn al-Mutrān then became proud and arrogant, thinking himself even above kings. Saladin was aware of this trait in him, but did not cease to show him respect and esteem, because he admired him for his great knowledge. Ibn al-Mutrān converted to Islam during his reign.

Someone who knew well Ibn al-Mutrān's conceited nature and arrogance told me that he once accompanied the sultan on one of his military expeditions. In time of war, during campaigns, it was Saladin's habit to occupy a red pavilion, complete with a red outer tent and vestibule. One day, when Saladin was out riding, he saw a red tent with a red vestibule and privy. He contemplated it for a while and then asked whose it was. Upon being informed that it belonged to the physician Ibn al-Mutrān, he said, 'By God, I knew it was some stupid freak of Ibn al-Mutrān's!' He laughed, but then said, 'What would happen if a messenger were to ride by and think that it belonged to a king? If he must have his tent, he shall change the privy,' and he ordered it to be destroyed. When this was done, Ibn al-Mutrān took it very hard, keeping to himself for two days and not providing his usual services, but the sultan mollified him with a gift of a purse.

The same source also informed me that there was in the service of Saladin a Christian physician by the name of Abū l-Faraj al-Nasrānī who served the sultan for a time and frequently visited his palace. One day he told the sultan that he needed dowries for his daughters and asked him for his assistance in this matter. Saladin replied, 'Write down on paper everything that you require and bring it to me.' Abū l-Faraj left and listed on a piece of paper jewellery, fabrics, utensils, and other things to the value of 30,000 dirhams. When Saladin read the list, he ordered his treasurer to buy everything for Abū l-Faraj, leaving nothing out. No sooner had Ibn al-Mutrān heard about this than his attendances on his master became surly and sporadic. Saladin noticed that his physician's face had changed and understood the reason. Then and there he ordered his treasurer to calculate the price of everything that he had bought for Abū l-Faraj. When the treasurer had calculated the total amount, Saladin ordered him to pay Ibn al-Mutrān a similar sum, and that was duly done.

A man who knew Ibn al-Mutrān and was on intimate terms with him told me that the vanity and the arrogance that became characteristic of him later on in life were entirely absent during his days as a young man in search of knowledge. He said that he would come to the mosque after he had finished his work at the sultan's palace. He would arrive with an escort of horsemen, accompanied by numerous Turkish slaves and others. When he approached the mosque, he would dismount and continue on foot, holding his books in his hand or under his arm. He would let none of the servants accompany him, but would greet the shaykh and sit among the group, alert and receptive, until the lesson was over and he returned to his attendants.

According to the venerable and respected Ibn al-Qiftī, the physician Ibn al-Mutrān, a Christian, became a good Muslim after his conversion to Islam.* Saladin—may God sanctify his soul—presented him with one of his favourite ladies at the palace, named Jawzah, as a wife. Jawzah was a servant of Khwand Khātūn, the wife of Saladin. It was Jawzah who managed the household and was her mistress's favourite handmaid. Khwand Khātūn gave her many pieces of jewellery and other precious articles, making her a rich woman. Jawzah put Ibn al-Mutrān's affairs into proper order, taught him how to behave, improved his manner of dressing, and embellished both his outer appearance and his character.

Ibn al-Qiftī also said that a preacher told him:

I went to his house in Damascus, entered with his permission, and found him a pleasant and good-natured man and a good listener and talker. His house struck me as extremely beautiful with respect to its construction and furnishings. I saw water spouting from pipes in his pond that were made of pure gold and were of the most excellent craftsmanship. I also saw a young and exceptionally handsome lad, who waited on him hand and foot, called ʿUmar. There were also luxurious carpets, and I smelled fragrances of which the sweet scent filled me with a sense of awe.

Ibn al-Mutrān was a zealous collector of books. When he died, approximately ten thousand medical and other works were found in his library, apart from those he had copied. He took a keen interest in copying and revising books, and there were three copyists in his service who were constantly transcribing books, and who received payment from him. One of them was Jamāl al-Dīn, known as Ibn al-Jammālah ('son of the camel-drivers'), whose handwriting was well-proportioned and symmetrical. Ibn al-Mutrān copied many books in his own handwriting; I have seen several examples of these, and they were unsurpassable as to script and grammatical correctness. He spent most of his time reading. The majority of the books in his library contain corrections and precise revisions in his handwriting. Most of the small books and miscellaneous items in the field of medicine in his library had been combined into single volumes. He had them all copied on small-format paper, one-sixteenth the size of Baghdadi paper,* and bound. He would never leave his house without a book in his sleeve, which he would read at the gate of the sultan's palace or wherever else he might go. After his death, all his books were sold, because he did not leave behind offspring.

A physician told me that he had attended the sale of Ibn al-Mutrān's books and saw that there were many thousands of these small-format items, most of them in the handwriting of Ibn al-Jammālah. They fetched 3,000 dirhams at auction.

I—Ibn Abī Usaybiʿah—say: Ibn al-Mutrān possessed the complete ideal of manhood and was a noble soul. He was kind towards his disciples and gave them books as presents. When one of them began to practise medicine and heal the sick, Ibn al-Mutrān would give him a robe of honour and devote his attention to him.

A person also told me that he once accompanied Ibn al-Mutran to the 'Great Hospital' in Damascus founded by Nūr al-Dīn ibn Zangī, where he treated patients. Among them was a man who was suffering from such severe dropsy of the belly that he was nearly bursting. At that time, the surgeon Ibn Hamdān, who was quite skilful in the treatment of patients, was also at the hospital. He and Ibn al-Mutrān decided to lance the hydropic swelling.

I—Ibn Abī Usaybiʿah—say:

We were present at the operation. Ibn Hamdān lanced the swelling in the right place, and yellow fluid came out, while Ibn al-Mutrān watched the patient's pulse. When he realized that the patient was not strong enough to withstand the removal of more fluid, he had the site dressed and the patient laid on his bed, ordering that the dressing should not be disturbed. The patient then felt greatly relieved and was able to relax. The patient's wife was with him, and Ibn al-Mutrān urged her not to allow her husband to remove the dressing or to change it in any way until he could examine the patient the next day. We then left the hospital. When night came the man said to his wife: 'I am well now, there is nothing wrong with me; those physicians only intend to prolong my illness. So, undo the dressing so that the rest of the fluid comes out and I can return to work.' She reproached him and said it would be a mistake, but he repeated his request over and over again, not realizing that the doctors wanted to extract the fluid at a later stage, as they were concerned about his condition. Finally, she undid the dressing, all the fluid ran out, his strength gave out, and he perished.

Ibn al-Mutrān is the author of eight books including:

> *The gardens of physicians and meadows of the intelligent.** [Its author] attempted to collect all the witty sayings, anecdotes, and appropriate information that he had read or heard from his teachers, or that he copied from medical books. He did not finish this book. All I found of it were two parts, written in the hand of our teacher, the physician al-Dakhwār. The first of these had been [proof]read by Ibn al-Mutrān and contained his handwriting. In the second part, however, al-Dakhwār makes mention of the fact that Ibn al-Mutrān died before he was able to proofread it.

I was told by a relative of Ibn al-Mutrān that, when he died, he left behind several drafts of medical works and other books, as well as scattered explanatory notes. His sisters took those drafts and they have never been seen since. This relative also told me that in the home of one of those sisters he had seen a chest that the lady had lined by gluing some of Ibn al-Mutrān's manuscripts to the inside of it.

10. RIDWĀN IBN AL-SĀʿĀTĪ (*d. c.*1230)

Ridwān ibn al-Sāʿāti, 'Ridwān the son of the Clockmaker', was born and raised in Damascus. His father Muhammad (al-Sāʿāti) was originally from Khorasan, but settled in Damascus. The father was unequalled in his time for his knowledge of clocks and the science of astronomy. It was he who operated the clock at the gate of the Umayyad mosque in Damascus.* He had constructed that clock in the time of Nūr al-Dīn ibn Zangī, who paid him an allowance for operating the clocks.

He had two sons. One was Bahāʾ al-Dīn (ibn al-Sāʿātī), who was one of the most outstanding poets of his generation. The other was Ridwān, who was an eminent physician and a distinguished man of letters. He was bright and intelligent, was extremely well-versed in all matters in which he took an interest, and eagerly devoted himself to every scrap of knowledge. Ridwān had the utmost regard for the medical teachings of the venerable shaykh, Ibn Sīnā, and wrote a commentary on his treatise on colic. Ridwān died in Damascus—may God have mercy upon him—of jaundice.

11. RADĪ AL-DĪN AL-RAHBĪ (*d.* 1233)

The eminent physician and learned authority Radī al-Dīn al-Rahbī was one of the most prominent practitioners of the art of medicine. He was outstanding in the eyes of his peers, enjoyed great respect, and was well spoken of by elite and common people alike. Rulers and subjects honoured him greatly. Furthermore, his conduct was unimpeachable, and he loved the good in people. He exerted himself tirelessly in the treatment of the sick and was kind and merciful to all. He never used indelicate words, nor was he ever known to wrong others or to speak ill of anyone during his entire life.

Radī al-Dīn and his father settled in Damascus in 555/1160. After they had lived there for some years, Radī al-Dīn's father died and was buried on Mount Qāsiyūn. Radī al-Dīn decided to remain in Damascus, where he kept a practice for the treatment of the sick; he wrote many books there.

Radī al-Dīn said to me one day:

'All those who studied under me and associated with me helped and benefited the people,' and he named many prominent men who had won renown in the medical profession, including some who were already dead and some who were still alive. He had deemed it appropriate, he said, never to teach medical principles to non-Muslims, to persons who were not worthy, for he considered that he was thereby enhancing the profession and upholding its prestige. He told me that in all his life he had taught only two non-Muslims: a Jew and a Samaritan* and he had taught them only because they had pestered him and pleaded with him incessantly, until finally he felt he could not turn them away. Both men were exceptionally gifted and became outstanding physicians.

I—Ibn Abī Usaybiʿah—have heard that Radī al-Dīn al-Rahbī employed the very best cooks and instructed them to apply the rules that he himself followed. This was most beneficial to him during daytime, keeping his humours in balance during the whole day. When the food was ready, the cooks would inform al-Rahbī, and he would invite one or more of his friends to join him at dinner. When they arrived, the cooks would ask permission to serve, but he would tell them that they must wait, as the guests as yet had no appetite. When al-Rahbī called to them, the cooks would serve the meal. Only then would he eat.

One day, one of his friends asked him the reason for this habit. 'Eating with appetite is essential for the preservation of health,' he replied, 'for when the members of the body require compensation for what they have used up, they demand it of the stomach, and the stomach in turn summons it from the outside. And it is thus that man will attain his natural life span.' 'But,' objected the friend, 'you have reached an age which is little short of man's natural life span, so what is the need for this rigmarole?' 'So that during this short period I may stay above the ground,' said al-Rahbī, 'inhaling air and swallowing water.' He continued to follow this practice until his time came.

Radī al-Dīn always followed his own precept. On Saturdays, he would always go to the garden to rest and refrain from work. Thursday was the only day on which he went to the bathhouse. He made it a rule to do these things in a regular order. On Fridays, he used to go to see all the prominent people and the notables. He steadfastly refused to climb a ladder—when he needed to visit a patient, the patient had to be in a place where it was not necessary to climb a ladder—or even to go near one, describing the ladder as 'the saw which cuts off life'. He once made a particularly astonishing remark to my father. 'Since the time I bought this place, in which I have lived for more than twenty-five years,' he said, 'I do not remember ever having gone up to the room at the top of the house, except for the one time when I inspected the house before buying it. I have never been up there again.'

Among the things I was told is the following anecdote:

The vizier* of al-Malik al-ʿĀdil I always ate poultry, but hardly ever ate mutton. He complained to Radī al-Dīn al-Rahbī about his pale complexion, for which physicians had prescribed many different syrups and other medicaments. Radī al-Dīn went out, returning shortly with a piece of chicken breast and a piece of red mutton. 'You are accustomed to eat the meat of fowl,' he said to the vizier, 'but the blood produced by fowls does not have such a reddish hue as the blood of sheep, and you can see that the colour of sheep meat is very different. You should give up eating fowl and stick to eating mutton. That will make you better and there will be no need for treatment.' The vizier accepted Radī al-Dīn's advice, and ate what the physician had recommended. It was not long before his colour returned and the balance of his humours was restored.

This is a very convincing account, and I—Ibn Abī Usaybiʿah—would recommend it to everyone who wishes to be cured and seriously wants to ensure the preservation of health. The vizier was a robust, well-proportioned man who possessed a strong physique and had a good digestion, but the members of his body were afflicted by the weak blood from fowl's meat. He needed coarser and stronger blood. When he went over to eating mutton, he began to produce stronger blood that supplied the needs of his bodily members, so that his humours became balanced and his colour returned to normal.

Shaykh Radī al-Dīn al-Rahbī was born in Jumādā I 534/December 1139–January 1140. He died—may God have mercy upon him—on the morning of Sunday, 10 Muharram 631/16 October 1233, and was buried on Mount Qāsiyūn. He lived nearly a hundred years* without any weakening of his hearing or sight; only during his last years he suffered from forgetfulness with respect to recent matters, but he remembered past events perfectly clearly.

One of his relatives, who had been at his bedside during his final illness, told me that at the time of his death Radī al-Dīn felt the pulse of his right hand with his left hand, with a pensive and reflective air as he did so. He then clapped his hands, for he knew that his strength had failed. He straightened the cowl on his head with his hands, disposed himself for death, and died.

12. ʿABD AL-LATĪF AL-BAGHDĀDĪ (*d.* 1231)

The shaykh and distinguished authority ʿAbd al-Latīf al-Baghdādī is also known as Ibn al-Labbād, 'son of the feltmaker'. He was born in Baghdad and was renowned for his knowledge of the sciences and his personal virtues. A prolific writer, he was a master of literary style as well as Arabic grammar and lexicography. In addition, he had mastered theology and medicine. He had lived in Damascus for a time, and while there had devoted much attention to the art of medicine and acquired a great reputation. Students and physicians used to frequent his lectures and study under him.

ʿAbd al-Latīf al-Baghdādī was a highly industrious person. He never let a moment pass without devoting himself to the study and composition of books and the art of writing. The works that I have seen in his own handwriting are very many, since he used to make numerous copies of his own works and copied several books of earlier authors as well.

He was a friend of my grandfather's; a strong friendship had grown up between them while they were both residing in Egypt. My father and my grandfather used to study the literary arts under him. My uncle also studied the works of Aristotle with him. From Egypt he went to Damascus. I saw him when he was living in Damascus on his final visit to that city. He was an old man of fragile physique, of medium height, a good speaker, expressing himself very well; still, his written word was more impressive than his speech. At times—may God have mercy upon him—he would go too far in his talk: he had a high opinion of himself, and would find many shortcomings in the intellectuals of his time and in many of former times also. He frequently disparaged the learned men of Persia and their works, especially the distinguished master, Ibn Sīnā, and people like him.

I have taken the following account from an autobiography written in ʿAbd al-Latīf al-Baghdādī's own hand:*

I was born in 557/1162 in a house belonging to my grandfather, in a street called 'Sweetmeats Alley' without knowing anything of pleasure or leisure. Most of my time was devoted to listening to the Hadith.* I also procured certificates of audition* for myself from the shaykhs of Baghdad, Khorasan, Syria, and Egypt. One day my father said to me, 'I have made you listen to all the luminaries of Baghdad, and I even had you included in the chains of transmission of the old masters.'* During this period I had also learned calligraphy, and I had memorized the Qur'an, the *Fasīh*,* the *Maqāmāt*,* the collected poems of al-Mutanabbī,* an epitome on jurisprudence, and another on grammar. Later when I grew up, my father brought me to the shaykh Kamāl al-Dīn, the leading teacher at the Nizāmiyyah law college in Baghdad. Between him and my father there was a long-lasting friendship, going back to the time of their study there. I studied the preface to the *Fasīh* under his direction, but found that he talked a lot of nonsense. I could not understand one bit of his continuous and considerable jabbering, even though his students apparently admired him for it. In the end, he said, 'I loathe teaching young boys and instead pass them on to my disciple al-Wajīh to study under his guidance. When the boy is more advanced, I will allow him to study with me.'

Al-Wajīh, a blind man from a wealthy and virtuous family, and also a teacher at the Nizāmiyyah law college, was the teacher of some of the children of the grand vizier. He welcomed me with open arms and began to teach me from early morning to the end of the day, showing me kindness in many ways. I attended his study circle at the Zafariyyah mosque, where he would place a series of commentaries in front of me and discuss them with me. Finally, I would read my lesson and al-Wajīh would favour me with his own comments. Then we would leave the mosque, and on the way home he would help me to memorize what I had learned. When we reached his house, he would take out the books that he was studying himself. I would memorize them with his help and help him memorize them as well. Thereafter he would go to see his teacher, shaykh Kamāl al-Dīn, to whom he would recite his lesson and who would then comment on the lesson, while I listened. I became so highly trained that I began to outstrip him in powers of memory and understanding. I used to spend most of the night in memorizing and repeating. We continued thus for a while, with me as a disciple of both my master and my master's master. My memory increased and improved continually, my insight became deeper and more acute and my mind became sharper and more reliable.

The first work that I had memorized was the *Lumaʿ*,* which I completed in eight months. Every day, I listened to a commentary on the greater part of it as it was recited by others. On returning home, I studied the commentaries on it. I commented on it myself for a group of competent and dedicated students, until I reached the point where I began to use up a whole quire for every chapter, but even that was not enough for what I had to say.

I then thoroughly memorized the *Adab al-kātib** by Ibn Qutaybah, the first half in a few months and the other half in fourteen days, for it comprised fourteen quires. Afterwards I learned by heart two other works by the same author, both in a very short time. I then devoted myself to a treatise and learned it by heart over many months. I applied myself constantly to the study of commentaries on it and worked through it with the utmost care until I had studied it in depth and was able to summarize what the commentators said. As to the *Takmilah*,* I memorized it in a few days, a quire every day. I used to read both extensive works and compendia.

The shaykh Kamāl al-Dīn had composed 130 works. I managed to learn most of his works by listening, reading, and memorizing them. He had begun to write two large works, one on lexicology and the other on law, but he was not fortunate enough to be able to complete them. Under his guidance, I memorized two more works, which I came to master thoroughly. After his death, I devoted myself exclusively to *The book of Sībawayh** and the commentary on it. I, then, studied a great number of works. I also studied the law of inheritance and prosody. As for Ibn al-Khashshāb, I listened to his reading of the *Maʿānī al-Qurʾān** which he again had studied from the writings of Shuhdah bint al-Ibarī.*

ʿAbd al-Latīf al-Baghdādī further reports that among the teachers from whom he derived great benefit, as he claims, was the son of Ibn al-Tilmīdh;* he speaks of him at great length and praises him highly, but this is due only to his extreme partiality for Iraqis, for in fact the son of Ibn al-Tilmīdh was not of such high merit, nor even close to it.

ʿAbd al-Latīf continues:

There arrived in Baghdad a man from the Maghrib, tall, dressed in the garb of a Sufi;* he displayed proud and haughty manners, spoke eloquently, and had a pleasing appearance. He had the air of a religious man and looked like a traveller; those who saw him before getting to know him were struck by his appearance. He was known as Ibn Tātalī* and claimed to be among the descendants of the Almoravids.* He had left the West when the caliph, ʿAbd al-Muʾmin, took possession of the region. When he settled in Baghdad, a number of great scholars and notables gathered around him. I, too, was one of those who paid him a visit.

Ibn Tātalī had a peculiar way of teaching. Those who came to see him considered him immensely learned, but in fact he merely possessed strange and radical views. He had carefully studied works on alchemy, talismans, and similar subjects. He won the hearts of many with his appearance, his eloquence, and his ability to influence others, and he filled my heart with a desire to know all the sciences. When he met the caliph al-Nāsir, the Commander of the Faithful was delighted. Then Ibn Tātalī set off again on his travels.

I, for my part, engaged in study, buckling down quite seriously to the task and with great endeavour, giving up sleep and pleasures. I turned to the books of Ibn Sīnā, both the small and large works. I transcribed and studied many books on alchemy and astronomy and I began to practise the false art (of alchemy) and to make the frivolous and idle experiments of error. The most potent of the influences that led me astray was that of Ibn Sīnā, by his book on alchemy, which he supposed completed his philosophy. However, it adds nothing to philosophy, but rather derogates from it.

In the year 585/1189, since there was none left in Baghdad who was able to win my heart, satisfy me completely, and help me to resolve the difficulties which I felt, I went to Mosul, but I did not find what I desired there. However, I encountered a man, expert in mathematics and law, but only partially learned in the remaining branches of knowledge. His love of alchemy and its practice had so drowned his intellect and his time that he attached no importance to anything else but that art.

Large numbers of students gathered around me, and various positions were offered to me; I chose the second-storey law college of Ibn Muhājir and the school of Hadith on the ground floor below. I stayed in Mosul for one year, always working incessantly, day and night. The people of Mosul declared that they had never before seen anyone with such an expansive and rapid memory, quickness of wit, and seriousness.

I heard people say exciting things about the philosopher Shihāb al-Dīn al-Suhrawardī. They were convinced that he surpassed all ancient and contemporary authors. I had in mind to go and look for him, but good fortune intervened. I read some of his works and in them I found clear proof of the ignorance of my contemporaries, and I realized that many of my remarks, with which I was not yet satisfied, were better than the arguments of this idiot. In the midst of his discourse, he would insert detached letters, by which he made people like himself believe that they were to be considered divine mysteries.

ʿAbd al-Laṭīf al-Baghdādī continues:

When I entered Damascus, I found there a great number of notables from Baghdad and elsewhere, who had been brought together through Saladin's generous patronage. I met with the grammarian al-Kindī, with whom I had many debates. He was brilliant, intelligent, and wealthy and enjoyed the favour of the sultan, but was quite taken with himself. We had many debates and God—exalted be He—permitted me to surpass him in many of the issues that we discussed. Later I neglected to attend him, and my neglect offended him, even more than he offended people.

I produced a number of works there in Damascus. I also found again Ibn Tātalī who had taken up residence at the western minaret. He had attracted a large group of followers. People were divided into two camps, with one party for him and the other against him.

Later, Ibn Tātalī made a serious mistake, aiding his foes, for he began to speak about alchemy and philosophy. Disparaging remarks about him soon became more frequent. I met with him, and he began to question me concerning pursuits that I regarded as contemptible and trivial, though he attributed great importance to them, and wrote down all that I said about them. I saw through him and found that he was not the man I had imagined him to be and came to have a poor opinion of him. When I spoke about the sciences with him, I found that he had only a superficial knowledge of them. One day I said to him, 'If you had devoted the time you have wasted on alchemy to some of the Islamic or rational sciences, today you would be without equal, waited on hand and foot.'

I learned a lesson from his example and kept my distance from the evil that befell him: 'the fortunate one is he who is warned by the fate of another,'* and renounced the art, albeit not entirely. Afterwards, Ibn Tātalī went to see Saladin on the outskirts of Acre* to complain about a preacher opposed to him. He returned sick, and was conveyed to the hospital, where he died. His books were taken by the military commander of Damascus, who was himself infatuated with the art of alchemy.

I then set out for Jerusalem, and on to Saladin in his camp outside Acre, where I met the military judge there at the time. My reputation had reached him in Mosul, so he was delighted to meet me and gave me his attention. 'Let us join ʿImād al-Dīn, the secretary,' he said, so we rose and went to his tent, which was next to that of the judge. I found him writing a letter in *thuluth* script* to the chancery of al-Malik al-ʿAzīz ʿUthmān, Saladin's son, without first having made a rough draft. He proceeded to put me to the test on some matters of speculative theology and then suggested we call on al-Qādī al-Fādil, head of Saladin's chancery.

When we entered his lodging, I saw a thin man, puny, with a relatively big head and a lively mind, who was simultaneously writing and dictating; his face and his lips moved about in all sorts of expressions due to the intensity of his effort to pronounce the words correctly, as if he were writing with all of his limbs. Al-Qādī al-Fādil questioned me about some of the Almighty's words: 'Where is the apodosis of the particle 'when' in the Qur'anic verse «*Until, when they arrive there, and its gates will be opened and its keepers will say*»?'* He also questioned me on many other matters, and all the time he never stopped writing and dictating. Then he said to me, 'Return to Damascus, for there you will be given a salary.' I said that I preferred Egypt, to which he replied: 'The sultan is worried about the capture of Acre by the Franks* and the killing of the Muslims in that town.' 'It can only be Egypt,' I answered, whereupon he wrote me a brief letter addressed to his representative in Egypt.

When I entered Cairo, I was met by his agent, Ibn Sanāʾ al-Mulk,* who was an old man of great virtue and authority. He lodged me in a house that had been thoroughly renovated and supplied me with money and a grain allowance. He then went to the high-ranking state functionaries and said, 'This is the guest of al-Qādī al-Fādil,' whereupon presents and blessings were showered upon me from all directions.

Every ten days or so a memorandum would come to the administrative office of the Egyptian government from al-Qādī al-Fādil, outlining important matters of state. In it there would be a paragraph that clearly set forth his instructions concerning my privileges. I stayed in the mosque of the chamberlain—may God have mercy upon him.

My purpose in going to Egypt was to meet three persons: Yāsīn the magician, the Jewish scholar *al-Ra'īs* Mūsā (Maimonides), and Abū l-Qāsim al-Shāriʿī. All of them came to call on me.

Yāsīn I found to be a swindler, a liar, and a common juggler. It was said of him that he was able to do things that even the prophet Moses was unable to do, that he could produce minted gold whenever he wished, of any quantity and coinage that he wished, and that he could turn the waters of the Nile into a tent, under which he and his companions would be able to sit; yet he was in a sorry state.

Maimonides came to see me as well. I found him to be extremely learned, but he was overcome with the adulation of authority and service to those in important positions. In his treatise on medicine based on the sixteen books of Galen and others, he imposed upon himself the rule of not altering a single letter unless it was a conjunction or a connection, and he only copied sections of his choice. He has also composed a book for the Jews entitled *The guide for the perplexed*, and pronounced a curse on anyone who would transcribe it in any but Hebrew characters. I looked through it and found it to be an evil book that corrupted the foundations of law and faith with elements that he had imagined would benefit them.

One day I was in the mosque with a number of people gathered around me, when an old man dressed in shabby clothes entered. He was sharp-featured, with a pleasing appearance. The crowd stood in awe of him and showed him reverence. I finished what I had to say, and when the meeting was over, the imam of the mosque came to me and said, 'Do you know this old man? He is Abū l-Qāsim al-Shāriʿī.' I embraced him and cried, 'It is you I seek!' I brought him to my house, and after our meal we entered into conversation. I found him to be all that souls can desire and eyes delight in. His conduct was that of a man of wisdom and intelligence, his bearing likewise. He contented himself with the barest necessities of life, not involving himself with anything that would distract him from moral excellence. Following our initial encounter, he frequently sought my company, and I found that he was well versed in the works of the ancient philosophers. I did not have much confidence in any of them, thinking as I did that the whole of philosophy had been encompassed by Ibn Sīnā and was embodied in his books. When we engaged in discussion, I would surpass him in strength of disputation and refinement of language, but he would surpass me in the force of his argumentation and the clarity of his methods. I did not yield to his arguments, nor did I give up my passionate and stubborn resistance to his allusions. But then he presented me with work after work in an effort to tame my aversion and to soften my headstrong disposition, until I inclined towards his side, putting one foot forward and the other back.

News arrived that Saladin had concluded a truce with the Franks and had returned to Jerusalem, so it was necessary for me to go and see him there. I took with me as many of the books of the ancient philosophers as I could carry and set out for Jerusalem. There I saw a formidable king, who filled all eyes with respect and all hearts with love, who was approachable, tolerant, and generous. The first night I entered his presence, I found myself at a meeting attended by men of learning, discussing various sciences. Saladin listened attentively and took an active part in the conversation, taking up the subject of the manner of building walls and digging trenches. He had a good understanding of this matter, and came up with all kinds of original ideas. He was concerned about the construction of the walls of Jerusalem and about the digging of its trenches. He himself took part in the work of carrying stones on his shoulders. His example was followed by the whole population, poor and rich, strong and weak alike, even the secretary ʿImād al-Dīn and head of chancery, al-Qāḍī al-Fāḍil followed his example. For this purpose, he would ride out on horseback before sunrise. At the time of midday prayer he would return home, have a meal, and rest. He would mount his horse again at the time of the afternoon prayer, and would return home in the evening, then spend most of the night planning what he would do the next day. Saladin assigned to me in writing 30 dinars a month to be paid by the administrative office of the mosque. His sons granted me stipends as well, so that I had a regular monthly income of 100 dinars.

ʿAbd al-Laṭīf al-Baghdādī continues:

I then returned to Damascus and devoted myself to studying and teaching at the mosque. The more deeply I studied the books of the ancient philosophers, the more my desire for them increased.

Saladin subsequently returned to Damascus, but left the city again to bid farewell to the pilgrims leaving for Mecca. Upon his return, he became feverish and was bled by someone without skill.* Thereupon, his strength ebbed and he passed away before his illness had lasted a fortnight. The people were as afflicted with grief as if he had been a prophet. I have never seen a ruler whose death so saddened the people; for he was loved by the pious and the profligate alike, by Muslims and infidels.

Saladin's sons and companions divided themselves like the descendants of the queen of Sheba and were scattered to the four winds throughout the various countries. The greater number of them went to Egypt, on account of its fertility and prosperity. I stayed in Damascus, which was then under the rule of al-Malik al-Afdal, Saladin's eldest son, until al-Malik al-ʿAzīz came with the Egyptian army to besiege his brother in Damascus. However, he failed and withdrew stricken with colic. I went to see him after his recovery, and he allowed me to return with him to Cairo and assigned me a salary from the treasury, which was more than sufficient for my needs.

In Cairo, I stayed with the shaykh Abū l-Qāsim al-Shāriʿī. We were inseparable from morning to night, until he passed away from a pleurisy arising from a head cold.

My occupations during this period were as follows: I gave lectures at the al-Azhar mosque from early morning until approximately the fourth hour. At midday, those who wished to study medicine and other subjects would come to me. Then at the end of the day, I would return to the al-Azhar mosque to teach other students. At night I used to study for myself. In this manner I continued until the death of al-Malik al-ʿAzīz. He was a generous, courageous young man, modest and unable to say no. In spite of his youthfulness, he was wholly abstinent from worldly possessions and sexual pleasures.

I—Ibn Abī Usaybiʿah—resume: After this ʿAbd al-Latīf al-Baghdādī lived in Cairo for some time, enjoying great prestige and receiving stipends from the sons of Saladin. Egypt was then visited by a huge rise in food prices and many deaths,* as never had been seen before. ʿAbd al-Latīf al-Baghdādī wrote a book on this subject, in which he described things that he had seen himself or heard from eyewitnesses, which make the mind reel.*

When the sultan al-Malik al-ʿĀdil I took control of the land of Egypt, ʿAbd al-Laṭīf al-Baghdādī moved to Jerusalem and stayed there for a while. In 604/1207 he set out for Damascus and from there to Aleppo, and beyond, into Anatolia, the land of the Rūm Seljuqs,* staying there for several years in the service of the governor of Erzinjan. In Dhū l-Qaʿdah 626/October 1228 he resumed his travels and arrived in Aleppo on Friday 9 Shawwāl/31 August 1229.

When ʿAbd al-Laṭīf al-Baghdādī was living in Aleppo it was my intention to meet with him, but it did not happen. I received a steady stream of books and letters from him, including some of his works in his own handwriting.

Here follows the text of a letter that I—Ibn Abī Usaybiʿah—wrote to him when he was in Aleppo:

'The servant conveys his prayers, his praise, his gratitude, and his commitment to the adored, eminent, illustrious, magnificent, most virtuous excellency, chief of scholars in times past and present who has united in himself the sciences scattered among the inhabitants of the world, protector of the Commander of the Faithful. May God elucidate the paths of right guidance to him and illuminate the ways of knowledge for him in the life hereafter, and confirm his authority through the true meaning of his words. May his happiness continue to exist unendingly, his mastery ascend to lofty heights, and may his writings remain in all lands the model for the learned and the main source for all men of letters and philosophy. The servant renews his homage, offers his most courteous salutations and his most affectionate thanks and compliments. He makes known to you the pain from which he suffers in his endeavour to witness the lights of the illuminating sun, the joy that is provoked by the exciting vision of your noble and illustrious presence and the intensification of his anxiety and the graveness of his insomnia on hearing of the nearness of his visit.

'Were it not for the expected return of the noble traveller and the arrival of the honourable and exalted excellency, the servant would have hurried to come to him and hastened to appear before him, and would come to pay his respects and be successful in seeing his beautiful appearance. How blessed is he who has the fortune to gaze upon it, and how glad is he who stands before him! How fortunate is the person in which he shows an interest, who draws from the seas of his excellence and is irrigated by its wholesome water, who is illumined by the sun of his knowledge and travels in brilliant light! We ask God, exalted is He, that He will soon unite us and by His grace and bounty bring about the merger between the delight of the eyes and the pleasures of the hearing, if God, exalted be He, wills.'

Among the letters of ʿAbd al-Laṭīf al-Baghdādī is one that he sent to my father. At the beginning of the letter he said of me, 'The son of the son is dearer than the son. This Ibn Abī Usaybiʿah is the son of my son* and no one is dearer to me than he. His excellence has been clear to me ever since his early youth.' He then continues to speak appreciatively of me and praises me. He also says in this letter, 'If I could go to him in order to enable him to study under my guidance, I should do it.'

Then it came to his mind to go on pilgrimage, making his way via Baghdad. After reaching Baghdad he fell ill and died—may God have mercy upon him—on Sunday 12 Muharram 629/9 November 1231. He was buried next to his father in the al-Wardiyyah cemetery. This happened after an absence of forty-five years from the city of Baghdad. God, exalted be He, sent him back and decreed his fate there.

The following are some examples of the aphorisms of ʿAbd al-Laṭīf al-Baghdādī, which I have taken from works in his own handwriting:

> It behoves a man to read histories, to acquaint himself with the biographies and experiences of nations, so that he becomes thereby, as one who, in this short life, has yet caught up with vanished peoples, has been their contemporary and companion and knows all the good and evil of them.
>
> It is necessary for you always to doubt yourself, rather than to hold a high opinion of yourself. Submit your ideas to men of learning and their writings. Proceed with caution, do not be hasty, and avoid being vainglorious, for vainglory ends in a fall, and rashness causes one to make mistakes.
>
> All activities aimed at gaining worldly goods require that one should occupy oneself exclusively with them, that one should possess the necessary skills to acquire them, and that one is able to devote one's time to them. Nevertheless, when a man masters science and makes a reputation for himself in it, he is in demand from all sides and positions are offered to him. Then the world comes to him in a submissive way and he conquers it with his honour perfectly intact and his dignity and his faith well kept. Furthermore, the learned man is loved wherever he is and whatever his situation is.

Beware of vulgarity in your discourse and harshness in debate, for that destroys the beauty of the speech, deprives it of its usefulness, robs it of its charm, causes grudges, and blots out friendships. The speaker then becomes a boring person, whose silence is more desirable to the listeners than his speech, and the people will choose to point out his faults, will denigrate him and take away his dignity.

Memorize a wealth of proverbial poems, aphoristic anecdotes, and unusual expressions.

Among the approximately 170 works of ʿAbd al-Laṭīf al-Baghdādī are the following books:

On what is sufficient and evident in Indian arithmetic.

On history (*K. tārīkh*), which comprises his biography. It was written for his son Sharaf al-Dīn Yūsuf.

On thirst.

On water.

On the enumeration of the aims of those who write books, and what kind of advantages and disadvantages follow therefrom.

On treatment by opposites.

Diabetes and suitable medicaments for it.

On rhubarb.

On the Egyptian lizard.

On the weighing of medical drugs in compound formulations.

On respiration, the voice and speech.

Treatise which deals with replies to certain questions about the slaughtering and killing of animals, and whether this is admissible from a natural and rational point of view, as it is according to religious law.

Treatise in which the analogical estimations that are considered right by Ibn Sīnā are shown to be false.

On minerals and the invalidation of alchemy.

13. YAʿQŪB IBN SIQLĀB (*d.* 1228)

Yaʿqūb ibn Siqlāb, a Christian, was one of the foremost persons of his time in the domain of knowledge, understanding, and critical examination of Galen's works. Early in my studies of the art of medicine, I—Ibn Abi Usaybiʿah—read some of the texts of Hippocrates with him. At the time, we were staying in al-Malik al-Muʿazzam [Sharaf al-Dīn ʿĪsā]'s military encampment, where my father too was employed. I observed that Yaʿqūb could explain everything excellently, in such clear, concise, and complete language as no one else could. He mentioned what Galen had said in his commentary with regard to the chapter in question, even quoting many of the very words used by Galen. He was the only man of his time capable of doing this.

Yaʿqūb suffered from gout in both legs and was sometimes in so much pain that he was hardly able to move, so that when al-Muʿazzam took him along on his travels, he was carried in a litter. Al-Muʿazzam visited him regularly, honoured him greatly, paid him a generous salary, and did him many favours. One day he asked him, 'O physician, why don't you cure that ailment in your legs?' Yaʿqūb replied, 'O master, once wood has become worm-eaten, there is no remedy for it.'

14. RASHĪD AL-DĪN IBN AL-SŪRĪ (*d.* 1242)

Rashīd al-Dīn ibn al-Sūrī had a comprehensive knowledge of all aspects of medicine and keen insight into the obvious and the hidden merits of that art. His knowledge of simple drugs, their nature, different names and characteristics, and the precise determination of their properties and effects was incomparable. He was born in 573/1177 in the city of Tyre. He grew up there, but later resided in Jerusalem for several years, practising medicine in the local hospital.

Rashīd al-Dīn once presented me with one of his books containing useful lessons and instructions concerning the art of medicine. By way of thanks, I wrote him a letter in which I said the following:

The knowledge of Rashīd* al-Dīn, in every assembly, has a lighthouse
    of lofty qualities, taken as a lead by every seeker of guidance.
A sage who possesses all noble traits,
    inherited from master to master:
He collected excellence from his fathers and grandfathers;
    it is something of old in him, not newly made.
He is unique in this era, without anyone resembling him,
    with the best characteristics that cannot be fully listed.

His fine *Instructions* came to me, which contained,
in prose speech, every well-composed paragraph.
Thus he imparted joy to my heart; he never ceases
to confer favours with his beneficence to people like me.
I found in them what I hoped for, and I shall
forever follow them in whatever I attempt.
No wonder that Rashīd, with his knowledge and excellence,
is, after God, in knowledge my guide.

Rashīd al-Dīn ibn al-Sūrī is the author of three books including:

On simple drugs. The book gives a full account of simple drugs, and provides insight into medicinal substances of which the author had acquired knowledge, and which had not been mentioned by his predecessors. Rashīd al-Dīn would go to places in which particular plants grew, such as Mount Lebanon and other spots, taking along with him a painter who had at his disposal all kinds of dyes and brushes. Rashīd al-Dīn would observe and examine the plants, and then he would show them to the painter, who would look at their colour, measure their leaves, branches, and roots, and then paint them, doing his utmost to make them as realistic as possible. Rashīd al-Dīn had an instructive method for these illustrations: first he would show them to the painter at the time of sprouting and tenderness, and would have him paint them at that stage. Then, he would show them to him when they were fully grown and in full bloom, and the painter would depict them at that specific stage. Finally, he would show him the plants when they were withered and dried up, and the painter would sketch them at that stage. In this way, the reader of the book could see the plants as he would encounter them in the field, and this would enable him to obtain more perfect information and clearer notions.

15. IBN RAQĪQAH (*d.* 1238)

Ibn Raqīqah was a man endowed with a noble soul and perfect virtues. He gathered together the medical teachings of the ancient authors that had become scattered, stood out above all his peers, and surpassed his fellow physicians and healers. Moreover, he possessed an outstanding character, flawless diction, and a wonderful gift for composing poems of high stylistic quality, of which many have become proverbs and maxims. As for verse in *rajaz* metre,* I have never seen any physician in his time who was quicker in composing it than he. He could take any medical work and render it in the *rajaz* metre in an instant, remaining faithful to the content and doing justice to the beauty of the words.

Ibn Raqīqah was also familiar with the art of ophthalmology and surgery, and in treating diseases of the eye, performing many surgical operations. He also removed cataracts from the eyes of many persons, who, thanks to his skill, were able to see again. The instrument that he used for that purpose was hollow and curved,* so that during the operation, the fluid could be more efficiently extracted, with the result that the treatment was more effective. [See Appendix 5, Figure 7 for an illustration of instruments for operating on the eye, including cataract needles, from a treatise on ophthalmology probably copied in Ibn Abī Usaybiʿah's lifetime.] Ibn Raqīqah also studied the *Book of ingenious devices* by the Banū Mūsā, from which he learned to make unusual things.

On the third of Jumādā II 632/23 February 1235, Ibn Raqīqah arrived at the court of al-Malik al-Ashraf Mūsa in Damascus, where he was kindly received and greatly honoured. He was ordered to attend the sultan's household in the citadel and also to treat the sick at the 'Great Hospital'. At that time, I—Ibn Abī Usaybiʿah—was also receiving payments for treating the patients at that hospital. Ibn Raqīqah and I became great friends. What I was able to observe of his perfect virtues, noble origins, rich knowledge, and excellent skills in the domain of diseases and their treatment is beyond all description. He lived in Damascus, devoting himself to the art of medicine, until he died—may God have mercy upon him—in 635/1238.

The following are some lines of Ibn Raqīqah's own poetry that he recited to me:

When I saw that people with excellence and intelligence
were not in demand, whereas any fool is,
I resigned myself to despair, knowing that I have
a Lord who is generous and who will grant what I desire.
I stayed at home and took as my companion
a book that speaks of all kinds of virtues.
In it, whenever I take it up to leaf through what it contains,
I have a lush and pretty garden.

The following lines he wrote on a wine cup, in the middle of which there was a bird sitting on a perforated dome. If a drink was poured into the cup the bird would turn round quickly and whistle loudly. The person facing the bird when it stopped had to drink. If he drank but left some drink in the cup the bird would whistle; likewise, if he drank it in one hundred draughts. But if he drank all the contents in one, without leaving as much as one dram,* the whistling would stop.

I am a bird in the shape of a sparrow,
beautifully shaped and formed.
Now drink to my tune a choice wine,
undiluted, which illumes the gloomy night,
Yellow, shining in the cups as if it were
the fire of Moses that appeared on the top of Mount Sinai;
And when one dram of your drink is left
in the cup, my whistling will alert you to it.

And he said—it is good advice—:

Beware of eating your fill, shun it!
Digest one kind of food before eating another.
Do not have sex often, for by doing it
continually one invites illness.
Don't drink water straight after eating
and you will be safe from great harm,
Nor on an empty stomach and being hungry,
unless you have a light snack with it.
Take a little of it: that is useful 5
when you have an aching, burning thirst.
Make sure your digestion is sound, that is the basic principle.
Purge yourself with laxatives once a year.
Avoid venesection, except for someone with
an illness of a mature and hot nature.
Do not exercise yourself straight after eating
but make it happen after digestion,
Lest the chyle* descend uncooked
and block the passages and pores.
But do not rest continually, for this means that 10
every humour in you will be made unhealthy.
Drink as little water as possible after exercise
and abstain from drinking wine.
Balance the mixing of your wine with water, for this preserves
the innate heat that always burns in you.

But do not become inebriated, shun it forever,
for drunkenness is something for common people.
Keep your soul well away from its cravings,
and you will attain eternity in the Abode of Well-being.

He also said:

A slender youth with languid eyes: he led
his lovers with his flirtation to the watering-place of death (*radā*).
He wore his cheek-down as an ample coat of mail, which protects him
from a lover's eye, while the glance of his eye is a sword (*ridā*).
If he had let me drink the coolness of his saliva,
this painful disease had not become a cloak (*ridā*) to me.
If he walks swaying from side to side he outdoes, with his bending, any sapling;
when he comes in sight he mocks the new Moon when it appears.
Whenever I look at the mole on his cheek he attacks
with a sword from both his eyes and becomes quarrelsome;
Or whenever I want one day's respite from my love of him,
he says, 'You intend to beg the question!'

He also said:

O young gazelle, for whose sake my exposure and disgrace
are pleasant, after guarding my reputation:
The sickness* (*ʿillah*) of your eyelids is the cause (*ʿillah*) of my disease,
and my cure is sipping the wine of your mouth.

And he said, as an elegy on a son of his:

Dear son, you have left in my breast,
because of losing you, a fire with blazing heat,
And you have incited my eyelids, after their sleep,
to be sleepless and now they never cease being wakeful.
I do not care, since you departed, about those who stayed behind,
not seeing anyone I should fear for or care for.
People say that grief diminishes the more time
passes, but my grief forever grows and increases.
I used to be steadfast when any calamity struck;
now, since you perished, showing fortitude is hard.
You were perfect; then fateful death came to you. Likewise,
an eclipse may come to a moon when it is full.

Ibn Raqīqah is the author of seven treatises.

16. AMĪN AL-DAWLAH (*d.* 1250–1)

The scholar and practitioner, the respected chief, the most excellent vizier, the chief physician, the learned authority, Amīn al-Dawlah was a Samaritan who converted to Islam under the name Kamāl al-Dīn. This Amīn al-Dawlah was a man of unsurpassable intelligence, whose knowledge was unparalleled among his contemporaries. He was charitable and high-minded, performed many acts of kindness, and continually bestowed favours upon everyone. He acquired exhaustive knowledge of the art of medicine, to the point that he had few peers, for even the learned and the accomplished were not able to attain to his superior status.

He first served al-Malik al-Amjad [in Baalbek]. Subsequently, Amīn al-Dawlah served as a vizier under the rule of al-Malik al-Sālih Ismāʿīl, emir of Damascus, enjoying eminent status, powerful authority, unquestioning obedience, and supreme importance. During his vizierate, Amīn al-Dawlah loved to accumulate money. He obtained large sums for al-Malik al-Sālih Ismāʿīl at the expense of the citizens of Damascus and took possession of many of their properties. When the sultan's representative in Damascus and the notables of the realm became aware of this, they decided to arrest him and confiscate his property. His possessions were seized and he was detained. He was then escorted to Egypt and thrown into prison in the fortress of Cairo.

After the Ayyubid ruler in Egypt died in 647/1249, al-Malik al-Nāsir Yūsuf marched from Aleppo, took Damascus, and, accompanied by al-Malik al-Sālih Ismāʿīl and other Syrian princes, marched on Egypt. The Mamluk ruler, al-Malik ʿIzz al-Dīn al-Turkumānī, who had come to power, led his army out to meet them. The Egyptian forces were repulsed at first, but rallied and finally routed the Syrians, capturing and imprisoning many of the princes. But al-Malik al-Sālih Ismāʿīl was never heard of again. It was said that he had been strangled with a bowstring.

News was mistakenly brought that the Syrian princes had defeated the Egyptian troops. When the vizier Amīn al-Dawlah, imprisoned in the fortress of Cairo, heard this, he said to the commander of the fortress, 'Let us go free in the fortress, until the Syrian princes arrive, and then you will see how well we shall treat you.' The commander of the fortress set them free. Amīn al-Dawlah went out to various places all over the fortress, commanding and forbidding. However, after the battle, the Egyptian victor, ʿIzz al-Dīn al-Turkumānī, ascended the fortress and ordered Amīn al-Dawlah hanged. Someone who had witnessed the hanging told me that he was clad in a vest of green ʿAttābī cloth* with his legs in gaiters which he had never seen on a hanged man.

I—Ibn Abī Usaybiʿah—say: A most amazing account of a judgement of the stars relating to these events, was told to me:

When Amīn al-Dawlah was imprisoned, he sent for an Egyptian astrologer, who possessed extensive knowledge of astrology and whose horoscopes were almost invariably accurate. Amīn al-Dawlah asked the astrologer about his situation and whether he would be released from prison. The astrologer examined the altitude of the sun at that particular moment, studied the degree of the ascendant, the twelve houses and the positions of the planets, wrote it all down, cast a horoscope, and made his prediction in conformity with it. 'Amīn al-Dawlah will be released from prison,' he said, 'and will leave it cheerful and happy. He will be favoured by fortune and remain in a high position in Egypt, and his orders and commands will be obeyed by all the people.'

When this response reached Amīn al-Dawlah, he received it joyfully. Upon being mis-informed of the arrival of the Syrian princes and their victory, he went out quite sure that he would remain a vizier in Egypt. Thus, the astrologer's prediction of his release from prison, his happiness, the obedience to all his orders and commands, and his ending up in a high position came true that day. But Amīn al-Dawlah did not suspect what would happen to him later on, for God, mighty and glorious, was already preparing that which had been predestined for him and was written in the Book.

Amīn al-Dawlah had a virtuous soul. He took a keen interest in collecting and studying books, and purchased many outstanding works in all the sciences. Once he desired a copy of a work in eighty volumes in minute script. 'One copyist will never be able to cope with this book,' he said, and he divided it among ten copyists, each of whom worked on eight volumes. They finished the work in approximately two years, and the whole book came into his possession. This shows his boundless ambition.

When Amīn al-Dawlah—may God have mercy upon him—was occupying the office of vizier in Damascus in the days of al-Malik al-Sālih Ismāʿīl, he was a close friend of my father's. One day he said to him, 'I have heard that your son has composed a book on the classes of physicians that is unprecedented. All the physicians in my service praise him greatly for his highly valuable book. I have in my library more than twenty thousand volumes, but none in that particular domain. I would like you to send him a letter and ask him to have a copy of that book made for me.'

At that time I was in Sarkhad, at the court of its ruler, and subject to his orders. Upon receiving my father's letter, I went to Damascus, taking along with me the rough drafts of my book. There, I called upon the illustrious copyist Shams al-Dīn Muhammad al-Husaynī, who did a lot of copying work for us: his handwriting was excellent, and his mastery of the Arabic language was admirable. I gave him space at our home, where he copied the book in a fairly short time, putting it into four sections, in quarter Baghdadi format.*

Having had these bound, I composed a panegyrical poem for Amīn al-Dawlah and sent all these items to him by the hand of the chief qadi of Damascus, Rafīʿ al-Dīn al-Jīlī.* When Amīn al-Dawlah read the book and poem, he was greatly surprised and extremely happy. He sent me back a large sum of money and honorary robes, along with many expressions of gratitude. 'It is my desire that you notify me of every new book you write,' he said.

Here is part of the poem that I composed for him, at the beginning of 643/1245:

My situation (*hālī*) with the people of this time is not sweet (*hālī*);
    my innermost thought (*sirr*) is not mixed with joy (*surūr*).
But if I complain about the time, my treasure is
    the Sāhib* Amīn al-Dawlah, the vizier:
A generous man, liberal, giver of favours
    that are general as a dark cloud pouring its rain.
He has risen in the sky of glory until 15
    the ether was marked with his footsoles' traces.
Can any poetry express his lofty qualities
    when Sirius is located beneath him?
He has authority and justice, continuously;
    through him people's affairs are justly balanced.

In times of famine (*azamāt*) he is charitable (*mubirr*) to the petitioner;
in times of firm resolve (*ʿazamāt*) he is a destroyer (*mubīr*) to the aggressor.
He has surpassed the ancients in noble deeds;
and how many an ancient one was surpassed by a later one?

I have also copied the following verses from a secretary. He wrote them for Amīn al-Dawlah, asking him for a manuscript promised him in 627/1227:

You promised the manuscript, so send what you promised,
O you who bestows benefits continuously without condescension!
He who does a good deed reaps every honour
and buys, without paying a price, eulogies that will be recited.
A manuscript that will increase your good fortune,* as long as
a grey dove coos on a branch in the trees.

Amīn al-Dawlah composed *The clear path in medicine*, which is one of the best books ever written on the art of medicine.

17. AL-DAKHWĀR (*d.* 1230)

Our teacher, the great and eminent authority, the learned and excellent Muhadhdhab al-Dīn ʿAbd al-Rahīm ibn ʿAlī, who was also known as al-Dakhwār, was—may God have mercy upon him—the outstanding man of his period, unrivalled during his lifetime, the most learned scholar of his generation. He held a leading position in the art of medicine. There was no one who could match him in diligence or keep up with him in respect of knowledge. He drove himself unsparingly, exhausting his mind in order to attain knowledge, until he surpassed all his contemporaries. To the day of his death, he enjoyed the good graces of rulers and was presented by them with more wealth and honour than any physician had ever enjoyed before.

Al-Dakhwār was born and raised in Damascus. His father was a renowned oculist and his brother was also. Al-Dakhwār was initially an oculist as well, but also worked as a copyist. His calligraphy was of a high order of skill, and he transcribed many books, of which I have seen at least a hundred or more volumes. He studied the Arabic language with the grammarian al-Kindī and constantly persevered to attain more knowledge by reading and memorizing, even during his periods of service, until his middle age. At the beginning of his medical studies, he studied under Shaykh Radī al-Dīn al-Rahbī—may God have mercy upon him. Subsequently, he attached himself to Ibn al-Mutrān, learning the art of medicine from him until he became a skilled and proficient physician in his own right.

Al-Dakhwār entered the service of al-Malik al-ʿĀdil I as a physician. When the sultan became ill, he was attended by the best doctors, including al-Dakhwār, who advised bloodletting, but the attending physicians did not approve. 'By God,' said al-Dakhwār, 'If we do not let blood from him, he will bleed of his own accord.' Very soon thereafter the sultan experienced heavy nosebleeds. When he recovered, he knew that al-Dakhwār outshone all the other doctors.

A similar story has it that one day, when al-Dakhwār was standing at the palace gate with some of the court physicians, a servant came out with a phial of urine from one of the slave-girls, saying that the girl was complaining of pain. When the other physicians had examined the phial, they prescribed something that they had prepared. But when al-Dakhwār examined the phial, he said, 'It is not the pain of which she complains that has caused the colour of the contents of this phial,' suspecting that the source of the colour was the henna with which the girl had been dyed. The servant informed him that he was correct and reported back to al-Malik al-ʿĀdil; this increased the ruler's confidence in him.

The following account of one of al-Dakhwār's most generous actions, one that illustrates his great sense of honour and solidarity, was told to me by my father, who said:

Al-Malik al-ʿĀdil was once very angry at the chief judge of Damascus. He had him imprisoned in the citadel, and he ruled that the judge was to pay him the sum of 10,000 Egyptian dinars. He managed to pay some of it, but was unable to raise the balance. Al-Malik al-ʿĀdil took the matter very seriously: 'He must pay the rest of the money, for otherwise I shall have him tortured.' The judge was at his wit's end, he became so worried about it that he hardly ate or slept and was on the verge of killing himself. Then, his old friend, the physician al-Dakhwār, paid him a visit. The judge complained to him about his troubles and asked him for help. Al-Dakhwār thought it over for a while and then said, 'I shall think of something for you and hope that it will be of use, if God, the exalted, wills,' and took his leave.

It so happened that the concubine of al-Malik al-ʿĀdil, the mother of al-Malik al-Sālih Ismāʿīl, emir of Damascus, was feeling out of sorts at that time. She was of Turkish origin, an intelligent, pious, and devout woman, and was very kind and generous. The physician al-Dakhwār came to see her, accompanied by the chief eunuch, and he brought up the situation of the judge. He requested her to mediate, in the hope that she could persuade the sultan to show mercy towards the judge and be lenient towards him. But the concubine said, 'By God, how can I do anything for the judge, or even mention him to the sultan? I cannot do this, because he will say to me, "What makes you speak about the judge, and how is it that you know of him?" If he were, for instance, a doctor who visits us from time to time, or a merchant who sells us cloth, it would be possible for me to speak to the sultan and intervene; but it is impossible for me to speak about this person.'

When the physician heard this, he said, 'My lady! You do not have to intervene with the sultan at all.' 'How so?' she asked. 'When the sultan and you are sleeping together,' said the physician, 'say that you saw in a dream that a judge was being treated unjustly.' He told her what to say, and she replied, 'It can be done.'

When she was well again and al-Malik al-ʿĀdil was sleeping next to her, as the night was ending, she lay awake pretending to be frightened, clutching her heart, trembling and crying. The sultan, who loved her dearly, woke up and said, 'What is the matter?' but she did not tell him what the matter was. He then ordered some apple juice to be brought, had her drink some, sprinkled her face with rose water, and said, 'Why don't you want to tell me what has happened to you and what is on your mind?' 'O husband,' she replied, 'I have had a terrible dream, which almost frightened me to death. I dreamed that the day of final judgement had come and I saw a large crowd of people. In one place, where there was a great fire burning, people were saying, "This is for al-Malik al-ʿĀdil, because he treated the judge unjustly." Did you ever wrong a judge?' she asked.

He did not doubt her words, felt uneasy about it, then rose up, called his servants, and said, 'Go to the judge and delight his heart, give him my regards and apologies for what has happened to him, and inform him that all he has paid will be returned to him. I, for my part, will ask nothing of him.' So they went to him. The judge was delighted with their news, blessed the sultan, and announced that he accepted his apology. When morning came, the sultan ordered that he be given a full robe of honour and a mule. He restored him to his office and ordered that all the money he had paid should be reimbursed from the treasury, and that all the books and other possessions he had sold were to be redeemed from the purchasers for the same amount as they had paid. Thus, relief was brought to the judge after hardship, by minimum effort and the subtlest of measures.

In 610/1213, when he was in the East, al-Malik al-ʿĀdil became very ill, and the physician al-Dakhwār treated his illness until he was cured. During that illness, the sultan paid him approximately 7,000 Egyptian dinars. The children of al-Malik al-ʿĀdil, some rulers of the East, and others also sent him gold, robes of honour, and mules, together with golden necklaces and the like.

I—Ibn Abī Usaybiʿah—continue: When al-Malik al-Muʿazzam [Sharaf al-Dīn ʿĪsā] established his rule over Syria, he wished to employ a number of those who had served his father, al-Malik al-ʿĀdil, among them my father. The physician al-Dakhwār, for his part, was provided with an ample salary and instructed to return to the 'Great Hospital' in Damascus where he was to treat patients.

During his time in Damascus, al-Dakhwār began to teach the art of medicine, and many of the best physicians joined him, while others studied under him. I, too, stayed in Damascus to learn from his teaching, but I had first worked under him at the military camp where he and my father were serving the great sultan, al-Malik al-ʿĀdil. I would frequent his classes as one of a group, and I began to study the works of Galen. I accompanied him also during my period of training at the hospital while he was treating the patients, and thus I gained practice in the art of medicine.

The physician al-Dakhwār became one of the most able representatives of the art of medicine. He was a prodigy and would prescribe medicaments that could cure in almost no time. Hence the impression was given that it was magic. Once, I saw him perform a feat of that kind. A man had come to him with a burning fever and extremely dilated pupils. After estimating the patient's strength, al-Dakhwār ordered that a quantity of camphor seeds should be pounded in a drinking cup.

Writing this down in a recipe, he told the man to drink it and not to take anything else. When the morning came, we found that the patient's fever had broken and that his pupils were no longer dilated.

It also happened, whilst in the ward for bilious patients, he treated someone suffering from the disease called mania, which is rabies, by prescribing that an ample amount of opium should be added to his barley water at the time when he was given it to drink. The man became better and his condition improved at once.

Meanwhile, al-Malik al-Ashraf Mūsa resided in the East. The physician set out in that direction in Dhū l-Qaʿdah 622/November 1225. Al-Dakhwār was in the service of al-Malik al-Ashraf for some time, but then contracted a speech problem, was overcome with lassitude, and was unable to talk at any length. He moved to Damascus when al-Malik al-Ashraf gained control over that city in 626/1229 and was appointed chief physician. He held that post for quite a long time, and the ruler created a *majlis** for him for instruction in the art of medicine. As time went on, his speech problem grew worse; when he tried to speak it was very hard to understand him. His pupils would discuss issues in front of him. Whenever a meaning was difficult, he would respond with the shortest word that pointed to the essence of it. At times it was difficult for him to speak at all, and then he would write on a slate, and the pupils would read what he had written.

Al-Dakhwār tried hard to cure himself and cleansed his body with several types of purgatives. He also took many medicaments and hot electuaries, swallowing them constantly. Then he contracted a fever that became so intense that his strength failed, with the result that many diseases followed in succession. When one's term has been reached, effort is in vain.

Al-Dakhwār passed away—may God have mercy upon him—in the early morning of Monday 15 Safar 628/22 December 1230 and was buried on Mount Qāsiyūn. He left no offspring.

In 622/1225 (before he had left Damascus to enter the service of al-Malik al-Ashraf) he dedicated his house in Damascus, near the old goldsmith's quarter east of the great market, as a charitable trust and converted it into a college for the study of the art of medicine.* For its support, he did the same with several estates and other properties, the revenue from which was to be used for its upkeep, the pay of the teacher and stipends for students.

Ibn Kharūf was a poet from the Maghrib who frequently ridiculed al-Dakhwār. He met his end in Aleppo, where he had gone to praise its ruler, al-Malik al-Ẓāhir. After reciting his eulogy, he took a step back. There was a well there, into which he fell and died.

Among the poetry of al-Dakhwār are the following lines, which he wrote to my paternal uncle, the physician Rashīd al-Dīn ʿAlī ibn Khalīfah, when he had fallen ill.

> You, for whom I hope when any misfortune occurs
> and for whom I fear if he has any symptoms (*aʿrāḍ*):
> Far be it that you should be visited on account of an illness,
> and may you live as long as we are in good repute (*aʿrāḍ*)!
> We count you as the substance of our epoch,
> while others, if counted at all, are accidents (*aʿrāḍ*).

Al-Dakhwār is the author of seven works.

18. MY PATERNAL UNCLE, RASHĪD AL-DĪN ʿALĪ IBN KHALĪFAH (*d.* 1219)

Rashīd al-Dīn ʿAlī ibn Khalīfah was born in Aleppo in 579/1183. My father had been born before him in 575/1179, in Cairo. They both grew up and studied in that city.

My grandfather—may God have mercy upon him—was a high-minded person, who had a great liking for men of virtue and studied the sciences himself. He was also known as Ibn Abī Usaybiʿah. He had moved to Egypt when Saladin conquered it, and was in his service and that of his sons.

Among my grandfather's acquaintances and friends in Damascus had been Jamāl al-Dīn,* the physician, and Abū l-Hajjāj Yūsuf, the oculist, for my grandfather was born and bred in Damascus and resided there for many years. By the time he met them again in Egypt, my father and my paternal uncle were in the prime of life. My grandfather had in mind to teach them both the art of medicine, because he was well aware of its noble rank and the people's great need for physicians, and held that one who was committed to its truths would be honoured and favoured in this world and be given the highest rank in the world to come. Accordingly, he set my father and my uncle to study under the guidance of these two shaykhs.

My grandfather set my father to study the science of ophthalmology under Abū l-Hajjāj Yūsuf who was then serving as an oculist in the hospital in Cairo—that is, not the later hospital belonging to the fort, but the older one that was situated, at that time, near the flea markets of lower Cairo. My grandfather lived nearby, so that my father was able to attend frequently, until he became an expert in that domain.

My uncle, for his part, was set to study the art of medicine. After my uncle—may God have mercy upon him—had learned to memorize the Qur'an and had become acquainted with mathematics, he began to study the art of medicine thoroughly under Jamāl al-Dīn who was then the chief physician in Egypt. Accordingly, he engaged in discussions with the physicians, saw the patients in the hospital, and learned about the various maladies and the appropriate prescriptions (there was a group of very notable physicians at the hospital).

The shaykh ʿAbd al-Latīf al-Baghdādī, who was a close friend of my grandfather's, was then living in Cairo. My uncle studied a little Arabic and philosophy under his guidance. They used to discuss Aristotle's books, debating the difficult passages. From an early age, my uncle devoted all his spare time to studying the sciences and filling his soul with virtues.

My grandfather returned to Syria in 597/1200. My uncle was then no more than approximately 20 years old, but he immediately began to treat patients and improve his knowledge of the art of medicine. He visited patients in the hospital founded by Nūr al-Dīn ibn Zangī, where the physician al-Dakhwār was also working. At the same time, he studied philosophy under ʿAbd al-Latīf al-Baghdādī, who had returned to Syria.

In addition, there was in Damascus a group of literary scholars who were celebrated for their knowledge of the Arabic language, whom my uncle came to know and under whom he studied, and the grammarian, al-Kindī, who had been a good friend of my grandfather's. My uncle attended his teaching sessions, and studied the Arabic language under his guidance. Before my uncle had reached the age of 25, he had already mastered all these sciences and become a shaykh whose example was followed and who had his own students. He also composed poetry, kept up a correspondence, spoke Persian, knew Persian grammar, and even composed poetry in it. He spoke Turkish as well.

In 609/1213 an esteemed eunuch of the sultan, called Sulaytah, became afflicted with an eye disease. Both eyes were affected, and his condition deteriorated to such an extent that he despaired of recovery. The best physicians and oculists were unable to cure him; they decided unanimously that he must inevitably become blind. When my father saw him and examined his eyes, he said, 'I will treat this man's eyes and he will see with both of them, if God, exalted be He, so wills.' In response to his treatment, both Sulaytah's eyes steadily improved, until his recovery was complete. He became his former self and was able to ride a horse again, so that the people were astonished and regarded the treatment as an unrivalled miracle. As a result, al-Malik al-ʿĀdil I gained a good impression of my father and paid him the utmost honour by presenting him with special robes and other items.

Even before this achievement, my father had been accustomed to frequent the palace of the sultan in the citadel of Damascus, treating those who were afflicted with serious eye diseases and curing them in short order. This also came to the attention of al-Malik al-ʿĀdil. 'Such a man should go with me wherever my travels lead me!' he exclaimed, and asked him to enter his service. My father asked to be excused and permitted to remain in Damascus, but his request was not granted. The sultan offered him a salary and allowances, and my father finally enrolled in his service on 15 Dhū l-Hijjah 609/9 May 1213. The sultan and all his sons relied on him for medical treatment, and they treated him with great generosity, bestowing many favours upon him.

In addition, my father, Sadīd al-Dīn al-Qāsim, used to frequent the 'Great Hospital' of Nūr al-Dīn Ibn Zangī, where he also received a salary and allowances. People flocked to him from all sides, when they found out about his rapid cures. Diseases that required the use of surgery he treated by that means, and those that could be treated with drugs he treated by that means, sparing those patients the ordeal of surgery. This method was praised by Galen: 'If you see a physician administering drugs in case of maladies that are usually treated by means of surgery,' he says, 'you may conclude that such a doctor is learned, experienced, and skilled.'

One of his patients, who was cured by him, composed the following poem about him:

Sadīd al-Dīn's ability in medicine
always saves an eye from its sore:
From so many an eye has it cleared its darkness
and from so many eyelids it has removed harm!
Eye doctoring should never be practised among mankind
except by such a skilled practitioner.
O Christ of our time! So many, blind from birth,
became seeing again through you, this one, that one . . . !
Through your sound opinions there is a cure for the disease,
in your words there is food for the soul.
I have obligations to you, the least of which, if I were
to thank you, would be 'Bravo!'

My father remained in service in the citadel of Damascus and frequented the 'Great Hospital' founded by Nūr al-Dīn ibn Zangī, until he died—may God have mercy upon him—during the night of Thursday 22 Rabīʿ II 649/14 July 1251. He was buried outside the Paradise Gate on the way to Mount Qāsīyūn.

My uncle (Rashīd al-Dīn ʿAlī), for his part, was serving at the court of al-Malik al-Amjad in Baalbek when al-Malik al-Muʿazzam [Sharaf al-Dīn ʿĪsā] came to reinforce al-Malik al-Amjad and help him fight his adversaries, the Hospitallers.* When the two princes met with their respective suites, my uncle would join them. At that time, there was no one who had a better knowledge of music and the art of playing the lute than he, nor was there anyone with a better voice, so that the listeners found their souls touched with deep emotion. Al-Malik al-Muʿazzam was greatly impressed by my uncle and engaged him in his service, beginning on 1 Jumādā I 610/18 September 1213. The sultan granted him a salary and allowances, visited him frequently, and treated him most generously. He spent most of his time in the company of his physician and relied upon him in all matters relating to the art of medicine. The same can be said of his brothers, al-Malik al-Kāmil and al-Malik al-Ashraf, both of whom depended upon him. Whenever one of them came to visit al-Malik al-Muʿazzam my uncle would constantly be at their side, and he obtained many presents from both of them.

I know of one occasion, when al-Malik al-Kāmil came to visit his brother; they had a meeting in a friendly atmosphere, and my uncle sat with them. That same night, al-Malik al-Kāmil gave my uncle a complete robe of honour and 500 Egyptian dinars.

When al-Malik al-Muʿazzam was in Damascus, he appointed my uncle as military secretary, and insisted on his acceptance of the post. The only thing my uncle could do was to obey this order. He sat in the administrative office and received the common soldiers and the officers. He spent all his days in his secretarial post, but realized that most of his time was spent in correspondence and calculations, with no spare time at all to devote to the rational sciences and other matters. He appealed to the sultan to be released from his job, asking a group of his intimate friends to put in a good word for him, until the sultan acceded to his request.

In the year 611/1214 my uncle accompanied al-Malik al-Muʿazzam on the pilgrimage to Mecca. Following that after much travel, my uncle then fell ill in 614/1217; his illness continued for the rest of that year, and he found that travelling was harmful to him. He was, by nature, inclined to solitude and the study of books.

While in Damascus, my uncle also met the learned authority, the Shaykh of Shaykhs, Sadr al-Dīn ibn Hamawayh, who presented him with the attire of Sufis on 20 Ramadan 615/10 December 1218. The following is the text of the inscription that was attached to his Sufi garment: 'In the name of God, the Merciful, the Compassionate; the esteemed master and learned authority, Shaykh of Shaykhs, may God maintain his support forever . . . herewith endows his novice, Rashīd al-Dīn ʿAlī ibn Khalīfah, may God grant him success in his obedience, with a Sufi garment.'

While dressing him in it, the Shaykh told my uncle that he had received that robe from his father—may God have mercy upon him—and that his father received it from his father, the Shaykh of Islam,—may God have mercy upon him. He had been presented with it* by the prophet al-Khidr*—peace be upon him—, who in turn had received it from the Messenger of God himself—God bless him and keep him.

In 616/1219, my uncle received a message from al-Malik al-Sālih Ismāʿīl, emir of Damascus, in his own handwriting, asking him to come to the town of Bosra and treat his mother and other sufferers at the court. It happened that a great epidemic was raging in Bosra. My uncle went there and successfully treated the sultan's mother, who felt well again within the shortest possible time, whereupon he was presented with gold and honorary robes. Shortly thereafter, however, he was stricken with an acute fever, which grew steadily worse, even after his return to Damascus. The best and most venerated physicians tried to cure him, but his time had come. He passed away—may God have mercy upon him— in the second hour of Monday 17 Shaʿbān 616/28 October 1219, at the age of 38. He was buried near his father and brother outside the Paradise Gate.

The following are some of my uncle's many wise sayings, as I heard them from him—may God have mercy upon him:

> Exhortation for the beginning of the day: 'This day has come, in which you are prepared to do all kinds of things. Choose to perform the finest deeds, so that you will be able to reach the highest of ranks. You should do good, for that will bring you nearer to God and endear you to men. Beware of evil, for it will keep you away from God and make people hate you.'
>
> Exhortation for the beginning of the night: 'Your day has passed with all that happened in it. Now this night has come, in which you do not have a necessary physical task to fulfil, so turn towards the things that are beneficial for you, by studying the sciences, and by reflecting on the knowledge of the true sense of things; as long as you can stay awake, do this; when you are feeling sleepy, concentrate on the subject of your concern, so that your dreams may also be of the same nature. Do what will be creditable to you tomorrow.'
>
> Know that this day of yours is a piece of your life that will be gone forever, so spend it on what might benefit you later; if you have satisfied your bodily needs, finish the rest of your day by doing things that are beneficial to you. Do unto people as you would like them to do unto you. Beware of anger and the sudden impulse to take revenge on an angry man or to dissociate from him, for you may come to regret it; you should be patient, for patience is the principal part of all wisdom.
>
> Respect your teachers, even if they kept silent and did not answer your questions. Perhaps that was because they learned things long ago (and have forgotten them), or because of weariness, or because you asked something that is not of your concern, or because they believed that you would not understand the answer; know that the benefit you will derive from them is greater than all of this.
>
> Be truthful, for a lie makes a man feel inferior in his own eyes, let alone in the eyes of others.

Stay far away from the rulers of this world and you will spare yourself the company of evil persons.

Diseases have their own duration, and remedies need the help of fate. The art of medicine is largely mere conjecture and assessment, in which certainty is a rare occurrence. Its two parts are analogy and experience, not sophistry and love of dominance.

What a wonderful thing is unanimous opinion!

Solitude is the best time of life.

Solitude is the best way of life.

Do not strive for solitude as long as you still have the least spark of ambition.

Here is some of the poetry that I have heard from my uncle—may God have mercy upon him:

My two friends,* ask Passion and leave me!
    What do you want from a yearning, suffering man?
Don't ask him about parting and how it tastes:
    parting is another kind of death.
The camel drivers have called: 'Departure will be soon, so say farewell!'
    Thus I was bereaved of my heart and my friends.
Their camels set off when darkness had fallen,
    but light shone from those who travelled, carried on the camels.
I did not know that your being far would kill me,
    until I did and I was deluded in thinking myself consoled.
I cried from passion after that, to no avail.
    How else, since meeting has turned into wishing?

He said, describing a gathering:

May rain bless a day on which our joy
    was complete and a cup of cool wine brought us together;
When Fate's vagaries had turned away from us
    and we, in delight, attained our desires
In a gathering perfect in its loveliness: if al-Junayd*
    had been there he would have been charmed.
We had fun there and fruit,
    and a cup of wine, and leisure, and song,
Amid drinking companions like suns, men 5
    of learning, excellence, high standing, and brilliance,
Whose conversation does not bore the listener,

so nice that the eye would envy the ear;
Sincere friends, their minds pure,
chaste, harbouring no immoral thought,
Magnanimous men, always doing good things that
earn them praise among people.
We recited our love poems, turning them into riddles
on the name of a gazelle who came to flirt with us.

He also said:

I would give my life for him of the graceful figure, who has
no equal in beauty and beneficence,
Drowsy, though his lover's eyelids
cannot escape a visit from insomnia.
His saliva seems a vintage wine,
cooled with water and ambergris perfume.
But now he resists me,
abandoning me, turning away, rejecting me.
I shall have to bear with his being bored with me;
perhaps my endurance will help me.

He said, as a riddle on the name Aqish:*

You who ask me about him whom the moons resemble:
Not so fast! I will conceal him forever.
The name is composed* of T and A;
a sixth of its third is half of its second,
And the first of the name is a tenth of Y;* so pay attention
to what I say and conceal it. I shall not name him.

My paternal uncle Rashīd al-Dīn ʿAlī ibn Khalīfah is the author of eight works.

19. NAJM AL-DĪN IBN AL-MINFĀKH (*d.* 1254)

This great physician and noble scholar was known as 'the son of the Bellows' (Ibn al-Minfākh). His mother was a Damascene singer* named Bint Dahīn al-Lawz, 'the daughter of one anointed with almond-oil'. Born in 593/1196, he was brown-skinned, of slender build, with a sharp mind, highly intelligent, and eloquent. No one could equal him in research or match him in debate. He studied the art of medicine under the guidance of our master, the physician al-Dakhwār, and in due course became distinguished in that domain himself. His writings are witty and well composed. He composed epistles and poetry, knew how to play the lute, and had fine handwriting.

Ibn al-Minfākh served al-Masʿūd, as a physician. For a time he enjoyed his patron's favour and was appointed vizier, but eventually the ruler became hostile and confiscated all his belongings. As a result, he removed to Damascus and settled there.

He died—may God have mercy upon him—on 13 Dhū l-Qaʿdah 652/25 December 1254. The judge, his half-brother by his mother, told me that he had died of poisoning.

Najm al-Dīn ibn Minfākh is the author of the following work among others: *Disclosure of the distortions of al-Dakhwār.*

20. ʿIZZ AL-DĪN IBN AL-SUWAYDĪ (*d.* 1291)

The great physician and renowned scholar ʿIzz al-Dīn ibn al-Suwaydī was born in 600/1203 in Damascus, where he grew up to become the most erudite man of his time and the cynosure of his generation. ʿIzz al-Dīn unites in himself all the virtues: outstanding excellence, noble ancestry, perfect manliness, and boundless generosity, and he is a guardian of brotherliness. He studied the art of medicine until he reached the utmost perfection in it, such as was never attained by any other master.

ʿIzz al-Dīn's father—may God have mercy upon him—was a merchant from al-Suwaydāʾ in the Hawrān, a man with a fine character. Between him and my father there was a firm and praiseworthy friendship.

I myself studied with ʿIzz al-Dīn. Our long-standing friendship has remained the same throughout the years, and has even grown steadily with time. The physician ʿIzz al-Dīn is indeed the most illustrious physician of his generation with respect to his knowledge and memory, treatment and amicability, beneficial cures and precise methods. He is still practising as a doctor at the al-Nūrī hospital, granting patients their ultimate desire by taking away their maladies and according them the finest gift by supplying them with health. He has also served at the hospital in the district of Bāb al-Barīd, at the citadel of Damascus, and taught at the medical college established by al-Dakhwār, receiving salaries from all of these posts.

ʿIzz al-Dīn has copied in his own handwriting a great many books on medicine and other sciences. Some of these are written in accordance with the method of Ibn al-Bawwāb,* whereas others resemble *muwallad* Kufic script.* Each of these scripts is more radiant than the most sparkling stars, brighter than the most sumptuous jewels, finer than the prettiest gardens and filled with more light than the rising sun.

The following is a specimen of his poetry, which he recited to me himself. In it, he is preoccupied with and worried about the discomfort of dyeing his hair with *katam*:*

If changing the colour of my grey hair
could bring back my lost youth
It would not fully compensate to me what my spirit
suffers from the trouble of dyeing it.

The following verses are some that he recited to me concerning my book on the history of physicians entitled *The best accounts of the classes of physicians*.

Muwaffaq al-Dīn* (Ibn Abī Usaybiʿah), you have achieved what you desire
and have reached the highest of splendid ranks!
You have provided a fine history of those who have gone,
though their bones have now decayed.
May God single you out with His beneficence
in this world and the next.

APPENDIX 1

WEIGHTS AND MEASURES

ALL weights and measures varied with time and place. Only approximate values can be given today. The classic study of weights and measures in Islamic lands is that of Walther Hinz, *Islamische Masse*; available in English translation, *Measures and Weights in the Islamic World*, Kuala Lumpur, 2003. It implies, however, a greater precision than existed in earlier times.

cubit (*dhirāʿ*) — An ancient unit of length that may have originated in Egypt close to 5,000 years ago, approximately equivalent to the length of the human arm from elbow to fingertip. The Egyptian cubit (*c.*52 cm) was used to calibrate the nilometer at Roda, an island in the Nile near Cairo. However, the 'Egyptian linen' cubit was much larger, approximately 73 cm.

dāniq — A small coin, one-sixth of a dirham in value, as well as the corresponding weight (roughly half a gram). It is derived from Middle Persian *dānag*, 'seed, grain'.

dinar (*dīnār*) — A gold coin weighing about 4.25 grams. Its name derived from the Roman *denarius.* First shaped by the Umayyad coin reform in 696, 'dinar' is still the name of various national currencies.

dirham (*dirham*) — A basic unit of weight (slightly over 3 grams) as well as a silver coin. As a monetary unit, its value varied greatly. It dates back to the Greek drachma, which, borrowed as *darāhim*, sounded like an Arabic plural and thus gave rise to the singular form *dirham*. It formed the basis of Islamic weighing in the pre-modern era. Dirhams are still used today in a number of Arab countries.

faddān — A commonly agreed quantity of land, loosely reckoned as the quantity of land which a yoke of oxen can plough in one day.

mithqāl — A basic unit of weight usually equivalent to 1 + 3/7 dirhams, or nearly 4.4 grams, though its value varied greatly. It was also a gold coin of that weight.

parasang (*farsakh*) — An ancient Iranian unit of distance, roughly equal to 3.5 miles or 5.5 km. The Arabic *farsakh* (from the

Parthian *frasakh*, Old Persian *parāthanga*, cf. Greek *parasangēs*, modern Persian *farsang*) was most often defined as equal to 3 Arabic miles (*mīl*) or 12,000 cubits.

ratl — A measure of capacity as well as a unit of weight, both of which varied considerably depending on time and place. The Baghdad *ratl* weighed approximately 400 grams and as a measure of liquids or capacity it was equivalent to approximately 400 ml. The word is derived from Greek *litron* (cf. 'litre').

shibr — A variable unit of height or length, frequently translated as 'span' and often defined as the span of a person's hand, or the maximum distance between the tip of the thumb and little finger; as used here (Ch. 8, no. 5), however, more likely equivalent to about a foot or 32 cm.

tapia (*tābiyah*) — A unit of measure, used exceptionally (see Ch. 13, no. 12), roughly equivalent to 10 spans. The *tapia* originates from a Maghrib term for a panel of wall made with mud pressed and dried in a wooden-frame; in Spanish it is *tapia*, Portuguese *taipa*.

APPENDIX 2

GAZETTEER OF PLACE NAMES

al-ʿAdhrāwiyyah law college	A *madrasah* founded in Damascus by ʿAdhrāʾ (d. 1197), a daughter of Shāhanshāh ibn Ayyūb, Saladin's oldest brother.
al-ʿĀdiliyyah law college	A *madrasah* in Damascus named after al-Malik al-ʿĀdil ibn Ayyūb (r. 1196–1218), a brother of Saladin. It was begun, but not completed, by Nūr al-Dīn ibn Zangī, and completed only by al-ʿĀdil's son al-Malik al-Muʿazzam.
ʿAdudī hospital	In Baghdad, named after the Buyid ruler, ʿAdud al-Dawlah (r. 949–83). The hospital was built in 979–80, and lasted until its destruction in 1258 by the Mongols during the Siege of Baghdad.
al-Ahwāz	An early important commercial town situated halfway between Baghdad and Shiraz.
ʿAlth	A town north of Baghdad in Iraq.
al-Anbār	An important town on the left bank of the Euphrates approx. 12 parasangs from Baghdad.
al-Andalus	Islamic Spain and Portugal.
Asyūt	A city in Upper Egypt important for trade and agriculture.
al-Azhar	A famous mosque and *madrasah*, founded *c*.970 in Cairo.
Bāb al-Barīd	A gate and district of Damascus; a small hospital was situated there to the west of Great Mosque of the Umayyads and had been functioning for about half a century before Nūr al-Dīn ibn Zangī founded what came to be called the Nūrī hospital to the east of the Great Mosque.
Badhdh	A district and fortress of northern Azerbaijan.
Badr	A place south-west of Medina at the junction of a road from Medina with the caravan route from Mecca to Syria. The site of the first battle of the nascent Muslim community against the Meccans, in 624.
Bahrain (Ar. *al-Bahrayn*)	In the medieval period, the name for the mainland of Eastern Arabia.
Balat (or Balad)	An old town on the Tigris above Mosul.

Balkh (classical Bactria) — In the Balkh Province of modern Afghanistan, near modern Mazar-i Sharif; the birth place of Ibn Sīnā.

Bamyan — A town on the ancient silk route, now in modern-day Afghanistan. The town was at the crossroads between the East and West when all trade between China and the Middle East passed through it.

Banyas River — A branch of the Baradā River in the south of Damascus.

Baradā River — The Baradā flows through Damascus.

Barqah — The eastern coastal region of Libya, named after the city of Barca, believed to have been where the city of al-Merj now stands.

Basra — Medieval Basra, built in 636, is located at Zubayr, 20 km south-west of modern Basra, a city on the Shatt al-Arab waterway, south-east of Baghdad.

Baʿqūbā — A place 64 km north-east of Baghdad, famous for its date and fruit gardens.

Bistām (or Bastām) — A town in Persia.

Bosra — A town in southern Syria near the Jordan border, about 23 km west of Sarkhad, an important stopover on the ancient caravan route to Mecca.

Budandūn (mod. Bozanti or Pozanti) — A river and town in modern Turkey. The Abbasid caliph al-Maʾmūn died suddenly there at a fortress by the Cilician Gates (Gülek Pass), while on a campaign against the Byzantines.

Būrnūs — Unidentified; possibly Praesus, an ancient coastal town in eastern Crete, or Portus, the port of Rome.

Būshanj — A town in Khorasan, close to Herat in modern-day Afghanistan.

Cave of Afqah — Between Baalbek and Byblos, 71 km north-east of Beirut in modern-day Lebanon.

Chalcedon — An ancient Greek city in Asia Minor, located across the Bosporus from Byzantium, now the Kadıköy district of Istanbul. The Fourth Ecumenical Council was convened there in 451 by the Byzantine emperor Marcian (r. 450–7).

Croton — An ancient Greek city in southern Italy.

Damietta — A port in Egypt on an eastern distributary of the Nile; an important naval base during the Abbasid, Tulunid, and Fatimid periods.

Diyār Bakr — The medieval Arabic name of the northernmost of the three provinces of al-Jazīrah, it encompassed the region on both banks of the upper course of the river Tigris and was bordered by Armenia to the north. The area is now part of Turkey, with its main city being Diyarbakır.

Dorylaeum — A Greek city in Phrygia, now Turkey.

Erzinjan — A city in Anatolia, modern-day Turkey.

Faramā (al-) (Greek Pelusium) — A fortified town and strategic city located in the Sinai on the eastern Delta of the Nile in Egypt.

Fardajān — A fortress-castle about 88 km from Hamadan in the district of Jarra.

Fārs (the lands of) — Persia, the south-west of modern Iran.

Fustat — Old Cairo, south of modern-day Cairo. The Graeco-Coptic township of Babylon, or Bābalyūn, on the east bank of the Nile, was known in Arabic as Qasr al-Shamʿ, and the first settlement founded by Muslim conquerors, Fustat, was built alongside it. It was the first capital of Egypt under Muslim rule.

Gate of Deliverance (Ar. *Bāb al-Faraj*) — In Aleppo, located at the northern side of the ancient city, it was one of nine main gates and was ruined in 1904.

In Damascus, one of the gates of the old city, built in the northern wall by Nūr al-Dīn ibn Zangī in 1154.

Gate of Victory — In Seville, probably the Puerta Osario, outside which was a Muslim cemetery.

Ghallah Lane, Ghallah Gate — In Baghdad, unidentified.

Gondeshapur — An ancient city, now ruined, in the modern-day province of Khuzestan in south-west Iran. It had a possibly mythical reputation as a centre of medical learning, complete with a hospital and an academy. It fell to Muslim Arab armies in 638 but persisted for several centuries.

Great Hospital — *See* Nūrī hospital.

Gūn Gunbad — A quarter and gate in Isfahan, unidentified.

Gurgānj (or Kurkānj) — Now known as Urgench, it was the capital city of Khorezm from 995 to 1602 and is situated in north Turkmenistan, near the border with Uzbekistan.

Harrān (Carrhae)	A major ancient city in Upper Mesopotamia near the modern village of Altınbaşak, Turkey, in a district that is also named Harran.
Hawrān (or Houran)	A volcanic plateau in south-western Syria extending into the north-western corner of modern-day Jordan.
al-Hirah	A town on the Euphrates River, south-west of the modern Najaf, in Iraq.
Ibn Muhājir law college	The *madrasah* founded in Mosul by ʿUlwān ibn Muhājir, father of the jurist and professor, Sharaf al-Dīn Muhammad ibn ʿUlwān ibn Muhājir, who died in 1218.
Ifrīqiyah	An area of North Africa, corresponding more or less to modern Tunisia, western Libya, and eastern Algeria.
ʿĪsābādh palace	The residence of caliphs Hārūn al-Rashīd and al-Hādī in Baghdad.
Jaʿfarī canal	A canal built in 859 by the Abbasid caliph, al-Mutawakkil to supply water to his new caliphal city, al-Mutawakkiliyyah or al-Jaʿfariyyah, to the north of Samarra.
Jaʿbar	A hilltop fortress overlooking the Euphrates in modern-day Syria, between Raqqa and Aleppo.
al-Jazīrah	Literally 'the island'; the traditional Arabic name for Upper Mesopotamia, the area of land more or less enclosed within the Euphrates and Tigris rivers. Today this area includes the uplands and plains of north-western Iraq, north-eastern Syria, and south-eastern Turkey.
Jurjān (or Gorgan)	The capital city of Golestan Province, Iran, lying approximately 400 km to the north-east of Tehran.
Kairouan	In modern-day Tunisia, about 50 km from the east coast, the city was founded by the Umayyads around 670. It became an important centre for Sunni Islamic scholarship and Qur'anic learning, attracting large numbers of Muslim scholars.
Karak	A city/castle on a hilltop plateau, 140 km south of Amman in modern-day Jordan.
Kaskar district	Kaskar was an ancient town on the west bank of the Tigris, just opposite the town of Wāsit, founded by the Umayyad governor al-Hajjāj ibn Yūsuf. It was renowned for its plump chickens.
al-Khandaq	Literally 'the moat' or 'the trench', a village near Cairo just outside the gates.

al-Khayf	The name of several places in the Hijaz, one of them at Minā near Mecca.
Khorasan (or Khurāsān)	The eastern part of Persia, including parts of what are today Central Asia and Afghanistan.
Khusrawshāh (or Khusrūshāh, mod. Khosrowshah)	A small settlement 30 km south-west of the city of Tabriz in Persia.
Khwārazm (mod. Khiva)	A city in the Khorezm region of modern-day Uzbekistan.
Kufa (Ar. *al-Kūfah*)	A town in Iraq, near modern-day Najaf and 90 km west of al-Qādisiyyah.
Laribus (or Lorbeus; Ar. *al-Arīs*)	A city in Libya where the army of Ziyādat Allāh III had encamped in 905 to fight the Fatimids. The Aghlabids were defeated there in 907 and Ziyādat Allāh fled to Egypt.
Locris	Founded about 680 BC on the Italian shore of the Ionian Sea by the Locrians, it was one of the cities of Magna Graecia; the city was abandoned in the fifth century and destroyed by the Saracens in 915. Today Locri is a town and municipality in the province of Reggio Calabria in Italy.
Lyceum	A temple in Athens dedicated to Apollo Lyceus (the wolf god). The school Aristotle founded there continued to function until the Roman general Sulla destroyed it during his assault on Athens in 86 BC. Its remains were discovered in 1996 behind the Hellenic Parliament.
al-Qādisiyyah	A historical city in southern Mesopotamia, south-west of al-Hillah and Kufa in modern-day Iraq.
al-Madrasah al-Dakhwāriyyah	A medical school in Damascus established by al-Dakhwār.
Maghrib	The Mauretania of the ancients, it included Moorish Spain, the Atlas Mountains, and the coastal plain of Morocco, Algeria, Tunisia, and Libya.
Mahdia (Ar. *al-Mahdiyyah*)	The capital city of the Fatimid caliphs founded in 916–21, now in modern Tunisia.
Marāghah	A city on the river Sufi Chay in the Azerbaijan province of Iran, famous for an early observatory.
Mayyāfāriqīn (mod. Silvan; ancient Martyropolis)	A city in south-eastern Turkey.

Memphis — The capital of the Egyptian Old Kingdom, situated on the west bank of the Nile, opposite modern Hulwān.

Metapontum — An important city of Magna Graecia, situated on the gulf of Tarentum, about 20 km from Heraclea and 40 km from Tarentum.

Muqtadirī hospital — Built in Baghdad by the caliph al-Muqtadir (r. 908–32).

Mūrātīr (mod. Sagunto) — A city on the Mediterranean coast of al-Andalus also known as Murviedro.

The Museum — Literally ‘House of Knowledge’. According to Arabic sources the House came to exist in Alexandria as a result of Alexander’s provision for the city. The real Museum at Alexandria, a foundation of Ptolemy I (d. 282 BC), was one of the most famous institutions of learning in antiquity.

Mutbaq prison — A notorious prison in Baghdad. It was located at the southern fringe of the city, between the Basra Gate and the Kufa Gate. It seems to have consisted of a series of oubliettes, into which, in some cases, prisoners were lowered at the end of a rope and left in inky darkness.

Nahr al-Malik — The ‘king’s canal’, a district at Baghdad said to comprise over 350 villages, one for each day of the year.

Nāsirī hospital — Founded in 1171 in Old Cairo by Saladin (al-Malik al-Nāsir), after whom it was named. It was created by modifying a part of the palace built by the Fatimid caliph al-ʿAzīz in 994 so that it could serve as a hospital.

Nighyā — A settlement in the region of al-Anbār; some have wanted to identify it with Nicephorium further north at the confluence of the Euphrates and the Balikh (mod. al-Raqqah).

Nisibis (mod. Nusaybin) — A city in the Mardin province of south-eastern Turkey, on the Syrian border.

Nizāmiyyah law college — A *madrasah* founded in 1057 by the vizier, Nizām al-Mulk, in Baghdad. Devoted to Sunni learning, the Nizāmiyyah served as a model for the establishment of an extensive network of such institutions throughout the eastern Islamic World especially Cairo, Damascus, and Aleppo.

Nūrī hospital (or Great Hospital)	Founded by Nūr al-Dīn ibn Zangī in Damascus.
Paradise Gate (Ar. *Bāb al-Farādīs* or *Bāb al-Imārah*)	One of the seven gates of Old Damascus. The gate was given its name because of its proximity to numerous water sources and lush gardens. There were initially eight gates of Old Damascus, but one was destroyed in Ottoman times.
Pergamum (mod. Bergama)	An ancient Greek city 25 km from the Aegean Sea and birthplace of Galen.
al-Qābūn	Once a stop on the trade route to Iraq, now a part of greater Damascus.
Qasr al-Shamʿ	The Arabic name for the Greco-Coptic township of Babylon (Ar. *Bābalyūn*), on the east bank of the Nile. *See* Fustat.
Qirmisīn (or Kermanshah)	The capital of Kermanshah Province, located 180 km west of Hamadan, in the west of Iran.
Qurrah	A village near Smyrna (mod. Izmir), in Turkey.
Qutrabbul	A village not far from Baghdad that was celebrated for its good wine.
Rahmah Gate (Ar. *Bāb al-Rahmah*)	'The Door of Mercy', one of the doors of the Golden Gate on the east side of the Haram al-Sharīf in Jerusalem.
al-Ramlah	A town situated on the coastal plain 40 km west-north-west of Jerusalem.
Raqqādah	The palace city of the Aghlabids, located approximately 10 km south-west of Kairaouan in modern-day Tunisia. Raqqādah surrendered to the Fatimid armies after the defeat at Laribus, in 909. It was the capital of Ifrīqiyah at the time of the Fatimid conquest.
Rayy (or al-Rayy)	One of the great Middle Eastern cities of the Abbasid period, rivalled only by Damascus and Baghdad. Islamic writers described it as a city of extraordinary beauty, built largely of fired brick and ornamented with blue faience. It was famous for its decorated silks and for ceramics. Its remains lie on the eastern outskirts of the modern city of Shahr-e Rey, a few kilometres south-east of Tehran.
Sābūr Khwāst	A town between Khuzestan and Isfahan.

Sarkhad (or Salkhad)	A stronghold in southern Syria near the border of modern-day Jordan.
Sawad (the) (Ar. *al-Sawād*)	The fertile alluvial plains of southern Iraq.
Siʿird (or Siʿirt, Isʿird; mod. Siirt)	A town in south-eastern Anatolia.
Smyrna (mod. Izmir)	A Greek city of Asia Minor.
Stagira	An ancient Greek city in Chalcidice, northern Greece.
Sūrā	A city east of the Euphrates in southern Iraq close to al-Hirah, known for its agricultural produce and as a centre of Torah scholarship.
Sus (ancient Susa; mod. Shūsh)	A town in the south-western province of Khuzestan, Iran.
al-Suwaydāʾ	A town in southern Syria, about 30 km north of Sarkhad.
Syria (Ar. *al-Shām*)	Referring to 'greater Syria', including modern-day Lebanon, Jordan, and Palestine; *al-Shām* may also refer to Damascus.
Tabarān	Unclear, possibly a corruption of either Tabarak, a citadel in Isfahan, or Tihrān, a district in the city.
Tabaristan	The name given to Mazandaran, a province now part of northern Iran bordering the southern shore of the Caspian Sea, also known as Tapuria. It was famous for its *Tabari* carpets and woollen furnishing cloth. The estate of Hārūn al-Rashīd (d. 809) included 1,000 cushions from Tabaristan and the caliph al-Maʾmūn apparently had 600 Tabaristan carpets delivered to him each year.
al-Tāʾif	A city about 80 km east of Mecca.
Tarentum	Founded as a Greek colony in 706 BC by the Spartans, it is now known as Taranto.
ʿUkbarā	A town located on the east bank of the Tigris River, approximately halfway between Baghdad and Samarra.
Uthayl (al-)	A place near Medina.
al-Wardiyyah cemetery	Lying beyond the Abraz Gate in eastern Baghdad, apparently the burial place of many persons of note.
Wāsit	A city built on the west bank of the Tigris by the Umayyad governor, al-Hajjāj, about 702 as his administrative centre for Iraq. He also died there.

Willow Tree Gate (Ar. *Bāb al-Gharabah*)	The uppermost of the gates in the Baghdad palace wall, near the Tigris, which took its name from a Gharabah or Babylonian willow tree.
Yabrūd	A city in Syria about 75 km north-east of Damascus.
al-Zafariyyah, Zafariyyah mosque	A neighbourhood in East Baghdad taking its name from the nearby Garden of Zafar.

APPENDIX 3

CONCORDANCE OF BIOGRAPHIES WITH THOSE IN THE FULL TEXT

OWC numbering	*Physician name*	*Brill no.*
Ch. 2	Asclepius	2.1
Ch. 4, no. 1	Hippocrates	4.1
Ch. 4, no. 2	Rufus of Ephesus	4.1
Ch. 4, no. 3	Dioscorides of Anazarbus	4.1
Ch. 4, no. 4	Pythagoras	4.3
Ch. 4, no. 5	Socrates	4.4
Ch. 4, no. 6	Plato	4.5
Ch. 4, no. 7	Aristotle	4.6
Ch. 5	Galen	5.1
Ch. 6	John the Grammarian (Yahyā al-Nahwī)	6.1
Ch. 7, no. 1	al-Hārith ibn Kaladah	7.1
Ch. 7, no. 2	al-Nadr son of al-Hārith	7.2
Ch. 7, no. 3	Ibn Uthāl	7.5
Ch. 7, no. 4	Hakam al-Dimashqī	7.7
Ch. 7, no. 5	ʿĪsā son of Hakam	7.8
Ch. 7, no. 6	Tayādhūq	7.9
Ch. 7, no. 7	Zaynab	7.10
Ch. 8, no. 1	Jūrjis	8.1
Ch. 8, no. 2	Bukhtīshūʿ son of Jūrjis	8.2
Ch. 8, no. 3	Jibrīl son of Bukhtīshūʿ	8.3
Ch. 8, no. 4	Bukhtīshuʿ son of Jibrīl	8.4
Ch. 8, no. 5	Jibrīl Ibn ʿUbayd Allāh	8.5
Ch. 8, no. 6	ʿUbayd Allāh son of Jibrīl	8.6
Ch. 8, no. 7	al-Tayfūrī	8.10
Ch. 8, no. 8	Zakariyyā son of al-Tayfūrī	8.11
Ch. 8, no. 9	Salmawayh	8.20

OWC numbering	*Physician name*	*Brill no.*
Ch. 8, no. 10	Māsawayh	8.25
Ch. 8, no. 11	Yūhannā son of Māsawayh	8.26
Ch. 8, no. 12	Hunayn ibn Is'hāq	8.29
Ch. 8, no. 13	Is'hāq son of Hunayn	8.30
Ch. 10, no. 1	Yaᶜqūb al-Kindī	10.1
Ch. 10, no. 2	Thābit ibn Qurrah	10.3
Ch. 10, no. 3	Sinān son of Thābit	10.4
Ch. 10, no. 4	Thābit son of Sinān	10.5
Ch. 10, no. 5	Sāᶜid ibn Bishr ibn ᶜAbdūs	10.13
Ch. 10, no. 6	Ibn Butlān	10.38
Ch. 10, no. 7	Ibn Shibl	10.51
Ch. 10, no. 8	Ibn Safiyyah	10.63
Ch. 10, no. 9	Ibn al-Tilmīdh	10.64
Ch. 10, no. 10	Awhad al-Zamān (Abū l-Barakāt al-Baghdādī)	10.66
Ch. 10, no. 11	Hibat Allāh ibn al-Fadl	10.68
Ch. 10, no. 12	Ibn Tūmā	10.77
Ch. 10, no. 13	Kamāl al-Dīn ibn Yūnus	10.83
Ch. 11, no. 1	Abū Bakr al-Rāzī (Rhazes)	11.5
Ch. 11, no. 2	Ibn Sīnā (Avicenna)	11.13
Ch. 11, no. 3	Fakhr al-Dīn al-Rāzī	11.19
Ch. 11, no. 4	al-Samawʾal	11.21
Ch. 12, no. 1	Shānāq (Cānakya)	12.3
Ch. 12, no. 2	Mankah al-Hindī	12.5
Ch. 12, no. 3	Sālih ibn Bahlah al-Hindī	12.6
Ch. 13, no. 1	Is'hāq ibn ᶜImrān	13.1
Ch. 13, no. 2	Is'hāq al-Israʾīlī	13.2
Ch. 13, no. 3	Ibn Malūkah al-Nasrānī	13.15
Ch. 13, no. 4	Tumlūn	13.17
Ch. 13, no. 5	Yahyā ibn Is'hāq	13.21
Ch. 13, no. 6	Ibn Juljul	13.36
Ch. 13, no. 7	Abū l-Salt Umayyah	13.58
Ch. 13, no. 8	Abū Marwān ibn Zuhr	13.60

OWC numbering	*Physician name*	*Brill no.*
Ch. 13, no. 9	Abū l-ʿAlāʾ ibn Zuhr	13.61
Ch. 13, no. 10	ʿAbd al-Malik ibn Zuhr (Avenzoar)	13.62
Ch. 13, no. 11	Abū Bakr ibn Zuhr 'the grandson'	13.63
Ch. 13, no. 12	Abū Muhammad ibn Zuhr	13.64
Ch. 13, no. 13	Ibn Rushd (Averroes)	13.66
Ch. 13, no. 14	Abū l-Hajjāj ibn Mūrātīr	13.68
Ch. 13, no. 15	Abū Bakr son of Abū l-Hasan al-Zuhrī	13.80
Ch. 13, no. 16	Ibn al-Asamm	13.88
Ch. 14, no. 1	Politianus	14.1
Ch. 14, no. 2	al-Hasan ibn Zīrak	14.3
Ch. 14, no. 3	Saʿīd ibn Tawfīl	14.4
Ch. 14, no. 4	al-Tamīmī	14.14
Ch. 14, no. 5	Ibn al-Haytham (Alhazen)	14.22
Ch. 14, no. 6	al-Mubashshir ibn Fātik	14.23
Ch. 14, no. 7	Ibn Ridwān	14.25
Ch. 14, no. 8	Salāmah ibn Rahmūn	14.27
Ch. 14, no. 9	Ibn Jumayʿ	14.32
Ch. 14, no. 10	Ibn Shūʿah	14.36
Ch. 14, no. 11	Maimonides (al-Raʾīs Mūsā)	14.39
Ch. 14, no. 12	Ibrāhīm son of al-Raʾīs Mūsā	14.40
Ch. 14, no. 13	Ibn Abī l-Bayān	14.43
Ch. 14, no. 14	Abū Sulaymān Dāwūd	14.49
Ch. 14, no. 15	Abū Hulayqah	14.54
Ch. 14, no. 16	Abū Saʿīd Muhammad son of Abū Hulayqah	14.55
Ch. 14, no. 17	Asʿad al-Dīn	14.57
Ch. 15, no. 1	al-Yabrūdī	15.3
Ch. 15, no. 2	Abū l-Hakam	15.8
Ch. 15, no. 3	Sukkarah al-Halabī	15.15
Ch. 15, no. 4	Ibn al-Salāh	15.17
Ch. 15, no. 5	al-Suhrawardī	15.18
Ch. 15, no. 6	Rafīʿ al-Dīn al-Jīlī	15.20
Ch. 15, no. 7	Shams al-Dīn al-Khusrawshāhī	15.21

OWC numbering	*Physician name*	*Brill no.*
Ch. 15, no. 8	Sayf al-Dīn al-Āmidī	15.22
Ch. 15, no. 9	Ibn al-Mutrān	15.23
Ch. 15, no. 10	Ridwān ibn al-Sāʿātī	15.29
Ch. 15, no. 11	Radī al-Dīn al-Rahbī	15.36
Ch. 15, no. 12	ʿAbd al-Latīf al-Baghdādī	15.40
Ch. 15, no. 13	Yaʿqūb ibn Siqlāb	15.43
Ch. 15, no. 14	Rashīd al-Dīn ibn al-Sūrī	15.45
Ch. 15, no. 15	Ibn Raqīqah	15.46
Ch. 15, no. 16	Amīn al-Dawlah	15.49
Ch. 15, no. 17	al-Dakhwār	15.50
Ch. 15, no. 18	Rashīd al-Dīn ʿAlī ibn Khalīfah	15.51
Ch. 15, no. 19	Ibn al-Minfākh	15.56
Ch. 15, no. 20	ʿIzz al-Dīn ibn al-Suwaydī	15.57

APPENDIX 4

LIST OF SOURCES USED BY IBN ABĪ USAYBIʿAH

ʿAbd al-Latīf al-Baghdādī	(Ch. 15, no. 12) a philosopher and physician of Baghdad, also known as Ibn al-Labbād (d. 1231).
ʿAbd al-Malik ibn Zuhr	(Ch. 13, no. 10) a celebrated physician of the Ibn Zuhr family in al-Andalus, known in the West as Avenzoar.
Abū Marwān al-Bājī	Sevillian jurist and verbal informant of Ibn Abī Usaybiʿah.
Abū l-Salt Umayyah	(Ch. 13, no. 7) twelfth-century author, poet, and physician from al-Andalus.
Abū ʿUbayd al-Jūzjānī	Eleventh-century pupil, assistant, and secretary of Ibn Sīnā (Avicenna).
Andromachus the Younger	A first-century Greek physician, and author of a work on pharmacy much quoted by Galen.
al-Bīrūnī	Abū l-Rayhān al-Bīrūnī, eighth- to ninth-century polymath, perhaps the greatest of all medieval Islamic scholars; his biography is 11.15 in the full text.
al-Dakhwār	(Ch. 15, no. 17) Muhadhdhab al-Dīn ʿAbd al-Rahīm ibn ʿAlī ibn Hāmid al-Dakhwār, a high-ranking Syrian physician and founder of the first medical school in the Arabic world (d. 1230).
Eutychius	Saʿīd ibn al-Bitrīq, physician, patriarch of Alexandria, historian, and author (d. 940); his biography is 14.11 in the full text.
Galen	(Ch. 5) renowned Greek physician and philosopher in the Roman Empire.
Hunayn ibn Is'hāq	(Ch. 8, no. 12) Christian physician in Baghdad and a prolific translator (d. 873).
Ibn Butlān	(Ch. 10, no. 6) eleventh-century Christian physician of Baghdad.
Ibn al-Dāyah	Secretary to Ibrāhīm ibn al-Mahdī and father of the well-known Tulunid historian, Ahmad ibn Yūsuf, active ninth century.

Ibn Juljul	(Ch. 13, no. 6) an influential tenth-century Andalusian Arab physician and pharmacologist of possible Spanish extraction, also known by the Latinized 'Gilgil'.
Ibn Jumayᶜ	(Ch. 14, no. 9) twelfth-century Egyptian-Jewish physician who served Saladin.
Ibn al-Mutrān	(Ch. 15, no. 9) a prominent Syrian Melkite Christian physician who converted to Islam (d. 1191).
Ibn al-Nadīm	The scribe known as al-Warrāq al-Baghdādī, a bookseller and author (d. 995 or 998).
Ibn al-Qiftī	Vizier, biographer, and an older Egyptian contemporary of Ibn Abī Usaybiᶜah (d. 1248), cited in Ibn Abī Usaybiᶜah's last revision of his work.
Ibn Raqīqah	(Ch. 15, no. 15) a colleague of Ibn Abī Usaybiᶜah's, fellow ophthalmologist and teacher.
Ibn Ridwān	(Ch. 14, no. 7) Egyptian physician at the court of the Fatimid Caliph al-Hakam II.
Is'hāq ibn Hunayn	(Ch. 8, no. 13) celebrated philosopher, translator, and physician, son of the great Hunayn ibn Is'hāq.
John the Grammarian	(Ch. 6) (Ar. *Yahyā al-Nahwī*) a shadowy author of a medical history probably dating to late antiquity, incorporated into Is'hāq ibn Hunayn's *History*.
al-Mubashshir ibn Fātik	(Ch. 14, no. 6) eleventh-century scholar and writer from Damascus who spent most of his life in Egypt.
Pethion	The author of a lost ecclesiastical history, active in the 870s.
Plato	(Ch. 4, no. 6) renowned fourth-century BC Athenian philosopher.
Ptolemy	Ptolemy al-Gharīb (dates uncertain), author of an *Epistle to Gallus on the Life of Aristotle*.
Sāᶜid al-Andalusī	al-Qādī Sāᶜid, Andalusian author (d. 1070).
al-Sijistānī al-Mantiqi	A leading Neoplatonist philosopher in tenth-century Baghdad (d. 985).
Thābit ibn Qurrah	(Ch. 10, no. 2) ninth-century mathematician, scientist, and translator.

Thābit ibn Sinān	(Ch. 10, no. 4) physician, historian, and grandson of Thābit ibn Qurrah.
ʿUbayd Allāh ibn Jibrīl	(Ch. 8, no. 6) eleventh-century member of the prominent Bukhtīshūʿ family of physicians.

APPENDIX 5

ILLUSTRATIONS AND DIAGRAMS

العلوم فقرأ في العلوم الحكمية والاجزاء الفلسفية على الشيخ شمس الدين عبد الحميد الخسروشاهي
وعلى عز الدين الحسن الغنوي الضرير وقرأ ايضا في صناعة الطب على الحكيم نجم الدين بن المنفاخ وعلى
موفق الدين يعقوب السامري وقرأ ايضا كتاب اقليدس على مويد الدين العرضي وفهم هذا الكتاب
فهما فتح به مقفل اقواله وحل مشكل اشكاله وخدم ابو الفرج بن القف بصناعة الطب في قلعة
عجلون واقام بها عدة سنين ثم عاد الى دمشق وخدم في قلعة دمشق لمعالجة المرضى وهو محمود
في افعاله مشكور في سائر احواله ه وله من الكتب كتاب الشافي في الطب
شرح الكليات من كتاب القانون ست مجلدات ، شرح الفصول كتابين ، مقالة في حفظ الصحة
كتاب العمدة في صناعة الجراح عشرين مقالة ، مقالة علم وعمل يذكر فيه جميع ما يحتاج اليه الجرائحي
بحيث لا يحتاج الى غيره ، كتاب جامع الغرض مجلد واحد ، حواشي على ثالث القانون لم توجد
شرح الاشارات مسودة ولم يتم ، المباحث المغربية مسودة ولم يتم ه

تم كتاب عيون الانباء في طبقات الاطبا
على يد اقل عبيد الله المتمولين بنعمته والراجين
غفرانه بسعة رحمته المتوسلين اليه بعظم ربوبيته
ان يصلي على النبي المرسل لكافة خليقته وعلى آله
وصحبه ويسلم تسليما كثيرا عبد الهادي بن
ابي المفضل بن ابي الفرج وكان الفراغ من تعليقه
في يوم ه ٢٧ [illegible] من عام ٧٧٣ للهجرة النبوية
على صاحبها افضل الصلاة والسلام والتحية والاكرام
وذلك بلطف الله وكرمه وتطوله ونعمه
له الحمد كثيرا اولا واخرا ه

قال كاتبها اما بعد حمد الله
والصلوة على نبيه واله
اني قد اجتهدت وبالغت
في المقابلة على هذه النسخة
جهد الطاقة من نسخ المصنف
ووجدت في مسودة بخطه
[illegible]

FIGURE 1 Colophon from *ʿUyūn al-anbāʾ fī tabaqāt al-atibbāʾ* (*The best accounts of the classes of physicians*) by Ibn Abī Usaybiʿah, completed 27 Shaʿbān 773/4 March 1372 and copied by ʿAbd al-Hādī ibn Abī l-Mufaddal ibn Abī l-Faraj (Istanbul, Süleymaniye, Şehid Ali Paşa MS 1923, fol. 306a).

كتاب

عيون الانباء فى طبقات الاطباء

تأليف الطبيب الفاضل العالم الأديب

موفق الدين أبى العباس أحمد بن القاسم بن خليفة

ابن يونس السعدى الخزرجى

المعروف بابن أبى اصيبعة

رحمه الله

نقله من النسخ الموجودة فى بعض خزائن الكتب وصححه

العبد الفقير الى عون الله ورحمته

امرؤ القيس بن الطحان

(الطبعة الاولى بالمطبعة الوهبية)

سنة ١٢٩٩ هجريه الموافقة لسنة ١٨٨٢ ميلاديه

FIGURE 2 Title page from *ʿUyūn al-anbāʾ fī tabaqāt al-atibbāʾ* (*The best accounts of the classes of physicians*) by Ibn Abī Usaybiʿah, ed. A. Müller. 2 vols. Cairo: al-Matbaʿah al-Wahbiyyah/Königsberg: Selbstverlag, 1882.

FIGURE 3 Physician with two patients, illustration from an Arabic translation of Dioscorides' treatise on medicinal substances, copied in Baghdad, 1224 (Arthur M. Sackler Gallery, Smithsonian Institution: Unrestricted Trust Funds, Smithsonian Collections Acquisition Program, and Dr Arthur M. Sackler, S1986.97).

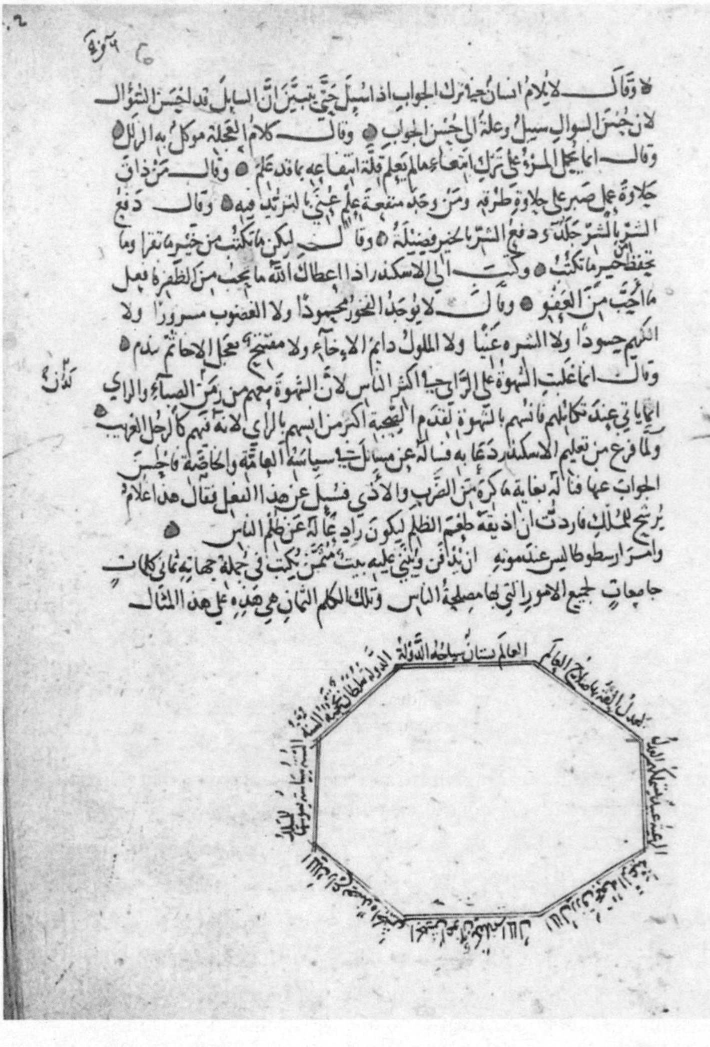

FIGURE 4 Folio with diagram showing placement of sayings on the octagonal mausoleum of Aristotle from *ʿUyūn al-anbāʾ fī tabaqāt al-atibbāʾ* (*The best accounts of the classes of physicians*) by Ibn Abī Usaybiʿah, completed 27 Shaʿbān 773/4 March 1372 and copied by ʿAbd al-Hādī ibn Abī l-Mufaddal ibn Abī l-Faraj (Istanbul, Süleymaniye, Şehid Ali Paşa MS 1923, fol. 45a).

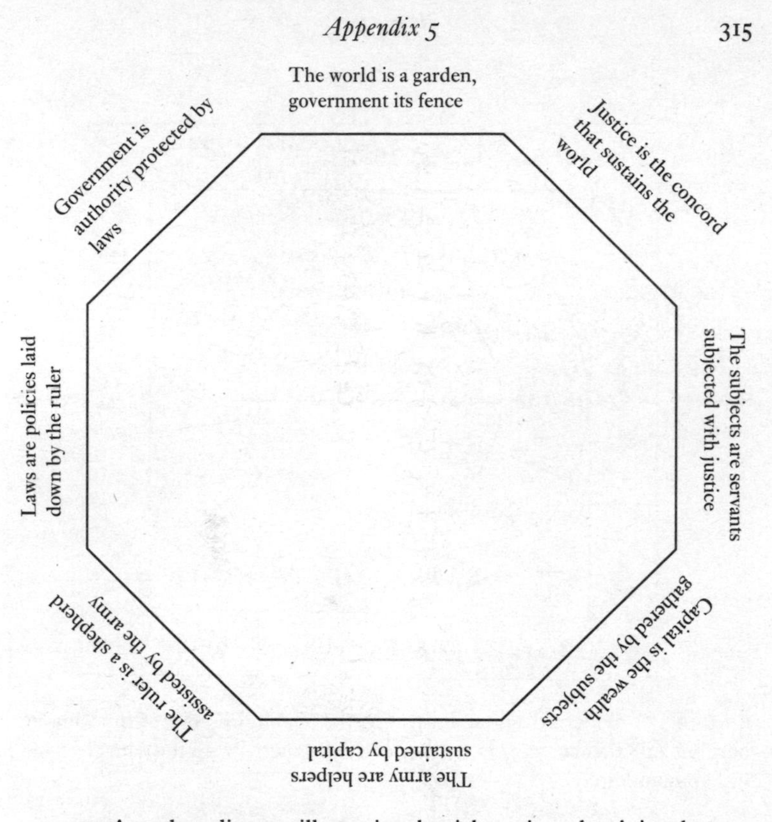

FIGURE 5 A modern diagram illustrating the eight sayings that Aristotle directed be placed on the sides of his octagonal mausoleum.

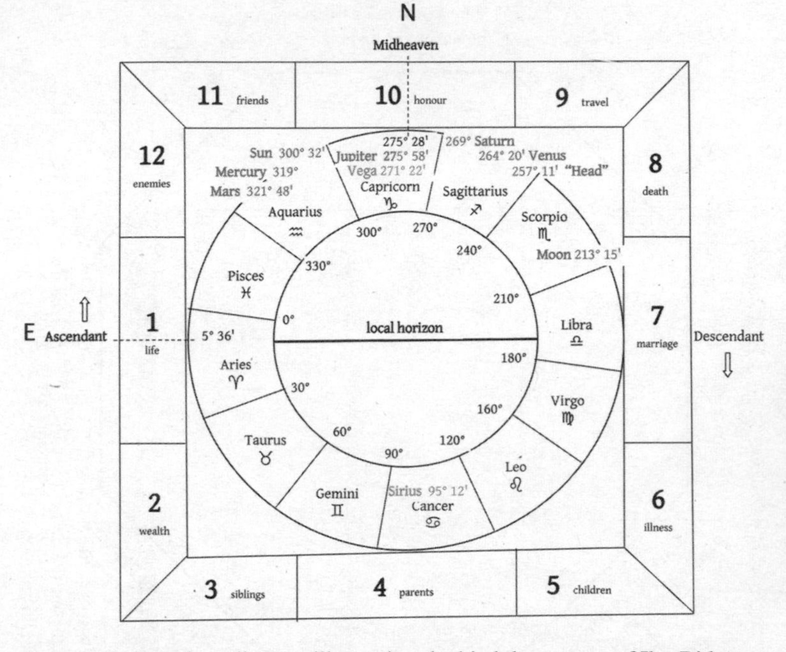

FIGURE 6 A modern diagram illustrating the birth horoscope of Ibn Ridwān, born on 22 Ramadan 377/15 January 988 constructed with the data given in his 'autobiography'.

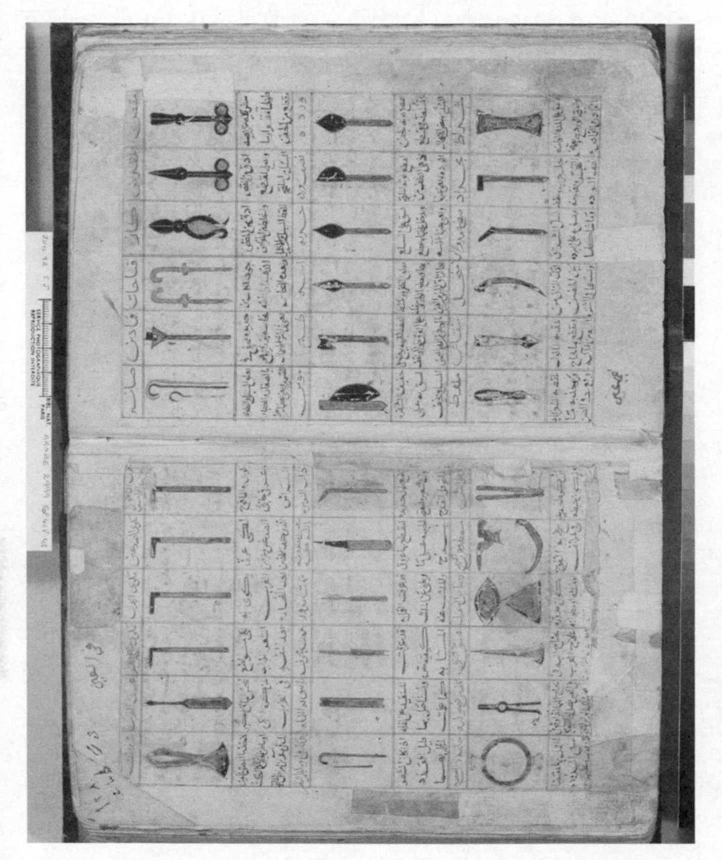

FIGURE 7 Instruments for operating on the eye, showing a standard and a hollow cataract needle in centre of the middle left row, illustration from *The Sufficient Book on Ophthalmology* by Khalīfa ibn Abī l-Mahāsin al-Halabī, copied in Syria *c.*1256–75 (Paris, BnF, MS arabe 1043 (de Slane no. 2999), fols 42b–43a).

EXPLANATORY NOTES

PREFACE

3 *ʿilm al-adyān*: al-Shāfiʿī (d. 820), the eponymous founder of one of the four authoritative Sunnite law schools, is credited with having said, 'Knowledge is of two kinds: knowledge of religion and knowledge of bodies,' quoted in Ibn ʿAbd Rabbih (d. 940), *al-ʿIqd al-farīd*, ed. Ahmad Amīn *et al.* (Cairo, 1948–56), vol. 6, 307.

5 *next*: this dedication to the physician and vizier was removed in copies made after his execution in 1250. The account of Amīn al-Dawlah's request for a copy of Ibn Abī Usaybiʿah's book is given in his entry (Ch. 15, no. 16).

CHAPTER 1

6 *Hippocrates' Book of oaths*: this is *Tafsīr li-Kitāb al-aymān li-Abuqrāt*, a commentary on the famous Hippocratic Oath. Only fragments survive of this commentary ascribed to Galen. The work is certainly Greek in origin and probably dates to the middle period of the Roman empire. Medieval authors had no doubts about its authorship.

7 *elecampane*: an herbaceous plant in the sunflower family found across much of Europe and Asia. The herb is of ancient medicinal repute, having been described by Dioscorides and Pliny and was considered a cure-all for a broad range of ailments. Helenin, a phytochemical occurring in the root, is now known to have anti-fungal and anti-microbial properties.

8 *Ibn al-Mutrān*: a trusted source used by Ibn Abī Usaybiʿah, he was a prominent Syrian Melkite Christian physician who converted to Islam (d. 1191); for his biography see Ch. 15, no. 9.

chicory: Ar. *hindabāʾ* or *hundabāʾ* is endive or chicory (*Cichorium sp*), an annual herbaceous plant, with many species, originally native to Europe. It has a long history of cultivation and some writers distinguished two basic varieties, wild (usually known as chicory) and cultivated (usually known as endive).

9 *a work . . . arts*: *De constitutione artium*, in three books, of which the book on medicine survives as *De constitutione artis medicae*.

Ibn ʿAbbās: ʿAbd Allāh ibn ʿAbbās, a seventh-century Qur'anic interpreter and soldier, a cousin of the Prophet Muhammad, and the eponymous ancestor of the Abbasid caliphs.

The Zoroastrians: (Ar. *al-Majūs*) followers of one of the world's oldest monotheistic religions, Zoroastrianism, founded by the Prophet Zoroaster in ancient Iran approximately 3,500 years ago. It was the official religion of Iran from 600 BC to AD 650.

On bloodletting: *K. fī l-fasd*, which combined several Galenic works; this story occurs in *De curandi ratione per venae sectionem*.

artery: bloodletting was usually performed only on veins. This passage in Galen's treatise occurs within a discussion of the dangers of cutting into arteries, with examples of how such treatment can occasionally be beneficial.

10 *Commander of the Faithful*: title of the caliph used especially when addressing him directly.

«From a . . . touched it»: Qur'an *Nūr* 24:35.

Ibn Ridwān: see Ch. 14, no. 7.

11 *theriac*: a type of drug originally formulated by the Greeks in the third century BC, primarily as an antidote to snakebite. In the Roman period it became known as a panacea for almost any ailment, as well as an antidote to any toxin, venom, or poison.

12 *mezereon*: Ar. *māzaryūn*, a variety of *Daphne mezereum* L, native to much of Europe and Western Asia and known in English as flowering spurge or spurge laurel. It has been used for a variety of medicinal purposes despite its very high toxicity that induces a choking sensation.

recovered: an almost identical tale is told in ch. 10 (though omitted in this edition) with boiled locusts. Ibn Abī Usaybiʿah notes there that it is an old story, already mentioned.

Abū Bakr al-Rāzī: tenth-century Persian physician; see Ch. 11, no. 1.

'jaundice stone': mentioned as early as Pliny the Elder (23–79), amulets made from stones supposedly taken from the nests of swallows were believed, through to early modern times, to relive many maladies including epilepsy and jaundice. Known specimens include small agate clasts, small fossilized marine organisms (such as so-called 'Pharaoh's Lentils'), and gastroliths (stones from a bird's crop or gizzard).

13 *eagles*: in early modern Europe, an 'eagle stone' was also called an aetites stone, which the *Oxford English Dictionary* defines as 'A hollow nodule or pebble of hydrated iron oxide containing a loose kernel that makes a noise when rattled, formerly regarded as having medicinal and magical properties'.

Oleander: a shrub or small tree in the dogbane family, toxic in all its parts. Found throughout much of southern Europe, the Middle East, and southern Asia, it typically occurs around stream beds, where it can tolerate both drought and flooding. It is hazardous for grazing animals, with as little as 100 g being enough to kill an adult horse.

antidote: Ar. *Bādzahr* (from the Persian *pād-zahr* meaning 'against poison') or 'bezoar', is frequently used, as here, in the general sense of any antidote against poison. More specifically, a bezoar is a stone-like concretion originating from the gastrointestinal system of certain animals, especially goats, and was highly prized in both medieval and early modern societies as a remedy for poisoning.

CHAPTER 2

15 *Asclepius*: god of medicine in ancient Greek mythology and son of Apollo; the rod of Asclepius, a snake-entwined staff, remains a symbol of medicine to this day. Though a mythological figure, Asclepius' existence was widely accepted by medieval Arab authors who considered him a real historical person. This was also the case for Hermes.

Annotations: though Ibn Abī Usaybiʿah refers to al-Sijistānī's work by a different title, the content he quotes is close to the kind of information found in the abridgement of al-Sijistānī's *Siwān al-hikmah*. It is possible that they are different versions of the same work.

son of Zeus: in Greek and Roman mythology Asclepius is generally considered to be the son of Apollo.

'*light*': the origin of the name Asclepius is uncertain. Ancient etymologies derived it from the roots *sklēr-* ('hard', 'dry') and *ēpi-* ('mild', 'soothing'). Another version, designed to dignify the great cult of Asclepius at Epidaurus, has Apollo name his son after his mother Aeglē, signifying 'brightness' or 'splendour'.

16 *streets*: quoted from a work by Ibn Juljul and corresponding to no particular statement in Galen.

al-Mubashshir ibn Fātik: an eleventh-century scholar and writer from Damascus who spent most of his life in Egypt; see Ch. 14, no. 6.

17 *taken Asclepius up*: the Greek poets in fact tell how Asclepius was blasted by a thunderbolt for restoring a man due to die.

two sons: the sons are Machaon and Podalirus.

Thābit ibn Qurrah: Ibn Abī Usaybiʿah quotes indirectly from an autograph by Thābit on the Baqāritah, i.e. the (several) individuals called Hippocrates; see Ch. 10, no. 2 for Thābit ibn Qurrah's biography.

18 *Galen*: from a *Commentary on the oath*, which Ibn Abī Usaybiʿah believed was written by Galen. See note to p. 6.

laurel tree: this refers to Bay Laurel or Sweet Bay, *Laurus nobilis* (not hedging laurel, *Prunus laurocerasus*), which has a long medicinal history. Today the plant is thought to have antiseptic and antioxidant properties, can be used as a mild insect repellent, and its oil is the main ingredient in Aleppo Soap.

castoreum: the exudate from the perineal castor sacs of beavers, extensively used in the perfume industry. From antiquity to as late as the nineteenth century, it was used to treat a variety of ailments, and it is one of the many ingredients of both mithridate and theriac. Effects, if any, have been credited to the accumulation of salicin from willow trees in the beaver's diet.

19 *chyme*: the semi-fluid mass of partly digested food that is expelled by the stomach into the small intestine after hours of mechanical and chemical digestion.

CHAPTER 4

23 *said*: Ibn Juljul (Ch. 13, no. 6) was an influential tenth-century Andalusian Arab physician and pharmacologist; this story is from his *Tabaqāt al-atibbāʾ* (*Classes of physicians [and sages]*).

Physiognomy: Polemon, courtier of the Roman emperor Hadrian, is the author of a manual of physiognomy that has not survived in its original Greek. The story represents the popularity of Polemon's book in the Islamic Middle Ages.

24 *witticisms*: Ibn Abī Usaybiʿah does not provide any information concerning the provenance of these sayings, but some of them were collected by the tenth-century scholar al-Sijistānī.

away: Hippocrates believed that semen came from all the humours of the body. The Qur'an, however, states that semen was produced in the backbones and the ribs; see Qur'an *Tāriq* 86:5–7.

25 *hellebore*: a plant in the Ranunculus family, and is also known as Christmas or Lenten Rose. All parts of it are toxic. It was used by Greek physicians, amongst other things, to induce abortion.

26 *Mālānā Ārsā*: the name of Hippocrates' daughter is not known in Greek sources.

quotes him: Rufus, an important Greek physician of the high Roman Empire, was famous for his work on melancholy.

28 *squill*: a plant in the lily family, it has a long history of medicinal use; it contains cardiac glycosides which can affect the heart and thin lung secretions and is used for edema and lung complaints; it is also poisonous and an effective rodenticide.

mallow leaves: this refers to many species in the Malva genus, most usually common mallow, all parts of which have long been known to be edible—it is mentioned by Horace; cooked leaves can be used as a thickener while dried leaves can be used for tea.

asphodel: a hardy herbaceous perennial; its roots can be eaten but were regarded by the ancient Greeks as food for the dead; its leaves are used to wrap Italian burrata cheese.

asūfā seeds: possibly hyssop seeds; hyssop, a herbaceous herb with a taste between mint and anise, has a long history of culinary, medicinal, and ritual use, though such use of the seeds is not recorded elsewhere.

purslane: common purslane, now usually regarded as a weed, can be eaten raw or cooked; it contains more omega-3 fatty acids than any other leafy vegetable.

35 *Epistle . . . Aristotle*: the original Greek of Ptolemy's *Vita Aristotelis* is lost but the Arabic translation has survived in the *unicum* MS Aya Sofya 4833. The identity of its author, referred to as Ptolemy al-Gharīb in Arabic sources, is unknown.

37 *Abū Nasr al-Fārābī*: a renowned ninth- to tenth-century Syrian physician, philosopher, and jurist (Ch. 15, no. 1).

38 *Herpyllis*: Diogenes Laertius 5.1: 'he also had a son called Nicomachus by his concubine Herpyllis.'

40 *Apellicon*: of Teos (d. *c.*84 BC), was a wealthy Greek book collector, who allegedly bought Aristotle's unpublished works from the descendants of Aristotle's pupil, Neleus of Scepsis.

CHAPTER 5

41 '*Seal*': recalling Qur'an *Ahzāb* 33:40, 'Seal of the Prophets', describing the Prophet Muhammad; i.e. having been found perfectly complete, beyond further addition, and thus 'sealed'; he is the last one, none can come after.

42 *works*: Ibn Abī Usaybiʿah relies on the chronology of physicians by John the Grammarian (lost in Greek) for many entries, using him through Is'hāq ibn Hunayn (Ch. 8, no. 13), though he was quite suspicious of its accuracy. Here he rejects it because John aligned Galen with Christ, thereby sectarianizing medicine; but the enticing coevality of Jesus and Galen, two great healers, was a popular theme. To fight this, Ibn Abī Usaybiʿah uses a trusted Christian source, ʿUbayd Allāh (see Ch. 8, no. 6), who calculates Galen's birth as 118, very close to the actual 129. It is unclear which work of ʿUbayd Allāh's the extracts here come from. For imaginative biographical information on John, see Ch. 6.

The . . . them: cf. the famous correspondence between Trajan and his special governor in Bithynia, Pliny the Younger, on what to do with the new religion of Christianity. See Pliny, *Letters* 10.96–7.

Antoninus Caesar: Galen in fact meant the emperor we call Marcus Aurelius, who began his reign in 161 and took the additional name Antoninus in honour of his adoptive father and predecessor, Antoninus Pius. Galen's first visit to Rome was in 162–6 and his second in 169.

Republic: Galen's summary of the *Republic*, apparently in six books, was known in Arabic.

43 *deeds*: Galen's remarks about Christians are important because they are an early witness to pagan views of the new religion.

Ibn al-Dāyah: a source much used by Ibn Abī Usaybiʿah, he was the secretary of Ibrāhīm ibn al-Mahdī and father of the well-known Tulunid historian, Ahmad ibn Yūsuf; in addition to his lost history of medicine, *Accounts of physicians* (*Akhbār al-atibbāʾ*) from which Ibn Abī Usaybiʿah frequently quotes, he probably provided much material for the works of his son.

Jibrīl ibn Bukhtīshūʿ: see Ch. 8, no. 3.

al-Anbār: for most of antiquity, including the late Roman world, Roman territory ended further north at the Khabur river in north-east Syria and the town of Circesium/Qarqīsiyā (modern al-Busayrah), which is located at its confluence with the Euphrates.

times: by Romans, Arabic authors meant both the Romans of the Roman Empire and its continuation in the East (the Byzantine Empire). Since Hārūn al-Rashīd is encamped in Asia Minor warring against the Byzantine emperor, the discussion over the old eastern border was of interest.

rug: *namat*, a decorative covering or rug placed over a carpet for sitting upon. Hārūn's treasury was well stocked with them at the time of his death according to the official inventory.

Smyrna: no other source mentions this, but Galen's education in the city under Pelops makes it likely enough that he owned property there.

47 *injuries*: Ibn Abī Usaybiʿah includes many of Galen's descriptions of his practice, since it was widely considered the model by which to judge all physicians. The work cited survives in Arabic only.

Arminus: a confusion between the Peripatetic philosopher Herminus, teacher of Alexander of Aphrodisias, and Albinus, a distinguished Platonist who taught Galen in Smyrna.

Cleopatra: though Galen could not have actually met or studied with Cleopatra, queen of Egypt, his extensive quotation in his *Composition of drugs by places* of a treatise on cosmetics supposedly written by her contributed to the legend of her career in medicine.

Chios: no voyage of Galen is attested to Chios, although he relates that he collected medicinal clays from Lemnos, another Greek island in the Northern Aegean and the usual source of *terra sigillata*.

terra sigillata: Lat. meaning 'sealed earth'; stamped tablets of so-called 'Lemnian earth' mixed with wine or juniper. The clay came from the island of Lemnos and was used as an adsorbent and protective medicinal clay from classical antiquity to the nineteenth century.

alum: in this case perhaps potassium nitrate (saltpetre), which can take sufficient heat out of water to cause freezing.

48 *Avoiding distress*: this work, recently rediscovered in Greek but lost in Arabic apart from quotations, develops from Galen's description of his composure following the loss of his books and possessions in the great fire that devastated central Rome in early 192, an event he refers to often.

50 *books*: Ibn Abī Usaybiʿah provides a list intended to be comprehensive and some 188 are listed in the full edition, though only a very small sample are included here. Further works are also mentioned on p. 56–7. Modern scholars recognize as genuine 123 of the 146 surviving works attributed to Galen.

Pinax: i.e. *On my own books* (*De libris propriis*).

51 *says*: from a short work by Hunayn on material 'not listed by Galen in the Catalogue of his books'.

52 *work*: Hunayn's statement that he wrote his main bibliographical work when he was 48 refers to the original, lost edition published in 856. Our two surviving recensions are revised versions of a second edition. Ibn Abī Usaybiʿah certainly used both.

CHAPTER 6

53 *Alexandria*: in late antiquity Alexandria could draw on the legacy of nearly a thousand years, and it attracted promising students and proficient masters of the medical art. The Library of Alexandria no longer existed, but there were numerous *akadēmia*s ('academies') and *museion*s (literally 'museums'), medico-philosophical schools-cum-libraries where both medicine and philosophy were taught side by side.

Gessius: a late fifth-century physician and philosopher from Alexandria, whose fame brought him great wealth.

54 *Jacobite sect*: the theologically opposed Melkite and Jacobite sects arose following a dispute over the nature of the divine and human aspects of Christ, which came to a head at the Council of Chalcedon in 451. This led to a schism between the Oriental Orthodox Churches and the Western and Eastern Orthodox Churches, which persists to this day. (Some of these events, involving Eutyches, are described in the chapter.)

Melkites were named after their fidelity to the Byzantine emperor (*malkā* in Syriac) and to the Council of Chalcedon. The vast majority of Christians now—Roman Catholic, Maronite, Eastern Orthodox, and traditional Protestant—belong to Chalcedonian Churches.

Jacobites, adherents of the Syrian Orthodox Church, were named after the sixth-century bishop of Edessa, Jacob Baradaeus. Oriental Orthodox Churches today include the Syriac Orthodox Church, the Coptic Church of Alexandria, the Armenian Apostolic Church, the Eritrean and Ethiopian Orthodox Churches and the Malankara Orthodox Church of India.

general: ʿAmr ibn al-ʿĀs had been one of the companions of the Prophet Muhammad; he conquered Alexandria in 641 and founded the Egyptian capital of Fustat (Old Cairo).

field: ʾUbayd Allāh's account is an imaginative and fictional one, based on historical events of the fifth century.

55 *Eutyches*: a Constantinopolitan archimandrite (one level below a bishop) who played a key role in the Christological battles of the first half of the fifth century; his confusion with John the Grammarian is puzzling.

anathema: a formal curse by a pope or a Council of the Church, excommunicating a person or denouncing a doctrine.

'lover of toil': the confusion with the great philosopher John Philoponus (d. 575) is understandable because Philoponus was also known as John the Grammarian.

Astīriyūs: Marcian's successor on the throne of Constantinople was actually Leo I (r. 457–74). Astīriyūs may be being confused with the later emperor Anastasius or the earlier Arcadius. In any case the events related refer to Marcian's predecessor, Theodosius II, and the temporary restoration of Eutyches at the Council of Ephesus.

bishop of Dorylaeum: Eusebius, bishop of Dorylaeum, was a fifth-century bishop who spoke out against heretical teachings, especially those of Eutyches during the period of Christological controversy.

appear: John of Antioch died in 441. His delaying is a reference to his involvement with the Council of Ephesus in 431.

Constantinople: having been deposed, the bishop took shelter in Rome, soon to return under the new regime and to achieve the removal of Eutyches himself.

doctrine: despite the slightly inaccurate facts and confused identities of the protagonist, complicated biographical stories like this were loved by readers of the time.

57 *Dogmatic doctrine*: Relating to the rationalist school of medicine, i.e. using logical theory to make sense of evidence gained through experience, which was the focus of empiricist physicians; Rationalism and Empiricism were often presented as opposing approaches to healing but most physicians combined them.

CHAPTER 7

59 *mawlā*: a person dependent (through formal contract) on a patron for legal status; usually a non-Arab Muslim convert, thus without tribal affiliation, sometimes a freed slave; cf. the Roman idea of the 'client'.

ibn Abīhi: the patronymic 'ibn Abīhi' ('son of his father') signifies he was not born in Abū Sufyān's house, not that his paternity was uncertain. Ziyād ibn Abīhi or Ziyād ibn Abī Sufyān became the governor of Iraq and the eastern provinces under the Umayyad caliph, Muʿāwiyah.

60 *Salsabīl*: see Qur'an *Insān* 76:18.

63 *nūrah*: this strictly means quicklime and, combined with other ingredients, is used as a depilatory paste also called *nūrah*. It is also an ingredient in compounds used as remedies for various skin complaints. Though its medical application is dubious, there are traditions of the Prophet using and recommending it, saying, 'Muslims! Use *nūrah*, for it is pleasant and purifying. With it, God removes your dirt and your hair.' In Arabic lore, it was invented by Solomon, for the Queen of Sheba, who had hairy legs.

64 *al-Hasan . . . ʿUmayr*: this is typical of the chains of narrators given to prove the authenticity of a report or story. Some run up to twelve or fourteen people, but further ones have been omitted in the interest of brevity.

64 *two*: an example of the ancient convention of addressing two unnamed companions at the beginning of a poem. See also note to p. 289.

65 *water-loaded*: Ar. *rayyā*, a play on words as the brother's wife's name was also Rayyā, according to some accounts.

SON OF AL-Hārith: the person here described was not actually the son of al-Hārith ibn Khaladah but rather the son of another man called al-Hārith. Because of this mis-identification, the following biography is actually that of al-Nadr ibn al-Hārith ibn ʿAlqamah, killed personally by the Prophet in 624.

67 *ʿAbd al-Rahmān*: ʿAbd al-Rahmān ibn Khālid ibn al-Walīd, the son of the famous general, Khālid ibn al-Walīd (d. 642), was a governor and general under the caliph Muʿāwiyah and was perceived as a rival to Muʿāwiyah's son, Yazīd.

Hāshimite: i.e. a supporter of the Prophet's family, the Banū Hāshim, and hence of ʿAlī ibn Abī Tālib's faction (later Shiites), contrary to the caliph Muʿāwiyah who was of the opposing faction (later Sunnis).

death: i.e. if Khālid had killed a Muslim, not an unbeliever, his punishment would have been greater.

68 *imprisonment*: Muʿāwiyah was renowned for his *hilm* (approximately = 'clemency'). He is reported to have said, 'I do not employ my sword when my whip suffices me, nor my whip when my tongue suffices me; and were there but a single hair binding me to my fellow-men, I do not allow it to break: when they pull, I slacken, and when they slacken, I pull.'

'*Fallen . . . mouth*': the Arabic phrase *li-l-yadayn wa-l-fam* is said to express pleasure at seeing someone fall, reminiscent of our expression 'to bite the dust'. ʿAlī ibn Abī Tālib had had enough of al-Ashtar by this time; he had sent him to Egypt to have him well out of the way. However, another school of historians attributes this remark to Muʿāwiyah.

69 *half of history*: Hakam's age was 105 *lunar* years, thus approximately 102 solar years. 'History' here means 'Islamic history', i.e. the 210 lunar years that had elapsed since the Hijrah.

basilic vein: a large superficial vein of the arm.

70 *Masīh*: the Arabic term for 'Messiah'. The nickname was probably meant as praise for one whose name was ʿĪsā (Jesus) and who was renowned for curing the sick.

umm walad: a slave or concubine who had borne a child (to her master) which had been formally acknowledged; Ghadīd ('Luscious'), also known as Umm Jaʿfar, was one of Hārūn al-Rashīd's favourite concubines.

al-Abahh and al-Tabarī: a note in one of the extant manuscripts reads: These were two leading astrologers. Both of them are the authors of works on the subject. Al-Abahh was the nickname of al-Hasan ibn Muhammad al-Tūsī, and means 'the hoarse one, one with a raucous voice'.

71 *until then*: according to ʿIsā ibn al-Hakam's astrological section in *al-Risālah al-Hārūniyyah*, certain medical treatments such as cupping were to be avoided when the moon was approaching Saturn. In this anecdote, it is al-Tabarī who follows this rule.

Salmawayh: see Ch. 8, no. 9.

73 *Banū Awd*: literally 'the sons of Awd' (Awd ibn Saʿb); a small branch of the Madhhij tribe of South Arabia.

CHAPTER 8

74 '*Syriac*': it is still used as a liturgical language. The modern Arabic name of Syria is Sūriyā, 'Syrian' is *Sūrī*.

family: the Bukhtīshūʿ family were a distinguished family of Christian physicians, who often have distinctly Christian names; Jūrjis, for example, is the equivalent of George, and Jibrīl is Gabriel. Other names predominantly used by Christians include Hunayn and Yūhannā (John).

76 *Jesus*: in fact, *bukht* does not mean 'servant', and it is not a Syriac word. It seems to be related to Middle Persian *bōkhtan*, 'to save', and the name means 'saved by Jesus'. The alternative form Bakhtīshūʿ is also used.

78 *Jaʿfar*: Jaʿfar ibn Yahyā ibn Khālid ibn Barmak from the celebrated family of viziers, he married Hārūn al-Rashīd's sister, al-ʿAbbāsah. Their names are connected with the fall of the Barmakids though the reasons are unclear.

79 *mawlā*: a person, usually a non-Arab Muslim convert, dependent on a patron for legal status (see also note to p. 59).

80 *logs*: teak was extensively used in Baghdad for house construction, among other things.

81 *oxymel*: essentially, nothing but a mixture of honey and vinegar.

caliph: Jibrīl had taken his revenge on the governor for not giving him the teak logs by advising the caliph, Hārūn al-Rashīd, not to eat or drink any of the expensive gifts of food and drink that the governor had brought.

82 *Masrūr the Elder*: a eunuch and executioner of Hārūn al-Rashīd's, he is often quoted as a transmitter of reports in literature and is a recurrent character in *The thousand and one nights*.

Harthamah: Harthamah ibn Aʿyan; for an account of his loyalty to Hārūn al-Rashīd, see Ch. 8, no. 7.

in front of him: the execution of Jaʿfar ibn Yahyā, who had been foster-brother and extremely close to Hārun al-Rashīd, presaged the sudden and dramatic fall in 803 of the renowned Barmakid family. They were an influential Iranian family from Balkh in Bactria where they were originally hereditary Buddhist leaders, before converting to Islam. They were highly educated, respected, and influential, coming to great political power under the Abbasid caliphs of Baghdad, as secretaries and viziers. The causes of their fall remain somewhat obscure and gave rise to much discussion and embroidery in Arab history and literature.

83 *Mansūrī silk*: possibly named after either the second Abbasid caliph, al-Mansūr, or al-Mansūrah, the principal city of Hind (in modern Pakistan).

ʿId al-Fitr: festival marking the end of the Ramadān fast.

Barmakids: see note to p. 82.

84 *90,600,000 dirhams*: it is impossible to say with certainty that these figures are accurate, but that enormous sums were spent and owned by the powerful and rich is without doubt. Modern equivalents are difficult to determine—a dirham was a silver coin, a dinar a gold coin worth 12 or often more dirhams. For comparison, Jibrīl ibn ʿUbayd Allāh (Ch. 8, no. 5) in tenth-century Baghdad is said to have earned 7,200 dirhams a year working as physician to the caliph and at the hospital, presumably a good wage, while in twelfth-century Jerusalem, ʿAbd al-Latīf al-Baghdādī (Ch. 15, no. 12) reports he was given 100 dinars a month from Saladin and his sons. In contrast, Ibn al-Haytham (Ch. 14, no. 5) in the eleventh century charged 150 Egyptian dinars for a year's work of copying which 'was his means of subsistence for the year'.

85 *punkah*: a large swinging fan, fixed to the ceiling pulled with a cord. The material used could range from utilitarian rattan to expensive fabrics, which could be quilted or stuffed. The date of this invention is not known, but it was familiar to the Arabs as early as the eighth century.

86 *tamarisk*: a variety of euphorbia, proverbial for the heat that it yielded when used as firewood.

87 *mandrake*: mandrake fruits have a very unique fragrance likened to strong red apples.

88 *sīnīzī linen*: the port of Siniz on the Persian Gulf was famed for its production of fine Egyptian-style linens; hence this description from the tenth or eleventh century, 'Sinizi gowns, like the very air in thinness, like a mirage' (al-Mutahhar al-Azdī's *Hikāyat Abī l-Qāsim al-Baghdādī*, trans. Geert Jan van Gelder and Emily Selove, manuscript submitted for publication).

relates: in Abū al-Rayhān al-Bīrūnī's *Collected information on precious stones*.

Zubaydah: *Amat al-ʿAzīz* or *Umm Jaʿfar Zubaydah* (*Mother of Jaʿfar, 'Little butter lump'*) (d. 831); granddaughter of the caliph al-Mansūr and mother of the caliph al-Amīn, she was particularly remembered for the series of wells and reservoirs that provided water for pilgrims on the route from Baghdad to Mecca and Medina; the exploits of her and her husband, Hārūn al-Rashīd, form part of the basis for *The thousand and one nights*.

89 *Muʿizz al-Dawlah*: the first of the Buyid emirs of Iraq (d. 967).

Daylamites: an Iranian people originally from the Gīlān region. The Buyid dynasty was of Daylamite origin.

90 *siglaton*: a heavy fabric of damask silk, widely used for upmarket clothing, bedding, and hangings.

ʿAdud al-Dawlah: the Buyid emir of Fars and Iraq (r. 949–83) under the Abbasid caliphate; at his height he ruled an empire stretching from the Mediterranean to the Indian Ocean and he is widely regarded as the greatest monarch of the dynasty; he founded the ʿAdudi hospital in Baghdad.

shujāʿī dirhams: bearing the inscription ʿAdud al-Dawlah Abū Shujāʿ. Specimens are still available from dealers.

91 *The merits of physicians*: The work, now lost, is often cited by Ibn Abī Usaybiʿah (see p. 54 and notes to p. 42) and Ibn al-Qiftī.

93 *al-Afshīn*: a title traditionally borne by the native princes of Usrūshanah, the mountainous district between Samarqand and Khujanda; here the prince Haydar ibn Kāwūs, who served the caliphs al-Maʾmūn and al-Muʿtasim as a talented and successful general.

Bābak: head of the Khurramī sect, a religious and social movement, Bābak led a persistent rebellion in the eastern part of the Islamic empire during the reigns of the Abbasid caliphs al-Maʾmūn and al-Muʿtasim, which was finally put down by al-Afshīn in 837.

camp: this camp was a substantial affair of some months' duration; al-Afshīn had established it as a base from which to attack Bābak's fortress of Badhdh, in northern Azerbaijan, which was finally stormed on 15 August 837.

95 *Ibn al-Dāyah*: from his *History of medicine* (*Akhbār al-atibbāʾ*).

97 *collyrium*: a term for an unguent, lotion, or liquid wash for the eyes particularly in diseases of the eye. The word comes from the Greek *kollyrion*, 'eye-salve'.

99 *Catholicos*: head of the Nestorian Christian Church and the official representative of the Christian community to the Abbasid caliphs.

Timothy I: (d. 823), patriarch of the Church of the East 780–823; a prolific theologian and translator of Aristotle, he is also known for his theological debate with the caliph al-Mahdī.

umm walad: See note to p. 70.

100 *Mīkhāʾīl*: his biography is Ch. 8.27 in the full text.

101 *two garments*: this seems to come from Matthew 10:9–11, 'Take no gold, or silver, or copper in your belts, no bag for your journey, or two tunics, or sandals, or a staff; for labourers deserve their food', while the clearest biblical prohibition to polygamy may be I Corinthians 7:2, 'But because of cases of sexual immorality, each man should have his own wife and each woman her own husband' [New Revised Standard Version].

102 *motherfucker*: literally, 'one who bites his mother's clitoris'.

103 *vagina*: Sahl the Beardless, by way of a merry prank, had apparently claimed to have seduced Yūhannā's mother and to be Yūhannā's biological father.

104 *Azadh dates*: the Persian word *āzād* or *āzādh* means 'free, faultless, solitary . . .' and is used in Arabic for high-quality dates.

courier system: the early Abbasid rulers employed a courier system using special horses or mules kept at a network of stations along major routes, to convey missives, intelligence, and sometimes goods.

106 *orpiment*: an arsenic sulphide mineral (As_2S_3), a deep yellow in colour, formerly widely used as a pigment, and also of interest to alchemists because of its resemblance to gold.

coins: copper coinage was a token currency used for petty commercial transactions. Its issue was at the discretion of local authorities and as a result, the coins varied greatly in weight, size, and probably value from one district to another.

107 *ʿIbādī*: this indicates that a person hailed from the Christian Arab community in al-Hīrah, the old city of the Lakhmid kings in Southern Iraq, being thus a member of the Syrian 'Apostolic Church of the East', often called Nestorian.

Sergius: Jibrīl alludes to Sergius of Resh ʿAynā, the first to translate a number of Roman scientific works into the Syriac language.

108 *himself*: many modern scholars doubt the authenticity of this autobiography.

Zarāfah: Zarāfah means giraffe; perhaps he had a long neck.

Hunayn ibn Balūʿ al-ʿIbādī: a composer-singer who was long dead by then.

109 *Bukhtīshūʿ ibn Jibrīl*: the son of Jibrīl whom Hunayn was found translating Galen with earlier; his biography is no. 4 in this chapter.

113 *pulp*: purging cassia fruit is the fruit (with laxative properties) of a tree native to the Indian subcontinent (*Cassia fistula* of the family Fabaceae).

manna: an exudate from the stems and leaves of camel thorn (*Alhagi persarum*), also with laxative properties.

114 *shake hands*: '*Musāfahah* is the Arab fashion of shaking hands. They apply the palms of the right hands flat to each other, without squeezing the fingers, and then raise the hand to the forehead' (R. Burton, *A Personal Narrative of a Pilgrimage to El Medinah and Meccah*, 2nd edn, London, 1857).

115 *Kufic script*: the earliest extant calligraphic form of Arabic script; the angular and often monumental writing probably developed around the end of the seventh century and takes its name from Kufa, Iraq, where it was believed to have originated. *Muwallad* means a later or hybrid form of a script; it can also mean a script that is not as well formed as the classical form or even a forged script.

Baghdadi paper: paper made in Baghdad was said to be of the best quality. The standard 'full' Baghdadi sheet of paper, considered large, was one cubit in width and one-and-one-half cubits in length (approx. 73 × 110 cm). A sheet would then be folded a number of times to produce the desired size of codex.

116 *On my own books*: referred to on p. 50.

umm walad: see note to p. 70.

Stories . . . old: this is the work from which Ibn Abī Usaybiʿah quotes on p. 39 and p. 49. It survives in an epitome, of which the authorship of the text, as existing at present, has been questioned.

On his trials and tribulations: the purported autobiography of Hunayn from which Ibn Abī Usaybiʿah quotes so extensively.

Empty Abode: an allusion to the euphemistic use of the word 'open space' for the place where one relieves oneself.

117 *electuary*: Ar. *maʿjūn*, a medicinal paste made with honey or syrup mixed with other ingredients.

On . . . medicine: this is his *Taʾrīkh al-atibbāʾ* (*History of physicians*), an important source for Ibn Abī Usaybiʿah in Chs. 1–6 that survives today mainly in the *Siwān al-hikmah*, falsely attributed to al-Sijistānī, and in an epitome. Is'hāq relied partly on John the Grammarian, whom we have met as a source in the earlier chapters.

CHAPTER 10

120 *Kindah*: the Kindah tribe, whose existence dates back to the third century BC, established a tribal kingdom in central Arabia in the fifth to sixth centuries. Al-Hārith ibn ʿAmr, the most famous of their kings, invaded Iraq and briefly captured al-Hirah. During Muslim times, descendants of the tribe continued to hold prominent court positions.

Basran: the consensus is that he was actually, as above, from Kufa.

Hashemites: the descendants of Hāshim ibn ʿAbd Manāf, great-grandfather of the Prophet Muhammad.

121 *Mūsā ibn Shākir*: employed by the caliph al-Maʾmūn as an astronomer and astrologer, he was the father of three sons, Muhammad, Ahmad (both mentioned here), and al-Hasan, who comprised the 'Banū Mūsā' (literally 'the sons of Mūsā'); the sons were patrons and sponsors of translation and intellectual endeavour in ninth-century Baghdad, as well as being scholars in their own right.

Baghdad: presumably from Samarra, the Abbasid capital between the years 836 and 892.

automata: the invention/construction of automata was regarded as an independent science with its own specific technological expertise, combining engineering and fine mechanics. The Banū Mūsā (sons of Mūsā ibn Shākir) were extremely knowledgeable and influential in this field and their *Book of ingenious devices* was a seminal work that, building on texts from late antiquity, described over one hundred devices, both entertaining and useful. Though reports testify to rivalry between Byzantine and Arab powers in installing impressive automata in the areas used for receiving foreign ambassadors, no specific information is known regarding the caliph al-Mutawakkil. Presumably, however, the Banū Mūsā's extraordinary prowess in this field, for which the caliph had a passion, gave them enough power and prestige to act with impunity.

121 *Jaʿfarī canal*: in 859 the Abbasid caliph, al-Mutawakkil, began to build a new caliphal city in the north of Samarra. Though called on coinage al-Mutawakkiliyyah, it was often referred to in written sources as al-Jaʿfariyyah, Jaʿfar being the caliph's given name. A canal was dug from a point 62 km north to supply the new city, crossing by aqueduct over the existing Qātūl al-Kisrawī canal, but the levelling was badly calculated, and little water flowed.

122 *commission*: incorporated in Muhammad and Ahmad's plea are two ideas on mercy: firstly, the great and noble virtue of *al-ʿafw ʿind al-maqdirah*, meaning that when one has power over another one should forgive them ('the power of a freeman dispels his grudges'); and, secondly, a paraphrase of a saying attributed to ʿAlī ibn Abī Tālib (*c.*600–61) that the confession of a wrongdoing effaces the doing of that wrong ('confession effaces the commission').

later: al-Mutawakkil was assassinated in December 861.

taste: literally 'the taste of your saliva'.

123 *blessings*: in the Arabic, this saying is in rhymed prose.

ornithomancy: divination by the behaviour of birds.

great phenomenon: the comet now known as Halley's Comet came within an exceptionally close proximity to the earth in 837 and was observed in Iraq.

124 *Sabians*: the Sabians mentioned here, to whom Thābit belonged, were the pagan, star-worshipping gnostics of Harrān (ancient Carrhae), which were tolerated by Muslims until the eleventh century because they were wrongly equated with the somewhat enigmatic Sabians mentioned in the Qur'an as being 'People of the Book'. Both are distinct from the pre-Islamic Sabaeans (inhabitants of Sabaʾ, the biblical Sheba) in south Arabia.

Badr: the son of one of the caliph al-Mutawakkil's freed slaves, he rose to prominence and power, and his daughter married one of al-Muʿtadid's sons, the future caliph al-Muqtadir; Badr was executed in the reign of al-Muktafi due to the machinations of the ambitious vizier, al-Qāsim ibn ʿUbayd Allah.

126 *stroke*: Ar. *saktah*, according to the early eleventh-century physician Ibn Hindū, 'occurs when there is loss of sensation and movement caused by an excess of blood or a cold thick humour filling the ventricles of the brain, causing the person to look as though he is asleep, although he is not.'

francolin: a partridge-like bird native to Asia and Africa.

Abū Hasan: Thābit ibn Qurrah's *kunyah* or nickname.

do not go far!: an old formula in elegies.

129 *chronicle*: though it was used by later historians, it has not survived. See also the following entry, Ch. 10, no. 4.

dhimmīs: non-Muslims, but members of a revealed religion (including Christianity and Judaism) that was given protected status under Islam. A *dhimmī* had fewer legal and social rights than Muslims, but more rights than other non-Muslim subjects.

133 *I wrote . . . twice*: Ibn Muqlah served as vizier three times and was also renowned for his calligraphy.

Bajkam: the Turkish military commander and official of the Abbasid caliphate, Abū l-Husayn Bajkam al-Mākānī, overthrew the regime that had imprisoned Ibn Muqlah. However, Ibn Muqlah was not released but remained imprisoned.

134 *wonders*: in the entry on Abū ʿAlī ibn Zurʿah (a Christian scholar and translator, not included in this edition) Ibn Abī Usaybiʿah, again quoting Ibn Butlān, describes how the scholar develops partial paralysis due to 'the combination of his inherent hot temperament, his unwholesome diet—of pungent and fried foods, salt fish, and all types of cold foods prepared with mustard—the fatigue of his mind from writing' and his business troubles. Ibn ʿAbdūs saw that the hemiplegia resulted from 'acute illness in a person of a hot constitution' and when the other physicians wearied, he changed the regimen to 'moisture-inducing foods, whereupon . . . a cure was in sight'.

135 *comet*: literally 'star leaving traces'. This refers to the supernova SN1054 which appeared in July of 1054.

1053: we know the comet appeared in July 1054, equivalent to Rabīʿ I–Rabīʿ II 446 in the Hijri calendar, so it seems there is an error here, either by Ibn Abī Usaybiʿah or in transmission.

136 *A tabular guide to health*: this is the manuscript that Ibn Abī Usaybiʿah uses for reference in Chs. 6 and 14.

hatchling . . . pullet: this theme, which reoccurs in the works of Ibn Ridwān (Ch. 14, no. 7) and al-Yabrūdī (Ch. 15, no. 1), refers to the matter of innate heat versus heat caused by movement, originally discussed by Aristotle. Discussion hinged on the chick or hatching being understood to be immediately and independently active, whereas the young bird or pullet must be fed to produce life and warmth. Ibn Butlān and Ibn Ridwān use this question as a pretext for a heated debate to discredit each other, show off their knowledge of ancient philosophy, and thus attract patrons and students.

137 *Qutb al-Dīn Qāymāz*: an emir of Armenian origin (d. 1174).

138 *Persia*: Ar. *bilād al-ʿajam*, the land of the foreigners, non-Arabs, barbarians in speech. Particularly applied to Persia.

139 *Awhad al-Zamān*: full name Abū l-Barakāt al-Baghdādī; see following entry.

says: this is most likely the same manuscript (probably the *Kitāb tārīkh*, written for ʿAbd al-Latīf al-Baghdādī's son, that Ibn Abī Usaybiʿah uses for his biography of al-Baghdādī (Ch. 15, no. 12).

139 *aloeswood*: a dark, fragrant, resinous wood, formed in the heartwood of the evergreen Aquilaria tree, native to northern India, and parts of South-East Asia, which, when infected with the parasitic fungus, *Phialophora parasitica*, produces an aromatic resin in response. Aloeswood is also known as Jinko, Agarwood, Gharuwood, and Oud.

141 *sheaths*: the common metaphor, in Arabic poetry, of the 'killing' glances of the beloved is supported by the fact that *jafn* means 'sword-sheath' as well as 'eyelid'.

142 *cheek*: the image revolves on the shape of the Arabic letter N: a curve with a dot above.

thinner than yesterday: note that the Arabic does not seem to mean 'thinner than I was yesterday' but is metaphorical. This rather odd proverb-like phrase has not been found elsewhere.

'*eggs*': Ar. *baydah*, can also mean 'testicle'.

The medical formulary: this pharmacopeia became the standard pharmacological work in the hospitals of medieval Islamic civilization.

143 *Saʿīd ibn Hibat Allāh*: an eleventh-century teacher of medicine in Baghdad who served the caliphs al-Muqtadī and al-Mustazhir.

144 *whitlow*: an infection of or abscess on the tip of the finger.

dhimmī: see note to p. 129.

147 *al-Sharābī*: a powerful official at the caliphal court of al-Nāsir (d. 1218).

Ibn al-Qiftī: vizier, biographer, and an older Egyptian contemporary of Ibn Abī Usaybiʿah who wrote an alphabetically arranged book on physicians and philosophers (d. 1248).

148 *Franks*: From the Middle Ages, following the rule of the Carolingian Franks over most of Western Europe, the term 'Frank' was used as a generic term for Western Europeans and the Holy Roman Empire; here Crusaders.

CHAPTER 11

151 *sempervivum*: Ar. *hayy al-ʿālam*, 'life of the world'; Latin meaning 'always alive' referring to its tolerance of extreme temperature and drought. The plant *Sedum telephium*, known in English as Stone Crop, Live-Long, or Orpine, has been used for centuries to heal. Modern research has shown that the plant does indeed have wound healing and anti-inflammatory properties, perhaps because of the presence of polysaccharides.

Philo: of Tarsus, a first-century druggist known through Galen for an analgesic pain-killer he devised that contained poppy juice, henbane seeds, and other drugs.

ʿAdud al-Dawlah: see note to p. 90.

152 *broad beans*: favism, a haemolytic response which can be triggered by the consumption of broad (*fava*) beans and which has similar symptoms to

jaundice, is not uncommon in Middle Eastern populations among others. However, it does not generally cause cataracts and blindness.

153 *madīrah*: a special dish of meat cooked with sour milk until the meat is well-done and the milk becomes thick.

155 *couched*: couching is one of the oldest surgical procedures, first described in the Indian *Sushruta-samhita* in the last centuries BC. The procedure was commonly practised until the nineteenth century but is now rarely performed. A sharp, fine needle was used to pierce the eye either at the edge of the cornea or sclera and push the cloudy lens down to the bottom of the eye, allowing light to enter. Early practitioners did not, of course, know that cataracts were due to the lens becoming opaque, but rather thought it was a fluid that descended in front of the lens (hence the name *māʾ*, 'water', in Arabic and 'cataracts' in English). The patient is left without a lens, ideally requiring a powerful prescription lens to compensate. However, many are left almost totally blind.

156 *The comprehensive book*: known as *al-Kitāb al-hāwī*, it was assembled by his students after his death from his extensive reading notes.

Khorasan: in fact the book was composed in 903 for the Sāmānid governor of Rayy.

al-Balkhī: Abū Zayd Ahmad ibn Sahl al-Balkhī (d. 934) was a philosopher, geographer, and theologian.

157 *Isagoge*: the 'Introduction' to Aristotle's *Categories*, written by Porphyry in Greek and translated into Latin by Boethius, was the standard textbook on logic for at least a millennium after his death.

158 *neck*: the *taylasān* was a piece of material worn over the shoulders and hanging down from them like a hood thrown back; it could cover the head, shoulders, and back, and eventually became the special clothing of judges.

159 *ʿAlāʾ al-Dawlah*: ʿAlāʾ al-Dawlah Muhammad, emir and governor of Isfahan on behalf of the Buyids, he sheltered Ibn Sīnā at his court after the latter had left the Buyid court at Hamadan; Ibn Sīnā died in his service in 1037.

disguised as Sufis: clothing was an important outward symbol of Sufi life. Though no definitive indication of Sufi dress in this period can be found, it is likely to have been similar to that worn later. Ascetics and Sufis commonly wore woollen garments (from whence their name *sūf*, 'wool') or the *khirqah*, 'patched robe'.

160 *mithridate*: a celebrated cure-all, used as an antidote for poisoning, allegedly devised by Mithradates VI, king of Pontus (120–63 BC) with as many as sixty-five ingredients, the major component of which was rue.

163 *ʿAlāʾ al-Dawlah*: probably around 1026.

164 *Mamelukes*: slave soldiers.

Solomon: or Sulaymān, is famous not only for his wisdom but also because he had power over birds and beasts.

168 *al-Bahlawān*: the Ildegizids or Eldigüzids, rulers of Azerbaijan.

168 *Pentateuch*: the first five books, and oldest part of the Bible: Genesis, Exodus, Leviticus, Numbers, and Deuteronomy, which according to tradition were written by Moses.

CHAPTER 12

172 *Jaʿfar*: the Barmakid vizier, Jaʿfar ibn Yahya; see note to p. 78.

postmaster: the postal service under the early Abbasids was an institution employed by the ruler, using couriers with special horses or mules kept at a network of stations along major routes, to convey missives, intelligence, and sometimes goods.

Truly . . . return: Qur'an *Baqarah* 2:156.

173 *Masrūr the Elder*: see note to p. 82.

174 *sneezewort*: *Kundus*, a plant apparently unknown to Greek physicians, has been identified in different ways by modern scholars. Some have aligned it with Struthium or 'soapwort'. However, at least one author states that *kundus* is not *strūthiyūn* and was not used to wash wool. Here it is clearly a sternutatory—that is, a medicine that when applied to the mucous membranes of the nose increases natural secretions and produces sneezing—and hence 'sneezewort' (*Achillea ptarmica L.*).

al-ʿAbbāsah: a sister of Hārūn al-Rashīd, she first married the vizier, Jaʿfar ibn Yahyā, who was executed by Hārūn in 803 (see Ch. 8, no. 3); following this she married Ibrahim, her cousin.

CHAPTER 13

175 *Ziyādat Allāh*: Ziyādat Allāh III was the last Aghlabid emir in Ifrīqiyah (r. 903–7).

177 *Imam al-Mahdī*: the redeemer of Islam who will appear before the Day of Resurrection (literally 'the guided one'), here Abū Muhammad ʿUbayd (or ʿAbd) Allāh al-Mahdī bi-Allāh, a Shiite Ismaili from Syria and the founder of the Fatimid Caliphate (r. 909–34). The Ismailis developed their own theory of the Mahdi with select Ismaili imams taking the title at various times.

178 *Fatimid dynasty*: named for their claim of descent from Fatima, daughter of the Prophet, the Fatimids were an Ismaili-Shia, political and religious dynasty of Arab origin that ruled over a large area of North Africa and subsequently the Middle East from 909 to 1171. Ismailism arose in the eighth century, following division among the Shia over a spiritual successor. The Ismailis accepted Ismāʿīl ibn Jaʿfar (after whom they are named) as the true imam, whereas the 'Twelver' Shiites accepted his younger brother. The Fatimids, as Ismaili imams, were, in the eyes of their followers, the rightful caliphs and custodians of the true faith—both by descent and by divine choice.

Laribus: this anecdote should be placed between 905 and 907.

179 *missionary*: the missionary for the Fatimid Mahdi in Yemen and North Africa recruited the Kutāmah Berbers for the cause of the Mahdi.

Kutāmah Berbers: the sedentary Kutāmah were pious but unsophisticated Muslim Berbers living in small village communities. In the early tenth century the Kutāmah formed a coalition with the Shia Fatimids against the Sunni Aghlabids who ruled Ifrīqiyah and supported the Abbasids. The Kutāmah became fierce protectors of the new Fatimid state and constituted the mainstay of its army until well into the eleventh century. There is still a Kutāmah identity today though their language has largely been Arabized or diluted with other Berber dialects.

182 *948*: the date given by Ibn Abī Usaybiʿah does not correspond to the rule of Romanos II (959–63) but to that of his father Constantine VII. The two were closely associated and were sometimes shown together as co-regents in imperial seals. There is also some evidence of a box with a portrait of Constantine VII sent to the Umayyad caliph in 949.

Great Theriac: originally formulated by the Greeks in the third century BC, primarily as an antidote to snakebite, theriac quickly became known as a panacea for almost any ailment. The Great Theriac, also known as *al-Fārūq*, became one of the most highly valued of all compound general remedies. It was said to have been called *al-fārūq* (a term applied to something or someone that distinguishes between two things such as truth and falsity) because it makes a distinction between disease and health.

183 *Classes of physicians and sages*: this work is a frequent source. Ibn Juljul apparently finished the book by the beginning of 987.

al-Afdal ibn Amīr al-Juyūsh: his father Badr al-Jamālī (d. 1094) was the Commander of the Armies of the Fatimid caliph al-Hakam II; he is sometimes referred to as the 'king of Alexandria' (d. 1121).

184 *released*: the story of his imprisonment is quite different in other sources. One states that he was imprisoned for ten years in the library of Alexandria after he was captured in Mahdia (in modern Tunisia) when he travelled as part of a Zirid embassy. Another relates that his imprisonment was tied to the fate of his patron who had lost the favour of the emir.

May . . . lost: a virtually untranslatable line exploiting some of the many meanings of the word *ʿahd*. Invoking rain is a traditional way of expressing a blessing.

185 *Hadith*: Hadith in Sunni Islam denotes the words, actions, and silent approval, of the Prophet Muhammad. Within Sunni Islam the authority of Hadith officially ranks secondary to the Qur'an as a legal source. However, in both practice and legal theory, Hadith is the prime source of the Sharia and sometimes overrules the Qur'an.

186 *gradually*: i.e. with a gradual increase from cold to hot and back again.

Almoravids: the Veiled Men (*al-Mulaththamūn*) was a denomination given to the Almoravid Berbers (*al-Murābitūn*) in al-Andalus because their men used to cover their faces. The Almoravids ruled Morocco and al-Andalus from the beginning of the eleventh century to the middle of the twelfth century.

187 *sickness*: 'sick', i.e. languid, eyelids or eyes are a common trait of the beloved in Arabic poetry.

1132: after the death of Abū l-ʿAlāʾ his students recorded various medical treatments that he had administered. The treatise was quite important and has survived in several manuscripts.

189 *Abū Marwān al-Bājī*: he met Ibn Abī Usaybiʿah in Damascus in early summer of 1237 providing unique information about Andalusian physicians. This also shows that Ibn Abī Usaybiʿah had started collecting biographical information about physicians by this time.

naghlah: in several sources this is the cause of death not of ʿAbd al-Malik, but of his father, Abū l-ʿAlāʾ ibn Zuhr.

190 *muwashshahah*s: a strophic poetic genre (Arabic poems are normally non-strophic, having only one rhyme throughout) with often intricate rhyme schemes and lines of different length which developed in Muslim Spain in the eleventh to twelfth centuries, with lyrical themes and meant to be sung. The poem beginning 'O cupbearer' and the poem that follows it, beginning 'The speech of a reproacher', are both *muwashshahah*s.

al-Yanāqī family: a prominent family of Seville.

Almohad: from the Arabic, *al-Muwahhidūn* (those who affirm the unity of God), the Almohads were a Berber confederation that created an Islamic empire in North Africa and Spain (1130–1269), founded on the religious teachings of Ibn Tūmart (d. 1130).

192 *poem*: this poem is considered one of the best and most lyrical examples of a *muwashshahah* (see note to p. 190). It is still extremely popular today and modern performances can be heard on YouTube (such as https://www.youtube.com/watch?v=HBZTNY103gk and https://www.youtube.com/watch?v=yEcNfo15SbM).

193 *four in four*: four drinks in four cups, presumably.

heart: literally 'liver', the seat of passions.

flowers: i.e. his teeth are whiter than chamomile flowers.

195 *Book of generalities*: this treatise is the most famous of Ibn Rushd's works on medicine. It has survived in several manuscripts and has been edited several times.

Berbers: two continents (Ar. *al-barrayn*), i.e. the territories in the Iberian Peninsula and in North Africa; in Arabic script *barbar* and *barrayn* are easily confused.

196 *Mudawwanah*: short for '*Mudawwanat al-ahwāl al-shakhsiyyah*', it is the branch of law based on the Maliki school of Sunni Islamic jurisprudence, concerning issues related to the family, including the regulation of marriage, polygamy, divorce, inheritance, and child custody. Still existing in Moroccan law, it was codified after the country gained independence

from France in 1956, and is now known as the personal status code, or family code.

CHAPTER 14

199 *death*: the fourth year of the reign of al-Mansūr began in 757–8, and if Politianus served for forty-six years until his death, it would mean that he died in 802, a date which coincides with that given at the end of the entry. However, other sources suggest that Politianus was made Patriarch of Alexandria in 768 and that he died in 813.

200 *Copts*: an ethno-religious group primarily inhabiting the area of modern Egypt and one of the oldest Christian communities in the Middle East.

Jacobites: see note to p. 54.

order: Ibn Tūlūn conducted a campaign in Syria in 878 with the purpose of defending the borders with Asia Minor against the Byzantines. The result was the occupation of Syria, making Ibn Tūlūn the first Muslim governor of Egypt to annex Syria.

202 *meat*: *bazmāward* (from Persian *bazm-āward*, 'banquet-bringing') consists of round flat bread loaves, with the pith extracted and then stuffed with a minced meat preparation combining the meat with mint leaves, salted lemons, walnuts, and vinegar. It is unclear here whether the cold young goat's meat was to be combined with the fowl in the bread, or whether it was served separately.

203 *durrāʿah*: this is a loose outer garment, slit in the front.

204 *Abū l-ʿAshāʾir*: probably Abū l-ʿAshāʾir Jaysh ibn Khumārawayh, a grandson of Ibn Tūlūn and third ruler of the Tulunid dynasty; the eldest son, he succeeded his father, Khumārawayh, at age 14 in 896, but after a few months was declared deposed and killed in favour of his younger brother.

205 *Great Theriac*: see note to p. 182.

Ibn Tughj: Muhammad ibn Tughj al-Ikhshīd (r. 935–46), the founder of the Ikhshīdid dynasty in Egypt and Syria.

208 *al-Muʿizz's city*: this is a reference to the fact that the construction of New Cairo (al-Qāhirah) was begun in 970 under the reign of the Fatimid caliph, al-Muʿizz (r. 953–75).

civilization: i.e. the ancient Egyptians.

209 *1039*: at this point, following his recording of this quotation, Ibn al-Qiftī adds a comment not quoted by Ibn Abī Usaybiʿah: 'I have seen in his [Ibn al-Haytham's] handwriting a volume on geometry that he wrote in 432/1040 and it is in my possession, thanks be to God.' If this is correct, then Ibn al-Haytham died in or shortly after 1041 and not at the end of September 1039. Ibn al-Qiftī does not comment on the contradictory dates.

211 *dispersed*: no specific passage in Galen's writings has been identified. There are, however, passages where Galen says something to the effect that people got hold of copies of his works and distributed them.

213 *Choicest maxims and best sayings*: this, his only surviving treatise, was composed in 1048. It is devoted to biographical sketches of ancient sages and collections of wise sayings. It includes 125 sayings, more or less, attributed to the legendary Egyptian-Greek sage Hermes, and it was largely responsible for establishing Hermes Trismegistus as a source of wisdom, not only in Fatimid Egypt but in later Ismaili literature as well. It is a frequent source of reference for Ibn Abī Usaybiʿah mainly in the early chapters.

longitude: while the determination of latitude for a given locality was relatively consistent in the medieval geographical tables, the value for longitude could vary considerably, depending upon the position taken for the prime meridian. In forty-three preserved medieval Arabic tables of geographical coordinates recording a value for Fustat, the longitude varies between 51° and 65°. A latitude of 30° and a longitude of 55° was given in an important set of astronomical and geographical tables prepared for the Fatimid caliph, al-Hākim, by Ibn Yūnus, a contemporary of Ibn Ridwān also working in Egypt.

214 *The Lot of Fortune*: 'lots' are points along the ecliptic calculated by measuring the longitudinal distance between two planets and counting the same number of degrees from a third point, usually the ascendant; they were used to indicate length of life, success in an endeavour, and other matters.

The Dragon's Head: not an actual planet, but rather the point where the Moon crosses to the north of the ecliptic. It was often treated as a 'planet' by astrologers, along with the point where the Moon crosses to the south—the Dragon's Tail.

Vega: two stars are indicated on this horoscope: Vega (α Lyrae) in the sign of Capricorn and Sirius (α Canis Majoris) in the sign of Cancer. The indication of prominent stars on a horoscope is relatively unusual and may reflect a particular interest among eleventh-century Cairene astrologers in their significance.

5°12': this personal natal horoscope, prepared (retrospectively) by Ibn Ridwān, a practising astrologer and physician and recorded in his own autobiography, is unique amongst preserved documents. There are a number of horoscopes preserved today that were constructed at the start of the reign of a ruler or were constructed to indicate the occurrence of plagues and disasters or the 'crises' of a disease, but birth charts are much rarer. For different types of horoscopes, see the entry 'astrology' (C. Burnett) in *The Encyclopaedia of Islam Three* (Leiden: Brill, 2000 [available on-line]).

215 *the government . . . Earth*: Qur'an *Aʿrāf* 7:185. See also the following paragraph.

On management: *Fī l-tadbīr* (Greek, *Oikonomika*), a treatise on management of the estate from the school of Aristotle.

216 *to this day*: the 1250s.

fled the country: two major famines struck the Nile Valley in the eleventh century: the first in the mid-1050s and the second began in 1065. Other sources record that the drought of 1056 was so severe that many people fled Egypt and that the majority of people in Fustat died of hunger or 'ate each other'.

Ibn Butlān: see Ch. 10, no. 6.

ruler: the Fatimid caliph, al-Mustansir (r. 1036–94).

a teacher . . . followed: this refers to the fact that Ibn Ridwān did not serve as an apprentice as was customary during this period for the study of medicine and other disciplines.

seven qualities: for the *Testament* of Hippocrates, on which Ibn Ridwān's list is based, see Ch. 4, pp. 22–3.

217 *sound of heart*: Ar. *salīm al-qalb*; in the Qur'an the phrase «*bi-qalb salīm*» means with a heart free from unbelief or divested of corruptness (Qur'an *Shuʿarāʾ* 26:89).

The useful book on the method of medical learning: Ibn Abī Usaybiʿah makes extensive use of this work in Ch. 6 concerning the medical curriculum.

219 *the poet*: this is al-Mutanabbī (d. 965), widely regarded as one of the greatest and most influential poets of the Arabic language and arguably the most famous Arabic poet of Islamic times.

Abū l-Khayr's: Abū l-Khayr is the *kunyah* or nickname of Salāmah ibn Rahmūn.

220 *shop*: a *dukkān* is a common word for a pharmacy or place where medical care and medicines were dispensed.

'all the Jews': the sense is that his mother was promiscuous to the extent that any male Jew could have been his father.

221 *qīthārah*: an early string instrument of the lyre family.

Sufi: Sufism is a mystical dimension of Islam which emerged in or before the eighth century, where practitioners' ultimate goal was annihilation of the ego in God.

222 *former faith*: in 1160, when Maimonides was in his early 20s, his family and he moved to Fez, close to the capital of the Almohads, a Berber dynasty that deprived religious minorities of their traditional protected rights in Islam. This passage is one of only two sources for the 'forced conversion' of Maimonides to Islam, the other source being Ibn al-Qiftī, who says: 'when he proclaimed the credo of Islam, he was obligated by its particular tenets to undertake recitation of the Qur'an and prayer.' Ibn Abī Usaybiʿah may have obtained his information from Maimonides' son Abraham/Ibrahim, whom Ibn Abī Usaybiʿah worked with in the Nāsirī hospital in Cairo (see Ch. 14, no. 12), while Ibn al-Qiftī may have relied on a student of Maimonides named Joseph Ibn Shimʿon.

hospital: this is the Nāsirī hospital in Fustat.

223 *Karaite*: Karaite Judaism traces its origins to disputes over the validity of the Oral Law in the first to second centuries BC. It believes in (direct, independent, and critical study of) the Scriptures as the exclusive source

of religious law. The development of its customs and laws shows some Islamic influence, and it follows patrilineal descent. In its 'Golden Age' (900–1100) Karaite Jews obtained autonomy from Rabbinite Judaism in the Muslim world, established their own institutions and often held high social positions, such as doctors, within Muslim society.

224 *five sons*: Abū l-Khayr's four brothers were all physicians; for their biographies see Chs. 14.50–3 in the full text.

ʿĪsā: brother of Majd al-Dīn Abū Bakr ibn al-Dāyah, who was governor of the province of Aleppo, and foster brother of the Zangid ruler Nūr al-Dīn Mahmūd ibn Zangī (r. 1146–74).

Franks: a generic term for Western Europeans, i.e. Crusaders from the Holy Roman Empire.

225 *faith . . . adheres*: he was a Christian, probably of the Melkite sect, as were his relatives who were also prominent physicians.

8 years old: in 599/1202 when Abū Hulayqah would have been about 8, al-Malik al-Kāmil would have been 22–3 years old. He was the eldest son of al-Malik al-ʿĀdil and at this time served as his father's representative. It is unclear from this account if al-Malik al-ʿĀdil was in the eastern provinces in Diyārbakr or Syria or even in Egypt.

226 *Andarānī salt*: Andarānī (or Darānī) salt is obtained by the evaporation of sea water, and it was considered the best type of salt.

227 *Abū Shākir*: one of the four physician sons of Abū Sulaymān Dāwūd and brothers of Abū l-Khayr previously mentioned (see note to p. 222).

229 *June 1269*: this date gives a *terminus post quem* for what was possibly the last revision Ibn Abī Usaybiʿah made of his book.

Asmaʿian insight: so called after the famous philologist al-Asmaʿī (d. 828).

Akhzamite nature: from an old proverb referring to the story of a man named Akhzam.

And . . . love: this is in fact the opening line of a poem by another poet, addressed to Saladin.

CHAPTER 15

230 *lungs*: he did not, as is often assumed, discover the *circulation* of blood in the lungs. See P. Pormann and E. Savage-Smith, *Medieval Islamic Medicine*, 46–8; and see N. Fancy, *Science and Religion in Mamluk Egypt: Ibn al-Nafīs, Pulmonary Transit, and Bodily Resurrection* (London/New York: Routledge, 2013).

231 *Jacobite Christian*: see note to p. 54.

wormwood: various types of Artemisia, also known as Absinthe or Mugwort. A perennial native to temperate Eurasia and North Africa, often found on uncultivated arid ground, rocky slopes, and at the edge of footpaths and fields, it was historically associated with extreme bitterness; burning it produces poisonous fumes. Medicinally it was used as an antiseptic and vermifuge.

233 *mouth*: this refers to the mouth of the stomach or cardiac orifice.

234 *beast of prey*: it is unclear what animal was the object of this vivisection, for the term used can refer to any wild predatory beast, including a lion, wolf, lynx, or leopard (but not a fox or hyena).

241 *Alan*: an Iranian people of the western Eurasian steppes and the North Caucasus, the region that we would now designate as Georgia. Some of them were idolaters, and others followed the Islamic faith, but the majority were Christians. However, the Arabic spelling (*ʿAlān*) here is highly unusual (normally Alān or al-Lān, without *ʿayn*). Whether the majority of the Alans were Christian, as the text says, is uncertain but they are said to have adopted Christianity in the tenth century.

242 *Timurtāsh*: an Artuqid ruler of Diyār Bakr.

Sufis: Sufism is a mystical dimension of Islam which emerged in or before the eighth century. See also note to p. 221.

shamshak: an unknown type of Baghdadi shoe, very likely a kind of sandal; also written *tamshak/shamashk*; presumably from the Persian *chamshak* meaning shoe.

244 *Hūd*: a legendary Arab prophet, precursor of Muhammad; he is mentioned several times in the Qur'an. He preached among the people of ʿĀd, who were destroyed because of their unbelief. Dhū l-Kifl (possibly meaning 'the man with double recompense') is an obscure personage mentioned twice in the Qur'an.

Yāsīn . . . Luqmān: the three first-mentioned are suras nos. 36, 20, 19, respectively; Luqmān is no. 38, and The Ants is no. 27.

calf: the story of Moses and the golden calf is told in the Qur'an in several places.

245 *Ibn Lūqa*: Qustā ibn Lūqa, a ninth-century Syrian physician and translator, see Chs. 9.6 and 10.44 in the full text.

al-Fārābī: see note to p. 37.

rain-stars: in ancient and popular Arab meteorology some stars and constellations were associated with rain.

sibling: al-Khansāʾ composed numerous laments for her two brothers, Sakhr and Muʿāwiyah, who died of battle wounds shortly before the coming of Islam.

'*Our . . . plight*': the opening of an elegy by the famous tenth-century poet al-Mutanabbī (see note to p. 219).

sore . . . before: while a rhymed version is used here, a more literal and annotated version is to be found in the full text.

246 *al-Suhrawardī*: a mystic, philosopher, and contemporary of Ibn Abī Usaybiʿah, who was put to death because of his alleged heretical ideas in

religious and political matters; he should not be confused with Shihāb al-Dīn Abū Hafs ʿUmar al-Suhrawardī (1145–1234), one of the most important Sufis in Sunni Islam.

247 *this place*: the location of this event is unclear. There is a Gate of Deliverance in both Damascus and Aleppo. The exclamation praising the beauty of Damascus is also ambiguous; it could well be that al-Suhrawardī praised the beauty of Damascus while actually walking in Aleppo.

hand: possible explanations for the experiences described include a state of trance, hypnosis, or mentalism.

248 *I . . . him*: this refers to the first version of the book, clearly written before the death of Rafīʿ al-Dīn al-Jīlī in 1244, which gives a *terminus ante quem* for the first writing.

250 *Baghdadi script*: presumably a cursive script, possibly *naskh* or *thuluth*, which was developed in Baghdad by Ibn Muqlah. However, Baghdadi paper was a very large format and size and could accommodate a very large script, and the comparison here may be to the size rather than the style of the script.

251 *empty*: the truth of this is not certain but his biographer and contemporary Bahāʾ al-Dīn Ibn al-Shaddād says that at Saladin's death his treasury was found to contain a mere 47 silver dirhams and a piece of gold of unknown weight (*al-Nawādir al-sultāniyyah*, Cairo, 1964, 17, quoted also in Ibn Khallikān's *Wafayāt*, vii, 204).

252 *Islam*: in the version of Ibn al-Qiftī's *Taʾrīkh al-hukamāʾ* preserved today, Ibn al-Mutrān is not mentioned.

253 *Baghdadi paper*: Baghdadi paper appears to have been the 'gold standard' of paper, both in size and quality, across much of the Arab world. Even a book one-sixteenth the size of Baghdadi paper, which would have been approx. 18 × 27 cm., was still a very reasonable size. See also note to p. 115.

254 *The gardens of physicians and meadows of the intelligent*: this is a medical anthology containing quotations and extracts from a large number of early medical writings, some of which have been lost for posterity. Ibn Abī Usaybiʿah uses it as a source in Ch. 1.

255 *clock . . . Damascus*: this was a monumental water-clock, with mechanical devices that, at the passing of each hour during the daytime, dropped weights into the beaks of two large brass falcons, making a noise and causing a shutter to fall over a window marking off that hour. At night time the circles indicating hours were illuminated one-by-one by a rotating lamp. For further information see D. R. Hill, *Arabic Water-Clocks* (Aleppo: University of Aleppo, Institute for the History of Arabic Science, 1981), where he has analysed Ridwān's treatise describing this clock.

256 *Samaritan*: Samaritans are Israelites, close to Jews, who recognize only the Pentateuch (the first five books of the Hebrew Bible). Their religion goes back several centuries before Christ and still survives. The split with

the Karaites (see note to p. 223) occurred around the time of the coming of Islam.

257 *vizier*: Safī al-Dīn ibn Shukr, a notorious vizier of al-Malik al-ʿĀdil.

a hundred years: that is, ninety-four solar years.

259 *own hand*: this is most likely the *Kitāb tārīkh* (*On history*), attributed to, or possibly written for, ʿAbd al-Latīf's son Sharaf al-Dīn Yūsuf, which is lost.

Hadith: in Sunni Islam, the words, actions, and silent approval of the Prophet Muhammad; its authority ranks inferior only to the Qur'an.

certificates of audition: Ar. *ijāzā*, pl. *ijāzāt*; a licence authorizing its holder to transmit a certain text or subject, which is issued by someone already possessing such authority. It is particularly associated with transmission of Islamic religious knowledge. The licence usually implies that the student has acquired this knowledge from the issuer of the *ijāzā* through first-hand oral instruction.

chains . . . masters: the idea was of hearing Islamic teaching directly from a reliable shaykh and, through him, becoming a part of an unbroken *isnād*, or chain of learning.

Fasīh: this is the *al-Fasīh fī l-lughah*, a work on grammar, by Thaʿlab (d. 904).

Maqāmāt: a collection of tales in rhymed prose with short poetic passages, referring here to the famous *Maqāmāt* of al-Harīrī (d. 1122).

al-Mutanabbī: the famous tenth-century poet.

260 *Lumaʿ*: *al-Lumaʿ fī l-nahw* ('*Flashes on syntax*') was a very popular work on Arabic grammar by the linguist Ibn Jinnī (d. 1002).

Adab al-kātib: '*The chancery scribe's handbook*', a work on technical terms needed by civil servants, on penmanship, matters of taxation, law, etc., by the polymath Ibn Qutaybah (d. 889).

Takmilah: '*The supplement*', by the grammarian Abū ʿAlī al-Fārisī (d. 987), a sequel to his *al-Īdāh* ('*The elucidation*').

The Book of Sībawayh: the foundational, voluminous, extremely influential, and also earliest work on Arabic grammar; considered the principal textbook; simply entitled *Sībawayh's book*. Sībawayh died in 793.

Maʿānī al-Qurʾān: '*The meanings of the Qur'an*', philological commentary on the Qur'an, by Ibn al-Khashshāb (d. 1172).

Shuhdah bint al-Ibarī: the renowned female grammarian and a great authority on the Hadith, (d. 1178).

son of Ibn al-Tilmīdh: this refers to the 'bad' son of Ibn al-Tilmīdh, mentioned in Ch. 10, no. 9.

261 *the garb of a Sufi*: clothing was an important outward symbol of Sufi life. See note to p. 159.

Ibn Tātalī: the name Tātalī is obscure and not Arabic but presumably Berber.

261 *Almoravids*: rulers of Morocco and al-Andalus from the beginning of the eleventh to the middle of the twelfth century. See note to p. 186.

263 ‘*the fortunate . . . another*’: a saying attributed to the prophet Muhammad, found in several authoritative collections of Hadith.

Acre: he undoubtedly went to see Saladin in the army camp set up during Saladin’s siege of Acre (1189–92).

thuluth script: a particularly ornate and monumental Arabic script.

Until . . . say: Qur’an *Zumar* 39:73 and Qur’an *Raʿd* 13:30.

Franks: see note to p. 148.

Ibn Sanāʾ al-Mulk: (1155–1211) qadi and famous poet of Ayyubid Cairo, and author of a treatise on the *muwashshahah*. See also note to p. 190.

266 *without skill*: according to the historian Sibt ibn al-Jawzī, it was the famous physician Radī al-Dīn al-Rahbī (see Ch. 15, no. 11) who bled Saladin and caused his untimely death.

deaths: the phrase *al-ghalāʾ al-ʿazīm wa-l-mawtān* suggests a food shortage and possible famine, with resulting deaths.

reel: in his book *Information and details about events witnessed and incidents observed in the land of Egypt* ʿAbd al-Laṭīf al-Baghdādī described some of the skeletal remains that he found in the cemeteries and corrected some details in Galenic anatomy.

267 *Rūm Seljuqs*: a Sunni Muslim dynasty of Turkish origin that ruled much of Anatolia (Rum) *c.*1081–1308. It was descended from the Great Seljuqs who ruled Iraq, Iran, and Central Asia.

268 *son of my son*: Ibn Abī Usaybiʿah’s father had studied under al-Baghdādī, and ‘son’ is used here in an affectionate and academic sense, as well as in a literal sense initially.

270 *Rashīd*: this means ‘rightly guided, following the right course’, hence the references to guidance in the poem.

272 *rajaz metre*: the oldest and simplest of Arabic poetic metres and the medium par excellence for didactic poems and the versification of knowledge meant to be memorized. It is unique among classical metres because its lines are not divided into hemistichs.

hollow and curved: cataract surgery by couching was usually performed by pushing the cloudy lens downwards. Though today we know that cataracts are due to an opaque lens, in medieval literature it was believed that an opaque fluid was interposed between the lens and the pupil. This passage is important evidence that at least some physicians attempted to remove the cataract by suction through a hollow instrument. If such a procedure was successfully carried out, it could only have worked on a soft juvenile cataract. Both methods allow light to enter the previously darkened eye but leave the patient without the use of a lens.

273 *dram*: both the Arabic ‘dirham’ and the English ‘dram’ ultimately derive from the Greek ‘*drachmē*’.

chyle: a milky fluid consisting of lymph and emulsified fat droplets formed in the small intestine during digestion of fatty foods.

274 *sickness*: i.e. languidness, deemed attractive.

276 *ʿAttābī cloth*: a silk with an irregular wavy finish similar to moiré, named after the al-ʿAttābiyyah quarter in Baghdad which was famous for its silk-cotton cloth; it is also the origin of the word tabby, referring to the striped or mottled colouring of cats.

277 *Baghdadi format*: see note to p. 115.

Rafīʿ al-Dīn al-Jīlī: see Ch. 15, no. 6.

Sāhib: literally 'companion, owner, master, or lord'; a title since the tenth century often given to a vizier, as 'companion' of the ruler.

278 *fortune*: the words *khatt* ('manuscript') and *hazz* ('good fortune') differ only in the placement of the single dot.

282 *majlis*: Ar. 'place of sitting', most often used to describe royal audiences, councils, and similar gatherings; here in the sense of a lectureship.

medicine: Al-Dakhwār is generally considered to have founded the first medical school in the Arabic world, al-Madrasah al-Dakhwāriyyah.

283 *Jamāl al-Dīn*: Jamāl al-Dīn ibn Abī l-Hawāfir, chief physician in Egypt; his biography is Ch. 14.44 in the full text.

286 *Hospitallers*: a religious military order founded at Jerusalem in the eleventh century, originally to care for sick and poor pilgrims. It later acquired wealth and lands and took on an additional role as defender of the Crusader kingdom. Along with the Templars, it became the most formidable military order in the Holy Land.

287 *presented with it*: clothing was not only an important outward symbol of Sufi life but also had spiritual significance and symbolism. It was believed that clothes retained *barakah*, the blessing and energy of the owner, thus the custom of handing over vestments meant that their *barakah* was also given over to the recipient. An 'ancestry' such as the one this garment had would have made it highly revered, and its bestowal a huge honour. See also notes to p. 159 and p. 221.

al-Khidr: this is the mysterious, unnamed figure alluded to in the Qur'an as accompanying Mūsā (Moses). According to popular belief al-Khidr, sometimes identified as Elijah, is immortal and roams the earth, and thus could have received the robe from the Prophet Muhammad.

289 *My two friends*: an example of the ancient convention of addressing two unnamed companions at the beginning of a poem. See also note to p. 64.

al-Junayd: a famous tenth-century mystic and ascetic.

290 *Aqish*: apparently a Turkish name.

The name is composed: the riddle, one of many of its kind, uses the numerical values of the letters of the Arabic alphabet, for which the ancient order is used, not the one used in dictionaries. Here T (tāʾ, 400) plus A (alif, 1) equals 401, the letters A (1), Q (100), Sh (shīn, 300) also add up to 401. A sixth of Sh (300) is 50, which is half of Q (100).

Y: (yāʾ) is 10; one tenth is 1 (the letter alif).

290 *singer*: he is also known as 'son of the singer (or songstress)', *ibn al-ʿālimah*.

292 *Ibn al-Bawwāb*: Persian calligrapher and illuminator (d. Baghdad *c.*1022) who helped develop the early cursive calligraphic scripts (in contrast to the angular Kufic script). The only surviving Qur'an copied by him (Chester Beatty Ms.1431) is one of the earliest extant Arabic manuscripts copied in cursive script on paper.

muwallad Kufic script: Kufic is the earliest extant calligraphic form of Arabic script and takes its name from Kufa, Iraq. *Muwallad* generally means a later or hybrid form of a script. See also note to p. 115.

katam: (*Buxus dioica*) also known as black henna, is a variety of box from Yemen, the dried and powdered leaves of which are used, often combined with henna, to dye one's hair; it is still in use today. Some have also identified it as 'troëne', a kind of privet.

Muwaffaq al-Dīn: a complimentary and respectful epithet, *most fortunate in religion*, i.e. Ibn Abī Usaybiʿah.

MORE ABOUT **OXFORD WORLD'S CLASSICS**

A SELECTION OF **OXFORD WORLD'S CLASSICS**

	Classical Literary Criticism
	The First Philosophers: The Presocratics and the Sophists
	Greek Lyric Poetry
	Myths from Mesopotamia
APOLLODORUS	**The Library of Greek Mythology**
APOLLONIUS OF RHODES	**Jason and the Golden Fleece**
APULEIUS	**The Golden Ass**
ARISTOPHANES	**Birds and Other Plays**
ARISTOTLE	**The Nicomachean Ethics**
	Politics
ARRIAN	**Alexander the Great**
BOETHIUS	**The Consolation of Philosophy**
CAESAR	**The Civil War**
	The Gallic War
CATULLUS	**The Poems of Catullus**
CICERO	**Defence Speeches**
	The Nature of the Gods
	On Obligations
	Political Speeches
	The Republic and The Laws
EURIPIDES	**Bacchae and Other Plays**
	Heracles and Other Plays
	Medea and Other Plays
	Orestes and Other Plays
	The Trojan Women and Other Plays
HERODOTUS	**The Histories**
HOMER	**The Iliad**
	The Odyssey

A SELECTION OF **OXFORD WORLD'S CLASSICS**

Horace	**The Complete Odes and Epodes**
Juvenal	**The Satires**
Livy	**The Dawn of the Roman Empire**
	Hannibal's War
	The Rise of Rome
Marcus Aurelius	**The Meditations**
Ovid	**The Love Poems**
	Metamorphoses
Petronius	**The Satyricon**
Plato	**Defence of Socrates, Euthyphro, and Crito**
	Gorgias
	Meno and Other Dialogues
	Phaedo
	Republic
	Symposium
Plautus	**Four Comedies**
Plutarch	**Greek Lives**
	Roman Lives
	Selected Essays and Dialogues
Propertius	**The Poems**
Sophocles	**Antigone, Oedipus the King, and Electra**
Suetonius	**Lives of the Caesars**
Tacitus	**The Annals**
	The Histories
Thucydides	**The Peloponnesian War**
Virgil	**The Aeneid**
	The Eclogues and Georgics
Xenophon	**The Expedition of Cyrus**

A SELECTION OF **OXFORD WORLD'S CLASSICS**

Bhagavad Gita

The Bible Authorized King James Version
With Apocrypha

The Book of Common Prayer

Dhammapada

The Gospels

The Koran

The Pañcatantra

The Sauptikaparvan (from the Mahabharata)

The Tale of Sinuhe and Other Ancient Egyptian Poems

The Qur'an

Upanisads

ANSELM OF CANTERBURY — **The Major Works**

THOMAS AQUINAS — **Selected Philosophical Writings**

AUGUSTINE — **The Confessions**
On Christian Teaching

BEDE — **The Ecclesiastical History**

KĀLIDĀSA — **The Recognition of Śakuntalā**

LAOZI — **Daodejing**

RUMI — **The Masnavi**

ŚĀNTIDEVA — **The Bodhicaryāvatāra**

A SELECTION OF **OXFORD WORLD'S CLASSICS**

	The Anglo-Saxon World
	Beowulf
	Lancelot of the Lake
	The Paston Letters
	Sir Gawain and the Green Knight
	The Poetic Edda
	The Mabinogion
	Tales of the Elders of Ireland
	York Mystery Plays
GEOFFREY CHAUCER	**The Canterbury Tales** **Troilus and Criseyde**
GUILLAUME DE LORRIS and JEAN DE MEUN	**The Romance of the Rose**
HENRY OF HUNTINGDON	**The History of the English People 1000–1154**
JOCELIN OF BRAKELOND	**Chronicle of the Abbey of Bury St Edmunds**
WILLIAM LANGLAND	**Piers Plowman**
SIR JOHN MANDEVILLE	**The Book of Marvels and Travels**
SIR THOMAS MALORY	**Le Morte Darthur**

A SELECTION OF **OXFORD WORLD'S CLASSICS**

An Anthology of Elizabethan Prose Fiction

Early Modern Women's Writing

Three Early Modern Utopias (Utopia; New Atlantis; The Isle of Pines)

Francis Bacon
: **Essays**
: **The Major Works**

Aphra Behn
: **Oroonoko and Other Writings**
: **The Rover and Other Plays**

John Bunyan
: **Grace Abounding**
: **The Pilgrim's Progress**

John Donne
: **The Major Works**
: **Selected Poetry**

John Foxe
: **Book of Martyrs**

Ben Jonson
: **The Alchemist and Other Plays**
: **The Devil is an Ass and Other Plays**
: **Five Plays**

John Milton
: **The Major Works**
: **Paradise Lost**
: **Selected Poetry**

Earl of Rochester
: **Selected Poems**

Sir Philip Sidney
: **The Old Arcadia**
: **The Major Works**

Sir Philip and Mary Sidney
: **The Sidney Psalter**

Izaak Walton
: **The Compleat Angler**